NMR for Physical and Biological Scientists

NMR for Physical and Biological Scientists

Thomas C. Pochapsky and
Susan Sondej Pochapsky

Brandeis University
Waltham, Massachusetts

Taylor & Francis
Taylor & Francis Group

New York • London

Vice President	Denise Schanck
Senior Editor	Robert L. Rogers
Associate Editor	Summers Scholl
Senior Publisher	Jackie Harbor
Production Editor	Karin Henderson
Cover Designer	Aktiv
Typesetter	Phoenix Photosetting
Printer	RR Donnelly

ISBN 0 8153 4103 2

Library of Congress Cataloging-in-Publication Data
Pochapsky, Thomas C.
   NMR for physical and biological scientists/Thomas C. Pochapsky and Susan Sondej Pochapsky.
      p.   ;   cm.
   Includes bibliographical references and index.
   ISBN 0-8153-4103-2 (alk. paper)
1. Nuclear magnetic resonance spectroscopy. 2. Nuclear magnetic resonance. I.
Pochapsky, Susan Sondej. II. Title.
   [DNLM: 1. Magnetic Resonance Spectroscopy. QU 25 P739n 2006]

   QP 519.9.N83P63 2006
   543′.66–dc22

                                                                    2006019638

Published in 2007 by Garland Science, member of the Taylor & Francis Group, LLC,
270 Madison Avenue, New York, NY 10016, USA and
2 Park Square, Milton Park, Abingdon, Oxon, OX14 4RN, UK.

Printed in the United States of America on acid-free paper.

10   9   8   7   6   5   4   3   2   1

Taylor & Francis Group, an informa buisness   Visit our web site at http//www.garlandscience.com

To our children, Elizabeth and Theodore

and our parents, Eugene and Mary Pochapsky and Bob and Mary Ann Sondej

CONTENTS

PREFACE

Nuclear magnetic resonance (NMR) has become an essential tool for scientists of all persuasions investigating physical, chemical and biological problems. This book is designed as a primary text in a one-semester course for graduate students or advanced undergraduates who are likely to use NMR in their research and feel the need for a more complete background in the subject. Problems are included towards the end of each chapter and are designed to be practical as well as instructive. No assumptions about prior training or knowledge have been made other than that the student has had basic college courses in calculus and physics.

Magnetic resonance theory can be intimidating to students (and teachers) who lack a strong physics background. Both Susan and I learned NMR on the job rather than through formal training, and in so doing we came upon concepts and methodology for which the underlying logic was not always obvious. My research required me to understand NMR theory and practice in more detail than my formal training provided, while Susan's job as an applications scientist with Bruker acquainted her with modern NMR hardware and practical spectroscopy. In the process of demonstrating instrument capabilities to potential customers, implementing experiments for users, and teaching NMR owners and operators how to use the equipment, she gained valuable experience on how NMR is used in research for a variety of applications. She also learned the practical things that never quite make it into the instruction manuals. Together we gained a unique perspective on this powerful experimental method. We decided that we could write a text that incorporated both the theory and practice of NMR. The text and problems have their origins in class notes from an NMR special topics course that was first offered at Brandeis in 1991.

Before saying anything more about what this text is, it is worth saying what it is not. This is not a compendium of pulse sequences and experiments, and there are not large numbers of references. We have tried to reference only those papers that deal with original concepts or that provide helpful explanations, rather than NMR application or pulse sequence papers. Such reference lists become obsolete almost as soon as they are made, and we have tried to concentrate on the basics that are less likely to change. The ease with which literature searching can be accomplished via online databases makes compendia of references in textbooks less important than they once were. We have tried to use common acronyms for experiments and techniques so that specific references can be easily identified.

The problems form an integral and important part of this text. They are designed to challenge the student to think mathematically and/or conceptually about topics as they are introduced. While working on the answer keys, we found it useful to have a standard mathematical software package such as Mathcad®, MATLAB® or Mathematica® available as well. Starred problems, in particular, provide important supplements to the chapter content. Solutions will be available on the Garland Science website (www.garlandscience.com) to qualified instructors. Professor Jeffry D. Madura of Duquesne University has prepared demos that use Mathcad® 13 and are also available at the Garland Science website.

By their very nature, textbooks require a linear format. Nevertheless, comprehending a subject as inherently complex as NMR requires a wide range of basic information, and the challenge for us was to introduce topics in an order that made logical sense. To do this, we first introduce important principles intuitively, using analogy and example. Only after all of the basics are in place do we attempt to bring more rigor to the explanations. In this spirit, Chapter 1 deals with principles common to all forms of spectroscopy. The quantum mechanics that provide the underpinnings of spectroscopic transitions are discussed. Physical principles that are used extensively in later chapters are introduced, including idealized and damped harmonic oscillators, standing waves, the semi-classical approach to spectroscopy and the time-independent Schrödinger equation. We discuss the importance of ensembles of chromophores, and emphasize the difference between incoherent and coherent ensembles in spectroscopy. The uncertainty principle is introduced in relationship to line-width and line-shape, as is the connection between these spectral observables and the relaxation of excited states. Standard methods of exciting, detecting and analyzing spectroscopic data are introduced, including sequential and simultaneous acquisition methods. The problems for this chapter deal with standing waves, simple radio frequency (RF) circuitry such as that used in NMR probes, Lorentzian line-shape, the relationship between line-shape and relaxation, basic quantum mechanics and signal averaging.

In Chapter 2, nuclear magnetism and NMR are introduced in light of the concepts from Chapter 1, and reasons are given for the unique position that NMR occupies in the spectroscopic smorgasbord available to today's scientist. The classical and quantum descriptions of nuclear spin, spin angular momentum, gyromagnetic ratio and Larmor frequency are described and compared. We go to some effort to distinguish between the "spinning top" classical model and the quantum approach for nuclear spin, describing where each approach is appropriately used. Boltzmann and non-Boltzmann distributions of spin states are described, and the concepts of coherence and populations in NMR experiments are specifically discussed. The importance of relaxation in the NMR experiment is introduced in a general way, with T_1 and T_2 described as enthalpic and entropic processes, respectively. With these basics in place, a description of the types of information available from NMR experiments is now possible. Chemical shielding and chemical shift, scalar coupling and dipolar coupling are described in terms of their origins and observed effects on NMR spectra. The effects of molecular dynamic processes on NMR observables are mentioned for the first time, as are the different time scales that are accessible to measurement by NMR. The chapter ends with the description of a time-dependent NMR experiment, RF decoupling. This section includes the first quantitative description of how electromagnetic radiation and nuclear spins interact, an important concept that occupies much of Chapter 3. The problems for this chapter make use of the principles introduced in the chapter for spectral analysis, e.g., slow vs. fast exchange phenomena and the chemical shift time scale, the relationship between gyromagnetic ratio, spectrometer frequencies and the sensitivity of the NMR experiment, and spin quantum number and multiplicity of spectral lines. The concept of RF as an oscillating magnetic field capable of inducing Larmor precession is introduced in a problem here as well.

Chapter 3 focuses on the theory and implementation of simultaneous acquisition Fourier transform NMR. The rotating frame of reference is introduced, first in terms of the Larmor precession of a single spin and then as applied to an ensemble of spins and their interaction with an RF field. The mechanics of detecting, digitizing and analyzing the free induction decay (FID) are discussed. Quadrature detection is described, and the RF circuitry associated with quadrature detection (mixers, phase shifting, and digital frequency generation) is introduced. Fourier transformation (FT) is considered first intuitively as a tunable frequency filter for time-domain signals, then as a mathematical concept (the transformation between time-domain and frequency-domain spaces). The digital approximation for the FT provided by the Cooley–Tukey algorithm is discussed. In the context of this discussion, the concept of complex numbers and the real and imaginary portions of the FID arise naturally, as does spectral phasing, zero- and first-order phase corrections, and absorptive versus dispersive line shapes. RF pulses and non-selective spin excitation are introduced after FT, since the connection between time and frequency domains can now be made more intuitively. The concepts of pulse and receiver phase are placed in the context of a simple phase cycle, CYCLOPS. The remainder of the chapter is dedicated to experimental and processing parameters that affect spectral quality. Magnet shimming, including cryoshimming and gradient shimming, are discussed. The role of window functions such as sine bells and exponential and Gaussian multipliers in optimizing spectral appearance is demonstrated, and digital manipulation techniques such as zero-filling, oversampling, digital filtering and linear prediction are introduced. Problems for this chapter cover the mathematical description of RF filters and mixer circuits, pulse excitation bandwidth and tip angle, effective field and off-resonance effects of RF pulses. A number of concepts important for the next chapter, including saturation of NMR transitions and chemical exchange, make their first appearance in Chapter 3 problems.

The fundamental mechanisms of nuclear spin relaxation are introduced in Chapter 4 with an emphasis on the difference between the return to thermal equilibrium (T_1) and loss of coherence (T_2). The experiments used for measuring T_1 and T_2 relaxation, the inversion-recovery and spin-echo experiments provide entree to multiple pulse NMR methods. The refocusing of chemical shift by the spin-echo is described using vector models, and the central role of this experiment in the development of NMR pulse sequences is discussed. The fact that NMR transitions required for relaxation must be stimulated by environmental magnetic fluctuations ("white noise") introduces the concept of the spectral density function and its relationship to spin relaxation. The common molecular mechanisms for spin relaxation are introduced at this point, and the relationship between dipolar coupling and the nuclear Overhauser effect (NOE) is described. The Solomon equation is shown to be a means of determining the time-dependent behavior of dipolar coupled spin systems. Standard experiments for measuring the NOE are discussed, and are used to illustrate the relationship between molecular weight, correlation time and the spectral density function. The problems for Chapter 4 provide examples of analysis of relaxation data and how the Solomon equation is used practically for analyzing the results of NOE experiments. The effects of multiple relaxation mechanisms on observed relaxation behavior and NOE is also illustrated by several of the problems in this chapter.

By this point in the course, the student has done some of the problems from each chapter, old math skills have been rediscovered, and we can venture into more quantitative descriptions of NMR. The first section of Chapter 5 introduces the Bloch equations of motion for macroscopic magnetization in the presence of a perturbing force (an RF pulse) and a restoring force (relaxation). It is demonstrated that a simple NMR spectrum can be completely simulated using the Bloch equations without recourse to quantum mechanics. However, in order to incorporate the effects of coupling on spectral appearance adequately, it is necessary to calculate transition energies by solving the time-independent Schrödinger equation. The Hamiltonian operator for a simple two-spin system is described, and the basic concepts of perturbation theory are applied to calculate the positions and relative intensities of doublet lines for a coupled two-spin system in an isotropic liquid. This exposition introduces many of the concepts that are important for the remainder of the text, including eigenfunctions, eigenvectors and eigenvalues, Hilbert spaces, spin functions, spin operators and the assumption of weak coupling as a way of simplifying spectral simulations for multi-spin systems. Because the time-independent solution to the Schrödinger equation does not yield an expression for line shape, the time-dependent description of the two-spin system is discussed briefly, and is covered more completely in Appendix A for the interested reader. The problems for this chapter give the student some practice using the quantum mechanical concepts and mathematical techniques introduced here, including the determination of eigenvalues and commutators for a variety of operators.

The critical link between the simple one- and two-spin systems described in Chapter 5 and the ensembles observed in real NMR experiments is made in Chapter 6. This is probably the most difficult part of the course for most students, since few undergraduate physics or physical chemistry courses deal in a meaningful way with the density matrix as a tool for describing ensemble behavior. It is also where most books on NMR are weakest; some don't bother with the density matrix at all, while others assume a sufficient statistical mechanics background so that the density matrix is a given. We have tried to introduce statistical ensembles, populations and coherence in an intuitive fashion, and to make the link between one- and two-spin operators and eigenfunctions to the corresponding descriptors for macroscopic observables.

Once the density matrix is part of the vocabulary, the student is then led to the operator descriptions of spin angular moment by expansion of the density matrix. These include population and coherence operators that are useful for characterizing evolution of particular coherence orders under the influence of field gradients and phase cycling, and Cartesian operators, used primarily for determining the effects of RF pulses, J-coupling and applied magnetic fields on spin ensembles. The product operator formalism is introduced as a tool for describing the evolution of weakly coupled spins. This is a standard formalism used in the description of multidimensional NMR experiments, and the student needs to be aware that product operators represent a special case of the density matrix formalism for weakly coupled spins. The relationship between the rules for product operator evolution and density matrix manipulation is emphasized, so that the reader does not lose track of the matrix mathematics implicit in product operator evolution. Finally, the relationship between the terms that evolve from product operators and spectral appearance after Fourier transformation is described, that is, dispersive and absorptive line shape and anti-phase versus in-phase multiplet structure as a function of the Cartesian operators that give rise to observable coherences.

At this point, the student is ready to begin examining multiple-pulse and multidimensional NMR experiments in more detail. Chapter 7 makes use of the product operator formalism in describing common homonuclear two-dimensional NMR experiments such as COSY and multiple-quantum-filtered COSY. Phase cycling for coherence pathway selection is introduced and intuitive pictorial descriptions of simple phase cycles are provided. Quadrature detection methods for indirectly detected dimensions, including States, TPPI and variations thereof are discussed. The central role of the spin-echo sequence in multidimensional NMR is described in the context of chemical shift refocusing, and this leads into the use of pulsed field gradients for coherence order selection and artifact suppression.

In the last part of Chapter 7, the NOESY experiment is described, and the role of the correlation time in determining the sign of the NOE first described in Chapter 4 is re-examined. The problem that arises for molecules of intermediate molecular weight for which no NOE is observed in the standard NOESY experiment provides a way of logically introducing the concept of spin-locking in the context of the ROESY experiment. From ROESY, the idea of coherent versus incoherent magnetization transfer is developed, leading to the discussion of TOCSY and the Hartmann–Hahn effect at the end of the chapter.

Heteronuclear correlation experiments are introduced in the context of polarization transfer to low-gamma nuclei in Chapter 8. For each type of polarization transfer, the one-dimensional version of the experiment is described first, with the extension to indirectly detected dimensions following. In this way, INEPT leads to HETCOR, and then to the HSQC experiment. DEPT is used to introduce heteronuclear multiple-quantum coherence and the HMQC experiment. The phase-cycled versions of each experiment are described first, showing that they are basically difference experiments, with all signals detected in each scan, with the selection of the desired polarization transfer pathway arising from the co-addition of individual steps of the phase cycle. The benefits of pulsed field gradient coherence selection are then described, particularly, that each acquisition contains only the desired signals. The student should be familiar with phase-sensitive quadrature detection in 2D NMR methods, so the problems attendant with gradient coherence selection (P/N selection in the indirect dimension and loss of sensitivity) can be discussed. The solution to these problems provided by sensitivity enhancement and Rance–Kay acquisition schemes complete this chapter.

Chapter 9 focuses on three-dimensional NMR experiments in the context where they are most commonly used, for the characterization of biological macromolecules in aqueous solution. The common building blocks of such experiments, HSQC and HMQC polarization transfer steps, frequency labeling periods, spin echoes for chemical shift refocusing and J-evolution delays, spin-locking periods are described in the context of commonly used 3D experiments, including NOESY- and TOCSY-HSQC, HNCA, HN(CO)CA and HCCH-TOCSY. The constant-time modification commonly used to combine polarization transfer and frequency labeling steps in multidimensional NMR experiments is discussed in some detail. The problem of solvent suppression and the various methods used to obtain such suppression introduce the concept of selective pulses and how such pulses are generated. Off-resonance frequency generation by incremental phase-shifting of composite pulses is discussed at this point, and composite pulse schemes (e.g., WALTZ, MLEV, DIPSI) and their various applications for decoupling and spin-locking are described. Special problems that arise when dealing with very large molecules are discussed in terms of selective and non-selective deuteration of samples, direct detection of ^{13}C, and finally, the TROSY modification of heteronuclear correlation experiments.

We have found that by the time Chapter 9 is covered in the classroom, the semester is nearly over. For this reason, the last three chapters cover more specialized topics, solid-state NMR, relaxation and MRI, and can be taught independently of each other. All three chapters require the basics discussed in earlier chapters, but are not sequential. Depending upon the audience, an instructor might prefer to end with any one of the three. Solid-state NMR is at this point in time undergoing a remarkable renaissance, and we have made some effort to capture the highlights of developments in this area in Chapter 10, while providing a more complete description of chemical shielding, non-averaged dipolar coupling, and the basics of solid state NMR including cross-polarization, magic angle spinning and dilute spin correlations. Chapter 10 also deals with the measurement and analysis of residual dipolar couplings, an increasingly important tool for macromolecular structure determination. Chapter 11 provides a reprise of spin relaxation in the context of macromolecular dynamics, introducing spectral density mapping, the model-free analysis of spin relaxation and chemical exchange. Some of the experiments used to make such measurements are also described. The special but common case of hyperfine interactions and relaxation in paramagnetic molecules is discussed, and the origin of hyperfine shifts and observable perturbations that arise in NMR spectra due to paramagnetic centers are explained.

Chapter 12 makes the connection between NMR and magnetic resonance imaging (MRI). Despite their similar theoretical and experimental bases, much of the terminology of MRI differs from that of NMR, and a practitioner of one often does not have more than a cursory understanding of the other. We have tried to bridge this gap by introducing the basics of MRI, and demonstrating the connection between NMR and MRI concepts. We first consider diffusion measurement by NMR as a way of introducing gradient encoding of spatial relationships in a sample. Then, the use of phase and frequency encoding of spin position is described, as well as the combined use of gradients and RF pulses to obtain spatially selective excitation of spins. Reciprocal space, phase encoding and the connection between time-domain data and spatial relationships are discussed. Standard MRI pulse sequences, including one-shot methods for rapid MRI, are introduced. Finally, we consider contrast agents and methods for improving speed of MRI acquisition.

Two appendices are provided for students who wish to learn more about solutions to the time-dependent Schrödinger equation, time-dependent perturbation theory, and density matrix analysis of NMR spin systems. While neither appendix is essential to the successful completion of the course laid out in this text, they may help to fill in some of the blanks that necessarily occur when covering a large amount of material in a limited period of time.

We are indebted to many people who have contributed in one way or another to this book. Juliette LeComte (the Pennsylvania State University) introduced Tom to many of the basics of biomolecular NMR spectroscopy during visits to Gerd LaMar's lab in Davis, and she has contributed to this text both by her heroic reviewing efforts to correct the chapters as they were written and by the wisdom and support she has provided for the project. Tom also thanks Mark Rance, who answered his many questions and put up with his occasionally clumsy use of the spectrometers at Scripps during the time that they worked together. We thank Al Redfield, who sat patiently and politely through parts of the Brandeis course and waited until after class to correct errors. He has also graciously allowed us to use the many interesting and useful problems that he wrote for the course over the years. We also thank the following reviewers for their comments: Yael Balazs (Technion–Israel Institute of Technology), Cecil Dybowski (University of Delaware), Judith Herzfeld (Brandeis University), Gary Lorigan (Miami University of Ohio), C. James McKnight (Boston University Medical School), David Peyton (Portland State University), Alex Pines (UC Berkeley), Jeff Reimer (UC Berkeley), James Satterlee (Washington State University), Richard Shoemaker (UC Boulder), and David Wemmer (UC Berkeley).

Thomas Pochapsky
Susan Sondej Pochapsky
Arlington, Massachusetts

SYMBOLS AND FUNDAMENTAL CONSTANTS

Unless otherwise noted, values listed are from the *IUPAC Report on Quantities, Units and Symbols in Physical Chemistry*, K. H. Homann, 1993. SI units are used unless otherwise noted.

$\vec{B}$ – magnetic flux density (vector)

c – speed of light in a vacuum, 299 792 458 m s^{-1}

χ – bulk magnetic susceptibility

$\bar{\bar{\chi}}$ – magnetic susceptibility anisotropy tensor

δ – chemical shift relative to a standard (dimensionless)

e – base of natural logarithms, 2.718281828

E – energy (Joules, kg m^2 s^{-2})

$\hat{E}$ – Hamiltonian operator

γ – gyromagnetic (magnetogyric) ratio in rad s^{-1} T^{-1}

h – Planck's constant, 6.626 0755 $\times$ 10^{-34} J s

$\hbar$ – Planck's constant in radial units, $\hbar = h/2\pi$ = 1.054 571 68 $\times$ 10^{-34} J s

Hz – hertz, 1 Hz = 1 cycle s^{-1} = 2π radians s^{-1}

i – square root of -1

I – nuclear spin quantum number

k_B – Boltzmann's constant, 1.380 6505 $\times$ 10^{-23} J K^{-1}

λ – wavelength (m)

m – spin angular momentum quantum number

$\vec{M}$ – macroscopic magnetization

MHz – megahertz, 1 MHz = 1 $\times$ 10^6 Hz

N$_o$ – Avagadro's number, 6.022 1415 $\times$ 10^{23} mol^{-1}

$\vec{\mu}$ – magnetic dipole

μ_B – Bohr magneton, 9.274 0154(31) $\times$ 10^{-24} J T^{-1}

μ_N – nuclear magneton, 5.050 7866(17) $\times$ 10^{-27} J T^{-1}

μ_o – permeability (permissivity) of free space, 4π $\times$ 10^{-7} = 12.566 370 614... $\times$ 10^{-7} N A^{-2}

ν – frequency in Hz (cycles s^{-1})

ω – frequency in radial units, $\omega = 2\pi\nu$

Ω – chemical shift operator, usually in radial frequency units

R – molar gas constant, 8.314 472 J mol^{-1} K^{-1}

σ – shielding constant

$\bar{\bar{\sigma}}$ – shielding tensor

T – tesla, units of magnetic flux density, 1 T = 1 V s m^{-2} = 1 × 10^4 gauss

τ_c – correlation time (inverse of the decay constant for autocorrelation, units of s)

1

WHAT IS SPECTROSCOPY?

This may seem an odd question to pose for people who are trained in the physical sciences, and who use spectroscopic techniques all of the time. Still, answering the question "what is spectroscopy?" provides a starting point for an examination of *n*uclear *m*agnetic *r*esonance spectroscopy (**NMR**), and allows us to consider what it is that makes NMR so uniquely popular among scientists who deal with condensed matter of all types, from simple fluids to ceramics. It also provides an opportunity to review fundamental concepts that come up fairly often.

Spectroscopy is the branch of science that deals with the interaction between energy and matter. Usually, the energy is in the form of **electromagnetic radiation** (**EMR**), and the interaction between EMR and matter is interpreted in terms of the behavior of atoms and molecules. In the absence of any time-dependent changes in their environment (**perturbations**), atoms exist in well-defined nuclear and electronic states (**stationary states**) that have fixed (**quantized**) energies. When atoms are bonded to form molecules, these also exist in quantized electronic, vibrational and rotational states. In order to move from one quantum state to another (that is, to undergo a **transition** from one state to another), energy corresponding to the difference between the energies of the quantized states must be gained or lost (depending upon whether the system is going from a lower to a higher energy state or vice versa). Thus, if the system is going from a state i with energy E_i to a state j with energy E_j, the system changes in energy by an amount $\Delta E_{ij} = E_j - E_i$. In the types of spectroscopy that we will be discussing, the energy must be absorbed in one discrete packet, called a **photon**. Within some small uncertainty, only a photon of the correct energy ΔE_{ij} will cause the $i{\rightarrow}j$ transition to occur.

Although the photon has some particle-like characteristics, it is often convenient to think about the energy in a spectroscopic experiment (the EMR) as a wave, with the usual characteristics. These include **velocity** (speed and direction of the wave propagation), amplitude (displacement of the wave perpendicular to the velocity), **wavelength** (distance between crests, or points of maximum amplitude) and **frequency** (number of crests that pass a fixed point per unit time). However, we need to reconcile this view with the "packet" nature of the photon. The relationship between the energy E of a single photon of EMR of frequency v is $hv = E$, where h is Planck's constant. In turn, the frequency of EMR radiation v_{ij} which will be either **absorbed** or **emitted** as the quantized system goes from state i to state j is given by:

$$hv_{ij} = \Delta E_{ij} = E_j - E_i \qquad (1.1)$$

For our purposes, the photon should be considered indivisible (one cannot use only one-third of a photon, for example) and nonadditive. Two photons with half the energy required for a given transition cannot normally be used in place of a single photon with all of the required energy in order to **excite** that transition. In a spectroscopic experiment, the most efficient absorption of EMR occurs when EMR of frequency v_{ij} is used to excite the transition from state i to state j. This condition is called **resonance**.

What information does one get from a spectroscopic experiment? Assuming that absorption of EMR by the sample can be measured as a function of frequency, one gets direct information concerning the spacing of quantized energy levels in the molecules under observation. This information can then be interpreted in terms of models for molecular structure and dynamics. The better the model, the better the match between the results of the spectroscopic experiment and those predicted by the model.

A semiclassical description of spectroscopy

The most convenient and commonly used approach to discussing spectroscopy is the semiclassical description. This approach presumes that EMR can be described mathematically as a wave (a classical description), but the atoms and molecules with which the EMR interacts must be considered in terms of their quantized energy levels (for which there is no classical description). Mathematical descriptions of these quantized energy states are obtained by treating atoms and molecules as **quantum oscillators**, that is, oscillating systems of particles whose characteristic frequencies are quantized. Before going into any detail on quantum oscillators, we will first look at the EMR side of the story, beginning with a brief review of wave properties.

A simple mathematical description of a two-dimensional (**plane-polarized**) wave is given by:

$$A = A_0 \cos\left(\omega t + \phi\right) \qquad\qquad (1.2)$$

where A is the **amplitude**, or wave height, at time t, A_0 is the maximum amplitude, ω is the angular frequency of the wave (in radians/s) and ϕ is the **phase shift** (in radians). Plane-polarization simply means that the amplitude vector is restricted to a single axis, and all wave motion is restricted to the plane dictated by the amplitude axis and the time axis. 2π radians equals one cycle, or $360°$, so $\omega = 2\pi\nu$, where ν is the frequency in hertz (Hz), or cycles per second. The frequency in hertz is simply the number of wave crests (positions of maximum amplitude) that pass a fixed point per second. Equation 1.2 describes the amplitude observed at a fixed point in space for a wave with a frequency ω as a function of time (or equivalently, the amplitude as a function of displacement along the time axis). The phase, ϕ, determines the value of A at the origin ($t = 0$). When ϕ is 0, $A = A_0$ when $\omega t = 0$ (cos 0 = 1) and $A = 0$ when $\omega t = \pi/2$ (cos $\pi/2 = 0$). When $\phi = -\pi/2$ (a phase shift of $-\pi/2$), the opposite obtains: $A = 0$ when $\omega t = 0$ and $A = A_0$ when $\omega t = \pi/2$. These two cases are shown in Figures 1.1 and 1.2.

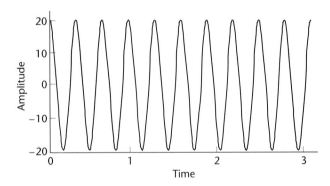

Figure 1.1 Plot of the amplitude of $A = A_0 \cos(\omega t + \phi)$ from $t = 0$ to $t = \pi$, with $\omega = 20$ rad/unit time (rad/time), $A_0 = 20$ and phase factor $\phi = 0$ rad.

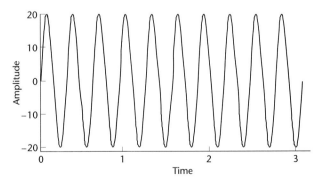

Figure 1.2 Plot of the amplitude of $A = A_0 \cos(\omega t + \phi)$ from $t = 0$ to $t = \pi$, with $\omega = 20$ rad/time, $A_0 = 20$ and phase factor $\phi = -\pi/2$ rad. This is equivalent to plotting a function $A = A_0 \sin(\omega t + \phi)$ with a phase factor of 0 rad.

EMR has a fixed speed $c = 2.998 \times 10^8$ m/s in a vacuum, so the wavelength λ of EMR is given by:

$$\frac{c}{v} = \lambda \qquad (1.3)$$

With these basic concepts in mind, one can start to think about the more complicated behavior exhibited by waves in the real world. Perhaps the most important phenomenon to be considered is that of **interference**. When waves are combined, the amplitude observed at a given point is the resultant of vector addition of the amplitudes expected at that point individually for the waves that are combined. The amplitude of the waves described by Equation 1.2 oscillates between A_0 and $-A_0$, and can be considered as a vector quantity perpendicular to the wave velocity vector, with origin at $A = 0$. If two waves both have amplitudes of the same sign at a given point, the amplitude vectors add and the resulting amplitude is greater than that for either wave individually. The two waves are said to **interfere constructively** with each other at that point. On the other hand, if one vector is negative and the other positive, the resulting vector will be the difference between the two original amplitudes, and the waves **interfere destructively** with each other. Note that the waves under consideration have their amplitude vectors pointing along the same axis, that is, they are plane-polarized in the same plane. If this were not the case, the problem would be more complicated.

Two waves with the same angular frequency ω and identical phase factors ϕ will interfere constructively at all points, and the resulting amplitude will be the sums of the amplitudes of the individual waves at all points. However, if two waves are combined that have the same frequency but have phase factors ϕ that differ by $\pi/2$ radians, the resulting wave has zero amplitude at all points. Note that a cosine wave with a phase factor ϕ of $-\pi/2$ radians is identical to a sine wave with a phase factor of zero:

$$A = A_0 \cos\left(\omega t - \pi/2\right) = A_0 \sin\left(\omega t\right) \qquad (1.4)$$

It is also important to understand what happens when waves of different frequencies interfere with each other. Consider two cosine waves (i.e. maximum amplitude at the origin) one with frequency ω_A and the other ω_B and both with maximum amplitude A_0. Assuming that there is no difference in phase factors ϕ and that both waves originate at the same point, the resulting amplitude will be at a maximum ($2A_0$) at the origin. However, frequency differences will result in the two waves usually not being at the same point in their cycle (referred to as **phase advance**) at any given point, and their amplitudes will exhibit either partially constructive or partially destructive interference at most points along t. The frequency of the resulting wave will be the average of the frequencies of the two components, $\omega_{int} = (\omega_A + \omega_B)/2$. Furthermore, the resulting wave will not again reach the maximum amplitude $2A_0$ until $t = 2\pi/(\omega_A - \omega_B)$. A graph of the resulting interference pattern shows an amplitude that oscillates at the average of the individual frequencies ω_{int}, and also shows a slower oscillation of the maximum amplitude envelope that occurs at a difference frequency $(\omega_A - \omega_B)$ (see Figure 1.3). This difference frequency is called the **beat frequency**. You have probably noticed beat oscillations in the amplitude of sound waves if you have ever heard two violins that are not quite in tune playing together (an excruciating sound!). As the sound waves produced by the violins go in and out of phase, the volume of the sound produced gets louder and softer at the frequency of the difference between the frequencies produced by the two individual instruments.

Damped harmonics

The wave equations shown in Figures 1.1, 1.2 and 1.3 are graphed over only a small region close to the origin, but their behavior would be identical in any region of the t axis where one cared to graph them. Using sound waves again as an example, such waves would be like a bell that, once rung, would continue to sound forever. Of course, such behavior is unrealistic. However hard one rings the bell, blows the horn or plucks the guitar string, eventually the sound will die out. A realistic mathematical description of a wave should account for this "dying out", which is known as **damped harmonic** oscillation.

The easiest way to describe damped harmonics mathematically is to multiply the wave equation by an exponential function of the independent variable t (assumed to be time for the present purpose) with a negative multiplier as shown here:

$$A = A_0 \left[\cos\left(\omega t + \phi\right) \right] \exp\left(-kt\right) \quad (1.5)$$

In the absence of the $\cos(\omega t + \phi)$ term, this would simply be an exponential decay function with a characteristic **decay constant** k, which has units of inverse time. In spectroscopy, it is common to talk about the inverse of the decay constant $k^{-1} = T$. T has units of time and is called the **relaxation time** for the decay process. The relaxation time is the time required for a function undergoing exponential decay to go from some starting value A_0 to a value of A_0/e, where e is the base of the natural logarithms. Figure 1.4 gives an example of this sort of damped harmonics.

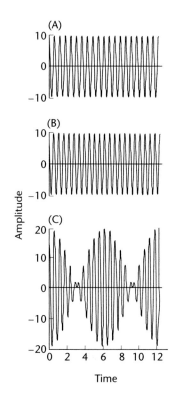

Figure 1.3 (A) Plot of the function $A = A_0 \cos(\omega t + \phi)$, with $\omega = 10$ rad/time, $A_0 = 10$ and phase factor $\phi = 0$. (B) Plot of the same function with $\omega = 11$ rad/time, $\phi = 0$. (C) Interference pattern resulting from superposition of the two functions in parts (A) and (B). Note that beat maxima occur at $t = 0$, 2π, 4π …

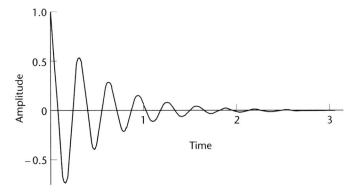

Figure 1.4 A plot of the damped harmonic function $A = A_0 \cos(\omega t + \phi)e^{-kt}$ as a function of time, with $\omega = 20$ rad/unit time, $A_0 = 1$, phase factor $\phi = 0$ and the decay time constant $k = 0.5$ inverse unit time.

Later on in this text, exponential decay will often be used to describe damped harmonics. But for the moment let us take a different tack. We have already seen that when two waves of slightly different frequencies interfere with each other, the resultant amplitude will be modulated by the difference in the individual frequencies. Now, one can ask what happens when multiple waves that are close but not exactly the same in frequency interfere with each other. Again, we will assume that the waves all start at the same origin with the same amplitude (i.e. they have the same phase factors). At first they exhibit almost completely constructive interference, but as time goes on they get more out of phase and begin to interfere destructively, until finally they cancel each other out almost completely. Figure 1.5 shows this graphically.

It is important to note that, if the function plotted in Figure 1.5 were to be graphed for longer periods of time, the individual frequency contributions would eventually come back into phase with each other and the amplitude would build up again. This is not true of the exponential decay function, which does not become nonzero at longer times. The intention of this exercise is to demonstrate that damping of harmonic oscillations can be mimicked over short time intervals solely by the use of a series of cosine oscillations with frequencies near to the frequency that describes the undamped oscillation.

The converse of the observation that multiple frequency components added together give rise to a damped oscillation is that *a damped harmonic oscillation cannot be completely described by a single frequency*. It *must* contain multiple frequency components that increasingly interfere with each other destructively with increasing time. This is a very important consideration in NMR.

Quantum oscillators

Now let us return to the matter side of the spectroscopy experiment. It has already been said that the atoms and molecules with which EMR interacts are best considered as oscillating systems of particles whose characteristic frequencies are

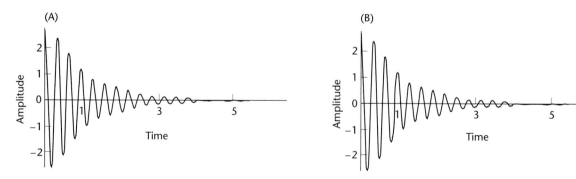

Figure 1.5 (A) A plot of $A = \sum A_{i0}\cos(\omega_i t)$, plotted for $\omega_i = 16$ to $\omega_i = 24$ rad/unit time with frequency increments of $\Delta\omega_i = 0.05$. The amplitude at each frequency ω_i is calculated by $A_{i0} = 1/(1 + (20 - \omega_i)^2)$, that gives a maximum at $\omega_i = 20$ rad/unit time. (B) A plot of the same function except that, instead of discrete summation, the function was integrated over the range $\omega_i = 16$ to $\omega_i = 24$.

quantized (quantum oscillators), and in order to go from one characteristic frequency to another, the appropriate amount of energy must be gained or lost. Now it is time to put a little more detail into this description.

The characteristic frequency of a quantum oscillator can be changed from v_i to v_j, only through the absorption or release of the appropriate amount of energy. Since each characteristic frequency can be associated with a characteristic energy by the relationship $E = hv$, the amount of energy involved in the transition ΔE_{ij} is the difference between the energies associated with the characteristic frequencies:

$$hv_{ij} = \Delta E_{ij} = E_j - E_i = h\left(v_j - v_i\right) \tag{1.6}$$

Note that the frequency v_{ij} is the beat frequency generated by interference between the frequencies characteristic of the two states i and j. Now we have two (related) descriptors for a given state i of a quantum oscillator, the energy E_i and the frequency of oscillation, v_i. Precisely what oscillates with a frequency v_i depends on the oscillator; it could be an electron moving in an orbital, a bond that vibrates, a molecular rotation or a nuclear precession. However, one can make a useful analogy to a quantum oscillator using a guitar string. When the guitar string is plucked, it vibrates back and forth, producing a **standing wave** with **nodes** (points at which the amplitude of the wave is always 0) at either end, with the maximum amplitude of the wave fixed at the center of the string. This vibration produces a sound with a well-defined frequency called the **fundamental**. The fundamental frequency is determined by the length of the string and the speed of sound according to the relationship $s = \lambda v$, where s is the speed of sound in air and λ is the wavelength. The wavelength corresponds to $2L$, where L is the length of the string. Every guitar player knows that if you put your finger lightly on the middle of the string before plucking it, you produce a higher-frequency note (called a **harmonic**), with a wavelength of L. The finger has created a new node at the center of the string, so the oscillation takes place as shown in Figure 1.6. Higher-order harmonics, with wavelengths $\lambda = 2L/n$, where n is an integer ($\lambda = 2L/3, L/2 \ldots$) are produced when the finger is placed at one-third of the length of the string, one-quarter of the length, etc.

The reason for the formation of a standing wave is simple; as the wave front travels along the string, reflecting back and forth from either end, any wave which has a wavelength that is not an integral fraction of $2L$ will interfere destructively with itself and quickly damp out.

The guitar string is not a quantum oscillator, and it can be described completely using classical physics. However, it does have many of the qualities of a quantum oscillator: the motion of the guitar string is described by oscillatory (wave) behavior, and the fundamental and harmonic frequencies are a function of the characteristics of the string and the medium in which the vibration occurs (air). Also, like the guitar string, it is useful to think of the stationary states of quantum oscillators in terms of standing waves, with characteristic frequencies determined by the types of particles involved and the boundary conditions of the system. Furthermore, it is possible to describe these stationary states mathematically as a wave function.

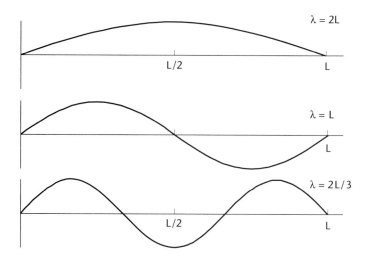

Figure 1.6 Standing waves formed along a guitar string of length L. The fundamental wave has a wavelength (λ) of 2L, the first harmonic, a wavelength of L, and the second harmonic a wavelength of 2L/3.

Fortunately, we do not need to know the exact mathematical form of such quantum wave functions in order to discuss many of their properties. The most important characteristic of the wave functions describing the stationary states of a system is that when treated with the appropriate mathematical operations, they will yield values for what are called **observables** of that state of the system. One of the most important observables obtainable in this fashion is the energy of a given state. For example, if a wave function ψ_i describing a state i of the system is treated with an appropriately constructed mathematical operator called a Hamiltonian operator $\hat{H}$, the result of this operation is the wave function ψ_i itself multiplied by a constant E_i that is the energy associated with the state i. This energy can in turn be related to the frequency of the harmonic oscillator using $h\nu_i = E_i$. This operation is described by the time-independent Schroedinger equation:

$$E_i\psi_i = \hat{H}\psi_i \tag{1.7}$$

Note that the Hamiltonian operator $\hat{E}$ is characteristic of the system (*not* just state i). As is the case for the wave function ψ_i, the exact form of the Hamiltonian operator is unimportant at the moment. It is constructed based upon consideration of the forces that act on particles in the system, and it will contain terms representing both the kinetic and potential energy of the system.

If Equation 1.7 is true (that is, operation of an appropriately constructed Hamiltonian operator on ψ_i yields ψ_i multiplied by a constant E_i), then ψ_i is an **eigenfunction** of the Hamiltonian operator with an associated **eigenvalue** E_i. Other observables can be obtained by the action of an appropriate operator on ψ_i if ψ_i is an eigenfunction of that operator. For example, ψ_i may be chosen so that it is simultaneously an eigenfunction of the energy (Hamiltonian) operator and of the angular momentum operator.

The spectroscopic experiment

Now we can begin to relate some of this to the spectroscopic experiment. Imagine a quantum oscillator that has two stationary states (two frequencies at which it may oscillate). We will describe the state that is lower in energy with the wave function α, and the higher energy state with the wave function β. To move the system from the α to the β state, energy must be absorbed. This can happen when the oscillator interacts with EMR of the appropriate frequency $\nu_{\alpha\beta}$, as determined by $h\nu_{\alpha\beta} = E_\beta - E_\alpha$ (see Figure 1.7). As already pointed out, **absorption** is most efficient when the frequency of the EMR ν is equal to $(E_\beta - E_\alpha)/h$. In order for the system to go from the higher energy β state to the lower energy α state, the same amount of energy must be lost. If the loss of energy is radiative, that is, involving the production of a photon of the appropriate energy, the process is called **emission**. However, this return to a lower energy level may also occur by nonradiative processes, which will be discussed in detail later. The combination of radiative and nonradiative processes that results in a return to the lower energy (**ground**) state is called **relaxation**.

If more than one distinguishable type of quantum oscillator is present in a molecule, the response of one oscillator to applied EMR may depend on the state of another oscillator in the same molecule, a phenomenon known as **coupling**. In the weak coupling case (with which we will be mostly concerned), the transition frequency for one oscillator changes slightly when a coupled oscillator occupies a different energy level. This is illustrated in Figure 1.8. Coupling provides a path over which energy (and information) may transmitted between oscillators.

Ensembles and coherence

Usually, the spectroscopist is not examining one quantum oscillator at a time, but a very large number of similar oscillators simultaneously. Such a collection of

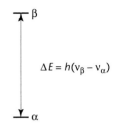

Figure 1.7 A spectroscopic transition between two quantum oscillator states with characteristic frequencies ν_α and ν_β occurs when EMR is applied at the frequency $(\nu_\beta - \nu_\alpha)$ corresponding to the energy difference between the states.

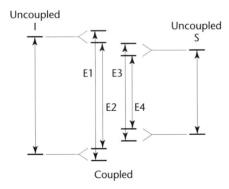

Figure 1.8 Uncoupled transitions for two oscillators, I and S (outside of figure), compared with coupled transitions of the same oscillators (inside of figure). The energy required for a given transition of I depends upon the state of S and vice versa. For example, $E1$ might be the energy required for an I transition when S is in the upper state, and $E2$ would be the same transition when S is in the lower state, and vice versa. Two distinct transitions would be observed for each oscillator.

oscillators of the same type is called an **ensemble**. Now even though an ensemble of quantum oscillators of the same type may all be oscillating at the same frequency, they may nonetheless have different phases and so their wave functions may interfere destructively with each other. If nothing is done to get the quantum oscillators into phase with each other, the distribution of phases at equilibrium will be completely random, forming what is called an **incoherent ensemble**. On the other hand, it is often possible to generate a nonrandom distribution of phases in an ensemble, so that there is constructive interference between individual oscillators. Ensembles in which the oscillators have not only the same frequency but also the same phase are called **coherent ensembles**.

The concept of coherence is extremely important in the theory of NMR (and many other forms of modern spectroscopy, particularly those involving lasers). A simple but useful analogy is that of marching soldiers. Soldiers in dress parade will march precisely in cadence (all at the same frequency, or steps per second), and legs will all come up and go down at the same time (all in phase). The soldiers form a coherent ensemble, driven by the drums and drill sergeant. As soon as the drums stop, however, the soldiers will begin to "dephase"; the beat will be lost, and the front of the parade will get out of step with the back. Eventually, no one will be in phase with anybody else, except accidentally, and now the soldiers form an incoherent ensemble (Figure 1.9).

The properties of a coherent ensemble are very different from those of an incoherent one. These are often quite dramatic: the difference between laser light (coherent radiation) and light emitted by a gas discharge tube is obvious, even though both sources produce radiation of uniform wavelength. Even the coherent ensemble of marching soldiers will have larger effects than the incoherent group. An army never marches across a bridge in cadence, because the effect of so many feet moving in unison can create vibrations of an amplitude sufficient to destroy the bridge.

Types of spectroscopy

During a spectroscopic experiment, a quantum oscillator is excited. The oscillator interacting with the EMR is called a **chromophore** and its nature depends on the energy carried by the EMR. The larger the forces holding the oscillator together, the greater the energy needed to excite that oscillator. Table 1.1 provides a list of

Figure 1.9 Marching soldiers as an example of a coherent ensemble.

Soldiers marching in step:
a coherent ensemble

Soldiers out of step:
an incoherent ensemble

Table 1.1 EMR used in spectroscopy, with approximate ranges of EMR frequencies and energies, types of transitions excited, and absorbing chromophores.

EMR	Approximate frequency (Hz) and energy (kJ/mol of photons)	Chromophore	Excited states
γ-Rays	$\geq 10^{20}$ Hz $\geq 10^{7}$ kJ/mol	Atomic nuclei	Nuclear vibrations
X-Rays	10^{18}–10^{20} Hz 10^{5}–10^{7} kJ/mol	Inner-shell electrons	Promotion to valence orbitals or ionization
Ultraviolet/ visible	10^{15}–10^{18} Hz 10^{2}–10^{5} kJ/mol	Valence shell electrons	Promotion to empty orbitals
Infrared	10^{12}–10^{15} Hz 10^{1}–10^{2} kJ/mol	Chemical bonds	Bond vibrations
Microwave	10^{10}–10^{12} Hz 10^{-2}–10^{1} kJ/mol	Molecules and functional groups, electron spins (ESR)	Molecular rotations, electronic precessions
Radio	10^{7}–10^{10} Hz 10^{-5}–10^{-2} kJ/mol	Nuclear spins (NMR), nuclear quadrupolar resonance (NQR)	Nuclear precessions

some of the more common types of spectroscopy along with their target chromophores, the approximate ranges of exciting EMR energies and frequencies, and the nature of the excited states produced. The highest energy EMR (γ-rays) excites transitions within the nucleus of the atom itself, which is held together by very strong short-range nuclear forces. The emission of γ-rays results from the relaxation of nuclei produced in excited states after nuclear reactions. X-ray spectroscopy excites the tightly held inner-shell electrons from low-lying inner shell orbitals to empty orbitals in higher-lying shells. The strong attractive electrostatic forces between the nucleus and low-lying electrons, modulated by electron–electron repulsion, provide the restoring force for this oscillating system, and require high energies for promotion of electrons to outer electronic shells. (It is important to distinguish X-ray absorption spectroscopy from X-ray diffraction, in which it is the scattering of X-rays by an ordered lattice and the resulting interference patterns that are of interest.) Promotion of valence shell and bonding electrons into unoccupied orbitals requires less energy than that of inner shell electrons since valence electrons are shielded from the nucleus by the inner shells, attenuating the electrostatic attraction. Hence, these transitions are usually excited by ultraviolet radiation, or, where the highest occupied and lowest unoccupied orbitals are closer in energy, visible light.

As one progresses to lower energy EMR (infrared and microwave), electronic transitions are no longer involved. Instead, changes in the relative oscillatory motions of nuclei (bond vibrations and rotations around bonds) within a molecule are the result of infrared or microwave excitation. The higher energy infrared causes distortions in molecular shape, whereas microwave transitions involve rotations around single bonds or rotations of the molecule as a whole. The restoring forces for these oscillations are still electrostatic in nature, but repulsive and attractive forces are more evenly balanced, and the energies involved in the quantum transitions are lower.

The lowest energy transitions that are commonly observed spectroscopically are electronic and nuclear spin transitions. These are at the root of *electron spin resonance* (ESR), NMR and *nuclear quadrupole resonance* (NQR) spectroscopy. Electrons and selected nuclei have a property known as spin, which for the present purposes can be thought of simply as a tendency to **precess** in a magnetic field, that is, they spin at a fixed frequency and angle with respect to the magnetic field. In the absence of an external magnetic field (or an electric field gradient, for NQR), the spin states of electrons and nuclei are usually **degenerate**, that is, equal in energy. It is only when an appropriate external field is applied that the states become non-degenerate and EMR can be absorbed or emitted as the spins "flip", or undergo transition from one spin state to another.

Practical considerations in spectroscopy

Despite their differences, all forms of spectroscopy have a few basic features and considerations in common. All **spectrometers** (instruments used to measure sample response as a function of EMR frequency) must have the same basic components. These are the EMR source, the EMR detector and the signal analyzer that provides the output spectrum. The sample cell is placed between the EMR source and the EMR detector, and the response of the sample to EMR is measured. The response is often recorded as the ratio of signal intensity in the presence of sample (I) with respect to that in the absence of sample (I_0) (Figure 1.10).

The response of the sample is analyzed as a function of input frequency (energy), and the result recorded as sample response versus frequency, or **spectrum**. There is a bewildering variety of ways in which these basic tasks are accomplished, and they vary according to the type of spectroscopy, but all of them address the same two basic concerns. These are **sensitivity** (how much sample do I need and how long will the experiment take?) and **resolution** (how close together in energy (or frequency) can two transitions be, and still be differentiated?). It turns out that in many cases the optimization of an experiment for sensitivity results in suboptimal resolution and vice versa, for reasons that we will discuss shortly. It is important that the spectroscopist remains aware of this trade-off (see Figure 1.12 below).

Sensitivity is usually expressed as the **signal-to-noise ratio** (**S/N**). Noise is the fluctuation of detector output around a zero-point (**baseline**) signal in the absence of

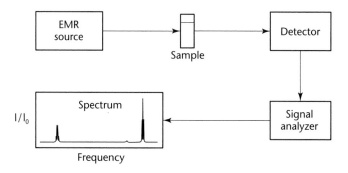

Figure 1.10 Basic components of a spectrometer.

sample absorption, and can come from a variety of sources. **Random noise** can come from thermal fluctuations in the environment and electronic noise in detector circuits. Nonrandom environmental influences, such as nearby equipment that produces EMR, can also be sources of noise. S/N usually refers to the ratio of the amplitude of a signal from a standard sample to the amplitude of the root-mean-square (**rms**) noise. The rms noise can be calculated by measuring the difference between the amplitudes of the highest positive and deepest negative noise fluctuations around the baseline (peak-to-peak noise) (see Figure 1.11).

Resolution is determined by how close in energy two transitions can be to each other and still be distinguished. As such, the **resolution limit** is usually expressed as a difference in energy or frequency units. Transitions that are closer to each other

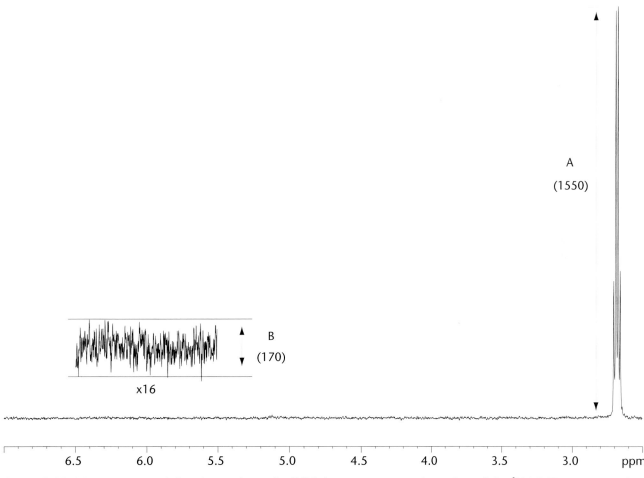

Figure 1.11 Measurement of signal-to-noise ratio (S/N) in spectroscopy. A portion of the ^{1}H NMR spectrum of ethyl benzene, a commonly used standard for S/N tests, is shown. Signal amplitude is taken as the height of the measured signal from baseline (A). Noise amplitude is measured peak-to-peak in an empty region of the spectrum (B). The noise region is shown with magnified (×16) vertical scale in the inset. The S/N = 16(2.5A)/B = 365, where the factor of 2.5 is used to convert peak-to-peak noise to rms noise, and the factor of 16 accounts for the scale magnification.

in energy than the resolution limit will be indistinguishable. In NMR, resolution is almost always expressed in hertz (Hz). Thus, in an experiment that provides 2 Hz resolution, two transitions (signals) must be at least 2 Hz apart in order to be distinguishable (Figure 1.12).

Acquiring a spectrum

The goal of the spectroscopist is almost always to acquire a spectrum of sample response versus frequency of EMR input. This implies that the EMR can be sorted by frequency in some fashion, either before or after passing through the sample. If the sorting is done prior to passing the EMR through the sample, it is necessary to scan through all the frequencies in the EMR region of interest in order to obtain a complete response spectrum. This type of experiment is called frequency sweep, or **sequential excitation**. In a sequential excitation experiment, the EMR spectrum is scanned by varying frequency with time, and detecting the response of the sample as a function of frequency. Sequential excitation requires that a **monochromator**, a device capable of separating EMR by frequency, be used to pass one frequency at a time through the sample cell. In fact, no monochromator can select only one

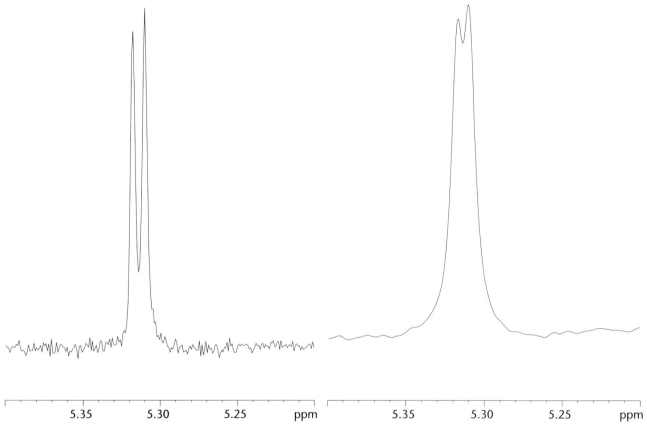

Figure 1.12 The trade-off of resolution for sensitivity. The figure on the left shows a doublet in an NMR spectrum "as acquired". The relaxation rate of the same signal was then artificially increased using an exponential decay function to generate the spectrum on the right. Note the increase in S/N at the expense of decreased resolution.

frequency, for both practical and theoretical reasons. A range of frequencies will always be present in the monochromator output, and the resolution available in the experiment is at least in part determined by how wide a range of frequencies (the **band width**) the monochromator allows to pass.

In the sequential excitation experiment, once the EMR is (nearly) monochromatic, response of the sample to the EMR is measured at the detector. Many spectrometers determine absorption by splitting the monochromatic EMR into two components. One component passes through a "blank", a cell similar to the sample cell except that it contains none of the sample of interest. The other component is passed through the sample cell. The intensity of the EMR after passing through the sample is compared at the detector with that passing through the blank, and the difference between them is recorded as a function of frequency, providing the spectrum. (Note that in solution NMR, this is not how background is removed: generally solvents are chosen so as to present little or no background, or else the signals from the solvent are removed using **solvent suppression** techniques.)

The resolution and sensitivity available in any spectroscopic experiment will depend upon experimental factors (the capabilities of the spectrometer), and theoretical limitations imposed by the type of spectroscopy and the type of sample. The theoretical limitations, especially as they pertain to NMR, will be discussed in detail later. The limitations imposed by the spectrometer in a sequential excitation experiment depend in large part upon the quality of the monochromator. It is generally true that for a given EMR source, as one attempts to improve resolution of the experiment by narrowing the bandwidth output from the monochromator (more monochromatic EMR), there is a loss in S/N because of lower EMR power output (photons/s) from the monochromator.

There are several drawbacks to sequential excitation methods. The first is the problem of making the EMR monochromatic. The particular type of monochromator used depends (obviously) on the type of spectroscopy being performed. Often, though, the monochromator contains delicate mechanical components, and as such may be sensitive to a variety of environmental factors to which purely electronic devices are immune. Furthermore, there is the inevitable loss of S/N in the experiment due to power losses at the monochromator, although this can be compensated for by using a more powerful EMR source. Perhaps the worst drawback of sequential excitation is the inefficient use of experiment time—much is wasted when sweeping regions of the spectrum in which the sample has no response. Simply increasing the speed at which the spectrum is scanned (increasing the **sweep rate**) can result in a distortion of the response spectrum, with a reduction in available resolution for the experiment. Clearly, it would be more time-efficient to use broad band polychromatic EMR to excite a response of all the chromophores in the sample simultaneously (**simultaneous excitation**), and then somehow sort the output signal by frequency, a process known as **deconvolution**. Until the advent of digital computers and the availability of appropriate algorithms for signal deconvolution, simultaneous excitation methods were impractical for most applications. Nowadays many different types of spectroscopy take advantage of simultaneous excitation methods in one form or another.

Another important advantage of simultaneous excitation is that it lends itself readily to **signal averaging**. In signal averaging, the same experiment is performed (acquired) many times, and the signal from each successive acquisition is added to those from previous acquisitions, and then stored in a computer memory. In the resulting spectrum, signal amplitude increases linearly with n, the number of acquisitions added together, while the noise increases as $\sqrt{n}$. Thus, the improvement of S/N over a single experiment from n averaged experiments is proportional to $\sqrt{n}$: a factor of two for four experiments, a factor of 10 for 100 experiments, and a factor of 100 for 10 000 averaged experiments (see Figure 1.13). Obviously, the same sort of thing can be (and is) applied to sequential excitation experiments, but the big advantage of simultaneous excitation for signal averaging is that the deconvolution need only be done on the finished average output signal, and not each time the experiment is performed. Thus, in the same time that is required for one sequential

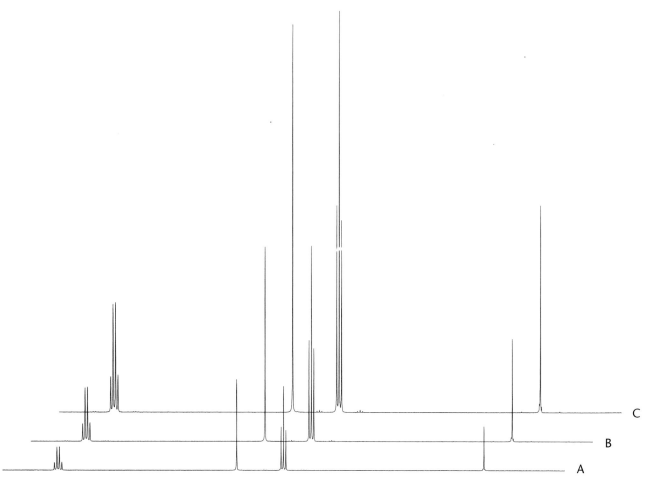

Figure 1.13 Improvement of S/N as a result of signal averaging in the ^{1}H NMR spectrum of ethyl benzene. The front spectrum (A) was obtained with a single experiment. The middle spectrum (B) was obtained by averaging four experiments, and is approximately twice the S/N of spectrum (A). Spectrum (C) was obtained as an average of 16 experiments, and is about four times the S/N of (A).

excitation experiment, many simultaneous acquisitions can be performed, with a concomitant improvement in S/N.

Resolution: the problem of line width

In addition to the limitations on spectral resolution imposed by the experimental setup, there are theoretical limitations to the extent to which two signals that are close in energy can be differentiated. Many of these limitations are imposed by the nature of the quantum oscillators themselves. One of the basic tenets of classical mechanics was that, given knowledge of the starting state of a system, it would be possible to determine the state of that system at a later time to whatever degree of accuracy is desired, if the forces acting on the system are known precisely enough. The unstated assumption was that the act of observation did not itself change the system, and that the observer could sit like a member of the audience at a trapeze act, not affecting the outcome in any way. However, imagine if our method of observation required us to throw rocks at the trapeze artist. How long would the artist continue to proceed along the expected path? Not long, one suspects, even for a Flying Wallenda. This is not a bad analogy to what spectroscopists do. The energies involved in many spectroscopic transitions are sufficient to result in physical changes in the observed system, including changes in molecular geometry, even breaking of chemical bonds, so it is not possible to think of such experiments as nonperturbing. It turns out that the more precisely we know the state of the system at the beginning of an observation (i.e. the position of particles in the oscillator or some related physical parameter), the less we can know about it at some later time, that is, the less precisely we know the momentum of the particles in the oscillator. This concept is mathematically expressed as the **Heisenberg uncertainty principle** (Equation 1.8).

$$\Delta z \Delta p \geq \hbar \qquad (1.8)$$

Here, Δz is the uncertainty in the measurement of position of a particle, and Δp the uncertainty in the measurement of momentum of the particle. The constant $\hbar$ is Planck's constant divided by 2π. Since $\hbar$ is a very small number, Equation 1.8 is really only important for quantum systems; the uncertainty introduced into macroscopic systems by the uncertainty principle is undetectable.

For spectroscopists, a more convenient and useful statement of the uncertainty principle is formulated in terms of uncertainty of the energy of a given state:

$$\Delta E_i \Delta t_i = \hbar \qquad (1.9)$$

where Δt_i is the **lifetime** of the ith excited state generated in the quantum oscillator by the absorption of EMR, and ΔE_i is the uncertainty in the energy of that state (note that the units of $\Delta z \Delta p$ and $\Delta E \Delta t$ are the same). One can consider the state lifetime as the time required for an ensemble of N oscillators in the ith excited state to relax sufficiently so that the number of oscillators remaining in the ith state has depleted to N/e, where e is the base of the natural logarithms. If the lifetime of the lowest energy state (the **ground state**) is very long, the energy of the ground state E_{gs} can be quite precisely defined, ΔE_i and will also be approximately equal to the uncertainty of the difference in energy between the ground state and the ith excited state, that is:

$$\Delta\left(\Delta E_{gs-i}\right) = \Delta E_{gs} + \Delta E_i \cong \Delta E_i \qquad (1.10)$$

The uncertainty in the transition energy implies that the frequency of EMR required to induce the transition will also carry an uncertainty, approximately equal to $\Delta E_i/h$. The effect of this relationship between the uncertainty in transition energy and the lifetime of the excited state on a spectroscopic experiment is simple. The shorter the lifetime of the excited state, the more imprecise the measurement of the transition energy, and the broader the range of EMR frequencies that will be absorbed (and the poorer the resolution of the experiment). This results in a phenomenon known as **uncertainty broadening** of spectral absorption lines.

Consider how uncertainty broadening affects the spectroscopic line corresponding to a transition between a ground state α and an excited state β. The uncertainty in the energy of the transition, $\Delta(\Delta E_{\alpha\beta})$, is given by the sum of the uncertainties in the energies of the individual states:

$$\Delta\left(\Delta E_{\alpha\beta}\right) = \Delta E_\alpha + \Delta E_\beta \qquad (1.11)$$

In turn, the uncertainty in the transition energy can be related to the uncertainty of the frequency of EMR exciting the transition by remembering that $\Delta E_{\alpha\beta} = h\nu_{\alpha\beta} = \hbar\omega_{\alpha\beta}$. Therefore:

$$\Delta\left(\Delta E_{\alpha\beta}\right) = \hbar\Delta\omega_{\alpha\beta} \qquad (1.12)$$

If we assume a long lifetime for the ground state α, the uncertainty in the ground state energy is small, and Equation 1.11 holds, and

$$\Delta\left(\Delta E_{\alpha\beta}\right) \cong \Delta E_\beta \qquad (1.13)$$

Substitution of the result of Equation 1.13 into Equation 1.12 gives the relationship $\Delta E_\beta = \hbar\Delta\omega_{\alpha\beta}$. Substituting this result into the uncertainty expression Equation 1.9 and rearranging gives:

$$\Delta\omega_{\alpha\beta} = 1 / \Delta t_\beta \qquad (1.14)$$

Equation 1.14 gives the relationship between Δt_β, the lifetime of the excited state β, and the spectral width of the absorption band in angular frequency units.

The trade-off that spectroscopists often make between S/N and resolution is clearly manifested by uncertainty broadening. On the one hand, if the excited state generated by absorption of EMR is short-lived, the system will return rapidly to the ground state and the experiment can be repeated quickly, improving the S/N of the experiment (more measurements per unit time). On the other hand, short excited-state lifetimes give rise to broader lines, thereby reducing the resolution attainable. Generally, excited state lifetimes in NMR tend to be much longer than in most other common forms of spectroscopy, and this is both bane and boon: long relaxation times limit the sensitivity of NMR, but also provide higher spectral resolution than most other spectroscopic methods. We will also see that the long excited-state lifetimes in NMR permit one to do "tricks" with excited states that give rise to many complicated and useful NMR experiments. It should also be noted that the situation is actually a little more complicated for NMR than is implied here, because there are other sources of line broadening besides uncertainty broadening, and so NMR line widths do not always reflect the rates of return to the ground state.

For convenience and relevance to later discussions, we now replace the excited state lifetime Δt_β with τ, the **relaxation time**, defined as the time required for a signal resulting from the excitation of a spectroscopic transition to decay from an amplitude A_0 to an amplitude of A_0/e. For a resonance with maximum signal amplitude A_{max} at frequency $v_{\alpha\beta}$, the **line width** of that signal at half-height $A_{max}/2$ is related to relaxation time by $\Delta v_{1/2} = (\pi\tau)^{-1}$. Since the resonance signal may decay for reasons other than return to the ground state, any experimental factor that causes time-dependent loss of signal contributes to the measured value of τ. If the decay of a signal with time is treated as a first-order exponential decay, the equations governing that decay are given by the following series of equations:

$$A = A_0 e^{-kt}$$
$$A_0 / e = A_0 e^{-k\tau}$$
$$k^{-1} = \tau \qquad\qquad (1.15)$$

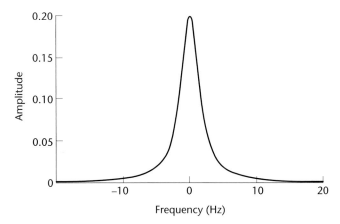

Figure 1.14 A Lorenztian line-shape function as shown in Equation 1.16, plotted over a 40 Hz range with a center frequency $v_{\alpha\beta} = 0$ and relaxation time $\tau = 0.1$ s.

where k is the relaxation rate constant. In the simple case of first-order decay, τ is the inverse of the rate constant for the decay.

Line shape

One last consideration in this general overview of spectroscopy is to introduce the concept of spectral line shape. Because of uncertainty broadening and other factors that contribute to relaxation, chromophores absorb EMR over a range of frequencies centered on their transition frequency, and we will spend a good deal of effort considering the factors that contribute to the shapes of the absorption versus frequency curves called spectra. Even prior to quantum theory, Lorentz succeeded in mathematically describing the characteristic shapes of resonance lines in electronic spectra as a function of frequency, and line shapes in NMR can often be described as Lorentzian. The equation for a normalized Lorentzian line for a transition from state α to state β as a function of frequency (v) is given by:

$$g(v) = \frac{2\tau}{1 + 4\pi^2\tau^2\left(v_{\alpha\beta} - v\right)^2} \tag{1.16}$$

where v_α is the frequency of the transition being observed, and τ is the relaxation time of the transition. Note that as the relaxation time τ decreases, the term $4\pi^2\tau^2(v_{\alpha\beta} - v)^2$ becomes increasingly less important at a given frequency, whereas the overall amplitude at any given point as a function of frequency is reduced. Decreasing the relaxation time therefore results in shorter, broader resonances with the same integrated intensity over all frequencies (for a normalized curve, integration of the area under the curve over all frequencies equals 1, see Figure 1.14). The practical implication of this relationship is that anything that enhances relaxation (i.e. leads to shorter relaxation times) broadens lines. For example, solution ultraviolet/visible spectra typically exhibit very broad lines, whereas in the gas phase a spectrum of the same species may give much sharper lines. This difference is due to more efficient collisional relaxation of excited states in solution versus the gas phase. In NMR, where relaxation times are often measured in seconds or tens of seconds, comparatively narrow line shapes are often observed. However, any factor that shortens relaxation times, such as the addition of a paramagnetic relaxation agent, will broaden lines and reduce resolution.

Problems

*1.1   The concept of standing waves is very useful for both the theory and practice of spectroscopy. In NMR, radio-frequency (RF) signals of a specific frequency range must be generated and detected. In order to efficiently detect radio frequencies, a detector circuit called a resonance circuit or LCR circuit (for inductor (L) capacitor (C) resistor (R) circuit) is used. The LCR circuit consists of a pickup coil (which acts as the inductor), a resistor and a capacitor:

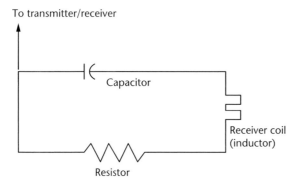

In a pure capacitor circuit, as the current amplitude changes, the voltage also changes. However, the capacitor resists changes in voltage, so that the rate of voltage change lags behind the rate of current change by $\pi/2$ radians. In a pure inductor circuit, the induction coil resists changes in current, so that the current change lags behind the change in voltage by $\pi/2$ radians. If an alternating current of frequency ω in radians (rad), where 1 cycle s^{-1} (1 Hz = 2π radians s^{-1}), is induced in a circuit that contains a resistor and both a capacitor and an induction coil, the *phase angle ϕ* between the current and the voltage is determined by the ratio of the difference between the *inductive* and *capacitive reactance $X_L - X_C$* of the circuit to the resistance R of the circuit:

$$\tan \phi = \left(X_L - X_C \right) / R$$

The *capacitive reactance* is given by $X_C = 1/\omega C$, where C is the capacitance in farads, and the inductive reactance is given by $X_L = \omega L$, where L is the inductance in henries.

(a) If the amplitude of the alternating current at a given frequency at a given point is a standing wave, the tendency of the inductor to lead the current change by the voltage change must balance the tendency of the capacitor to lead the voltage change by the current change, that is, the phase angle ϕ is zero. Given an inductance of 5 nanohenries in the detector coil of an NMR probe, what capacitor is needed in order to detect ^{1}H at 500 MHz? How about ^{13}C at 125 MHz? (Note that 1 MHz = 10^6 Hz).

(b) When connecting a detector circuit to the preamplifier (the first-stage amplifier of the NMR spectrometer, which boosts the very weak signal from the detector), it is common to use what is called a "quarter-wave cable", a coaxial cable that is approximately one-quarter of the wavelength of the EMR frequencies being detected. Suggest a reason for this.

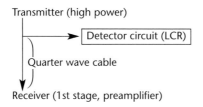

1.2   The simplest quantum mechanical example of a standing wave is the (in)famous particle in a one dimensional box. This example, which has some applications to ultraviolet/visible spectroscopy involving conjugated π-electron systems, consists of a particle of mass m in a one-dimensional path of length L blocked at both ends by walls of essentially infinite potential. The time-independent Schrödinger equation 1.7 for the particle is given by:

$$\left(-\frac{h^2}{2m}\frac{d^2}{ds^2} + V\right)\psi = E\psi$$

where the Hamiltonian operator is the term in parenthesis. The potential V of the particle is assumed to be zero between $s = 0$ and $s = L$, and infinite everywhere else. The kinetic energy is given by the first term within the parentheses. A general solution of the differential equation is given by $\psi = Ae^{iks} + Be^{-iks}$.
   (a)   Sketch the system.
   (b)   Using a standard relationship between complex exponentials and trigonometric functions, rephrase the general solution $\psi = Ae^{iks} + Be^{-iks}$ in terms of sin and cos of ks.
   (c)   Using the boundary conditions imposed by the potential walls, eliminate one of the two terms in the sin, cos form of ψ.
   (d)   Using the same boundary conditions, show that the allowed wavelengths of the particle fit with the requirements for a classical standing wave ($\lambda = 2L/n$).

*1.3   Determine the time dependence of the amplitude of a superposition of two cosine waves with frequencies ω_A and ω_B, and phase factors ϕ_A and ϕ_B, respectively. Assume that both waves have the same maximum amplitude M. Hint: use standard trigonometric identities in order to get the answer.

*1.4   Using the equation for the Lorentzian line shape (Equation 1.16), find the relationship between the range of frequencies $\Delta\omega$ at half of the maximum amplitude used to generate the damped harmonic decay shown in Figure 1.5 and the relaxation time τ of that harmonic decay.

*1.5   If a hypothetical student obtains a spectrum in 1 h with S/N of 50, how long will it be before the student has a spectrum with S/N of 200 using signal averaging? Does it matter whether the student is doing sequential or simultaneous acquisition experiments?

*1.6   What time decay constant for a particular signal will give rise to a line width of 3 Hz at half-height?

2

ELEMENTARY ASPECTS OF NMR: I. INTRODUCTION TO SPINS, ENSEMBLE BEHAVIOR AND COUPLING

Nuclear and electronic spin

Elementary particles such as protons, electrons and neutrons exhibit behavior that can be explained in terms of a spinning motion for which the angular momentum is quantized. This property is called spin. Spin was originally postulated based on evidence from fine structure in atomic spectra, and was predicted from quantum mechanics. Of particular interest for NMR is the phenomenon known as **Zeeman splitting**, that is, the splitting of energetically equivalent (**degenerate**) nuclear or electronic spin states into energetically nonequivalent (**nondegenerate**) states in the presence of a magnetic field. Electrons, protons and neutrons all exhibit two allowed spin states that are degenerate in the absence of an external magnetic field, but become non-degenerate in the presence of such a field. Depending on how many protons and neutrons are present in the nucleus and how they pair, the nucleus can also have a net spin. Any nucleus with a nonzero spin will exhibit Zeeman splitting in a magnetic field. Any such nucleus is a potential target for the spectroscopist, since an energy (and therefore a frequency) is associated with each spin state, and transitions between the quantized spin states require either absorption or emission of energy. Like many quantized phenomena, spin has a quantum number associated with it (the origins and importance of which we will look at later), as do the individual quantized spin states. The nuclear spin of a particle is usually designated by the quantum number I. The basic unit of nuclear spin, the spin on a proton, is $I = 1/2$. If a nucleus has $I > 1/2$, the nucleus is said to have a **quadrupole**

moment. A quadrupolar spin exhibits not only Zeeman splitting (splitting of spin energy levels in a magnetic field), but also splitting of spin energy levels in the presence of an electric field gradient. For most of this text, we will concentrate on nuclei with $\mathbf{I} = 1/2$, since these are the most commonly observed in solution NMR experiments.

There are two different ways of looking at nuclear spin. One way relies strictly on quantum mechanics, and the other on classical physics. The latter is referred to as the "spinning top" model. The two approaches are related, and are often used interchangeably, which is a little unfortunate, since some confusion can result. In this text, both approaches will be used, but we will try to keep them distinct, and to be clear which is being invoked to explain a given observation.

The quantum picture of nuclear spin

In the quantum approach, the spin quantum number $\mathbf{I}$ defines the number of stationary spin states that the nucleus may occupy in an imposed magnetic field as follows:

$$\# \, levels = (2\mathbf{I} + 1) \tag{2.1}$$

Thus, for $\mathbf{I} = 1/2$ nuclei (this group includes ^{1}H, ^{3}H, ^{13}C, ^{15}N, ^{19}F, ^{29}Si and ^{31}P, among others), the nucleus can have two spin states. For $\mathbf{I} = 1$ nuclei (^{2}H and ^{14}N, for example), the nucleus has three possible spin states, while $\mathbf{I} = 3/2$ nuclei (this includes ^{11}B, ^{7}Li and ^{23}Na) can have four. Besides the spin quantum number $\mathbf{I}$, each individual spin state also has a quantum number $\mathbf{m}$ associated with it. Allowed values of $\mathbf{m}$ are given by:

$$\mathbf{m} = -\mathbf{I}, -\mathbf{I} + 1, \ldots, \mathbf{I} - 1, \mathbf{I} \tag{2.2}$$

Therefore, the allowable spin states for $\mathbf{I} = 1/2$ nuclei are $\mathbf{m} = -1/2, +1/2$, for $\mathbf{I} = 1$ nuclei, $\mathbf{m} = -1, 0, 1$, and for $\mathbf{I} = 3/2$ nuclei, $\mathbf{m} = -3/2, -1/2, 1/2$ and $3/2$, and so on. As noted above, all spin states of the nucleus are degenerate in the absence of a magnetic field (or an electric field gradient for $\mathbf{I} > 1/2$). However, when an external magnetic field (usually designated as the vector quantity $\vec{B}$) is present, the various spin states become nondegenerate. [A note is required at this point concerning the use of $\vec{B}$ to designate magnetic field. In fact, what is being discussed is **magnetic flux density**, which is related to the magnetic field $\vec{H}$ in a vacuum by $\vec{B} = \mu_0 \vec{H}$, where μ_0 is the permeability constant. The magnetic field $\vec{B}$ is typically reported in units of gauss (G) or tesla (T) where $1 \, \text{T} = 10\,000 \, \text{G}$. One tesla is equal to 1 weber/m^2, and so is a measure of magnetic flux.]

Now this raises an important point: the Zeeman splitting (that is, the difference in the energies of the spin states) depends on the magnitude of $\vec{B}$, as shown by:

$$\Delta E_{\alpha\beta} = \hbar \gamma B \tag{2.3}$$

where γ is the gyromagnetic ratio of the nucleus [usually given in units of rad·$(s\cdot T)^{-1}$], and is a constant for a particular nucleus (see Table 2.1).

Figure 2.1 shows the extent of Zeeman splitting between the two spin quantum states $\mathbf{m} = -1/2$ and $\mathbf{m} = +1/2$ for an $\mathbf{I} = 1/2$ nucleus (^1H) as a function of magnetic field intensity B. Recall that the energy is directly proportional to the frequency of the transition, so:

$$2\pi v_{\alpha\beta} = \omega_{\alpha\beta} = \gamma B \qquad (2.4)$$

where $v_{\alpha\beta}$ is the transition frequency in Hz and $\omega_{\alpha\beta}$ is the frequency in angular units (radians s^{-1}). When the field strength of an NMR spectrometer's magnet is described as a frequency (as in the sentence "I have an 800 MHz NMR and you don't"), the reference is usually to the transition frequency of a ^1H nucleus in that

Table 2.1 Gyromagnetic ratios, spin quantum numbers and transition frequencies of some commonly observed NMR nuclei (values from the online Magnetic Resonance Periodic Table, Beckman Institute, University of Illinois, Urbana).

Nucleus	Spin (I)	$\gamma/10^{-7}$ rad·$(s\cdot T)^{-1}$	Transition frequency at 11.74 T (in MHz)
^1H	1/2	26.7510	500.0
^2H	1	4.1064	76.7
^3H	1/2	28.5335	533.3
^7Li	3/2	10.3964	194.3
^{11}B	3/2	8.5794	160.4
^{13}C	1/2	6.7263	125.7
^{14}N	1	1.9331	36.1
^{15}N	1/2	−2.7116	50.7
^{19}F	1/2	25.1665	470.6
^{31}P	1/2	10.8289	202.6

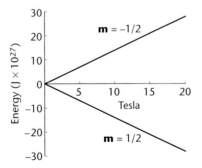

Figure 2.1 Energy splitting of nuclear spin levels for a ^1H nucleus as a function of magnetic field intensity. Magnetic field intensity is plotted along the horizontal axis in tesla (T), and the energy is reported along the vertical axis in joules (J). Splittings are calculated from Equation 2.3, using a value of 1.054592×10^{-34} J/s for $\hbar$ and 26.7519×10^7 rad·$(s\cdot T)^{-1}$ for γ of ^1H.

field. The more correct description would be in commonly accepted units of magnetic field strength such as tesla or gauss, since this does not vary with the nucleus being observed. Thus, a "500 MHz" NMR magnet is one in which protons resonate around 500 MHz (5×10^9 Hz), but ^{13}C will resonate at 125 MHz and ^{15}N at 50 MHz, the result of a magnetic field strength of ~11.74 T (see Table 2.1).

The "spinning top" model of nuclear spin

A semi-classical approach to a description of nuclear magnetic resonance (one that nevertheless requires the assumption of quantized angular momenta and energies) arises from the equations of motion used in analyzing the macroscopic behavior of magnetism in condensed matter. In this picture, the response of a sample to EMR in the presence of a laboratory magnetic field arises from coherent behavior of the individual nuclear spins that are treated as magnetic dipoles with nuclear magnetic moment $\vec{\mu}$. A simple picture of an **I** = 1/2 nucleus is of a charged sphere with an imposed magnetic dipole. The nuclear magnetic moment is proportional to the spin angular momentum, which is quantized. The gyromagnetic ratio is the proportionality constant between the magnetic moment and the spin angular momentum $\vec{P}$:

$$\vec{\mu} = \gamma \vec{P} \tag{2.5}$$

The angle that $\vec{\mu}$ makes with respect to the imposed magnetic field $\vec{B}$ determines the energy of the state (and since energy is a scalar quantity, it is the dot product of the two vectors that is important):

$$E = -\vec{\mu} \cdot \vec{B} = -\mu_z B \tag{2.6}$$

The quantity μ_z is the magnitude of the component of the nuclear magnetic dipole parallel to the magnetic field vector $\vec{B}$, which is usually taken as lying along the **z** axis in the Cartesian coordinate system (see Figure 2.2). Since we have already seen that the energies of the spin states are quantized, $\vec{\mu}$ may only make certain angles relative to $\vec{B}$, as determined by the spin state:

$$\mu_z = \gamma \hbar \mathbf{m} \tag{2.7}$$

By replacing μ_z in Equation 2.6 with the term $\gamma \hbar \mathbf{m}$, the energy of the individual spin states can be determined, and the difference in energy between the two states will give the same result as in Equation 2.3. For an **I** = 1/2 nucleus, two values of μ_z are allowed, one at $\mu_z = +\gamma\hbar/2$ and the other at $\mu_z = -\gamma\hbar/2$, with corresponding energies of $-\gamma\hbar B/2$ and $+\gamma\hbar B/2$, respectively. For nuclei with positive gyromagnetic ratios, the state with **m** = +1/2 is lower in energy than the state with **m** = −1/2 in the presence of a magnetic field.

Classically, a magnetic field $\vec{B}$ exerts a torque $\vec{N}$ on a magnetic dipole $\vec{\mu}$ determined by the cross-product $\vec{N} = \vec{\mu} \times \vec{B}$. The direction of the torque is determined by the

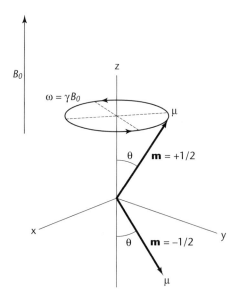

Figure 2.2 In the classical picture, the nuclear magnetic dipole $\bar{\mu}$ precesses around an imposed magnetic field $\bar{B}$ (which is parallel to the **z** axis) with a frequency ω determined using Equation 2.4. $\bar{\mu}$ is tipped with respect to the imposed field (and the **z** axis) by an angle θ as determined by Equation 2.8. The magnitude of the component of $\bar{\mu}$ parallel to the magnetic field, μ_z, is a scalar quantity that determines the energy of the nuclear spin orientation according to Equations 2.6 and 2.7. Because nuclear spin levels are quantized, only particular orientations are allowed. The example shown is for a spin 1/2 nucleus, and two orientations (**m** = +1/2 or **m** = –1/2) are possible, one aligned with and one aligned against the field. Note that the three-dimensional (**Cartesian**) axis system used is defined by a counterclockwise progression of +**x** to +**y** to +**z** unit vectors. This axis system will be used consistently throughout this text.

right-hand rule, as shown in Figure 2.3, and its magnitude is determined by the component of $\bar{\mu}$ perpendicular to the magnetic field $\bar{B}$ (usually the **x**–**y** component of $\bar{\mu}$ in the standard Cartesian frame of reference). The angle θ that $\bar{\mu}$ makes with the applied field is determined by the quantization of the spin angular momentum; without going into the derivation, it is given by:

$$-\cos\theta = \mathbf{m}[\mathbf{I}(\mathbf{I}+1)]^{-1/2} \tag{2.8}$$

The result of the applied torque is a top-like rotation of the magnetic moment around the applied field, called **precession** (Figure 2.2). For a nucleus with a positive gyromagnetic ratio, the direction of the torque results in counterclockwise precession when viewed in the direction of the applied field (see Figure 2.2). The precession frequency is proportional to the imposed field intensity B and the nuclear dipole moment, which in turn depends upon the gyromagnetic ratio γ. Not surprisingly, the precession frequency turns out to be the same as the transition frequency described in Equation 2.4, ω = γB. The precession of nuclear dipoles is referred to as **Larmor precession**, and the precession frequency as the **Larmor**

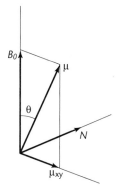

Figure 2.3 The magnitude and direction of the torque $\bar{N}$ on a magnetic dipole $\bar{\mu}$ due to the presence of a magnetic field $\bar{B}$ is determined by the cross product of $\bar{\mu} \times \bar{H}$. The torque causes the dipole to precess around the magnetic field $\bar{B}$ as shown in Figure 2.2.

frequency. For both the **m** = +1/2 and the **m** = –1/2 states, the direction of precession remains the same, as does the frequency, that is, both have the same magnitude for their **x**–**y** component of the magnetic dipole moment.

Spin-state populations in ensembles

The above descriptions, both quantum and classical, deal with a single nuclear spin. We are almost never concerned in NMR with only one nucleus; we are dealing with huge numbers of spins acting more or less in concert (coherent ensembles). It is now time to begin thinking in terms of these ensembles, which must be dealt with statistically. In any ensemble of quantum oscillators, the populations of the available quantized states are determined at thermal equilibrium (that is, after free exchange between all available states for a very long time) by the relative energy of those states, with lower energy states being more populated. In many forms of spectroscopy, (UV/visible, infrared, etc.) the energy difference between states is large enough so that the lowest energy state, or ground state is almost completely populated at thermal equilibrium, with negligible population of excited states. In contrast, the splitting between nuclear spin states is energetically very small, even at high magnetic fields. For example, the splitting between the lower energy (**m** = +1/2 for nuclei with positive γ) and higher energy (**m** = –1/2) states for ^{1}H at 11.74 T is only 0.239 J/mol, which is much smaller than RT (where R is the gas constant equal to 8.314472 J mol^{-1} K^{-1} and T is the temperature in Kelvin units) at temperatures of interest in NMR (RT = 2.48 kJ/mol at 298 K). In all further discussions, we will refer to the lower energy state as α and the higher energy state as β. The relative population of the α and β states at thermal equilibrium can be obtained from Boltzmann's equation:

$$\frac{N_\beta}{N_\alpha} = \exp\left(\frac{-\Delta E_{\alpha\beta}}{kT}\right) \tag{2.9}$$

where k is Boltzmann's constant (equal to R, the gas constant, divided by Avogadro's number, $k = 1.38 \times 10^{-23}$ J K^{-1}) and T is the absolute temperature. At 300 K, for protons at 11.74 T, Boltzmann's equation yields a ratio of N_β/N_α of 0.99992, corresponding to a very small population difference. As the magnetic field intensity decreases, this difference becomes even smaller. It also becomes smaller as the gyromagnetic ratio decreases, so that other nuclei have smaller population differences than ^{1}H at a given temperature and field strength (with the exception of ^{3}H, which has a γ of 28.535 $\times$ 10^7 rad T^{-1} s^{-1}).

An ensemble of nuclear spins can undergo net absorption or emission of EMR only as long as there is a *population difference* between the α and β states, (i.e. $\Delta N_{\alpha\beta}$ = $N_\alpha - N_\beta \neq 0$) and thus, although $\Delta N_{\alpha\beta}$ is small, it is an important characteristic of the system. For reasons that will be discussed later, the probability per unit time P of a single spin in the α state going to the β state in NMR transitions is about the same as that for the opposite transition, i.e. of a single spin in the β state going to the α state. The transition rate of α to β for the ensemble is given by PN_α, and the transition rate of β to α is given by PN_β. In units of power (energy/time), the maximum rate of EMR absorption is:

$$P\Delta N_{\alpha\beta}\Delta E_{\alpha\beta} \qquad (2.10)$$

With time, as EMR is applied to the system, the population difference $\Delta N_{\alpha\beta}$ approaches zero. Once this happens, the transition is said to be **saturated**, and no further absorption of EMR can occur until a population difference is re-established by relaxation.

Because the transition from the α to β state and the reverse transition from the β to the α state are almost equally likely, the re-establishment of thermal equilibrium after the absorption of EMR in the NMR experiment tends to be quite slow. (We will see later that the low probability of spontaneous emission in NMR transitions, that is, an unstimulated loss of energy from the excited state to return to the ground state, is an important reason why nuclear spin lifetimes tend to be long.) Long relaxation times and small equilibrium population differences $\Delta N_{\alpha\beta}$ conspire to reduce the sensitivity of the NMR experiment. Nevertheless, these same factors are also the basis for the success of NMR as a spectroscopic technique, since long relaxation times give narrow line widths. Even more importantly, *non-equilibrium coherent ensembles can be readily prepared and operated on during the course of an experiment*. Relaxation in NMR takes place on a time scale that is readily accessible to modern electronics, and one can perform multiple perturbations on a system of spins and observe the results of those perturbations. This is the basis for the huge number of multinuclear and multidimensional NMR experiments in use today.

The student should note that different relaxation rates often apply to processes that broaden lines and those that return the system to thermal equilibrium. The former are called T_2 processes and the latter T_1 processes. These will be discussed in detail later, but for the moment it is sufficient to know that T_1 processes control the repetition rate of the experiment, while T_2 processes determine how long coherent states can be maintained during an experiment.

Information available from NMR: 1. Nuclear shielding and chemical shift

A wide range of information concerning molecular structure, environment and dynamics is available from NMR, the most basic of which concerns the local environment of the reporting nucleus. The magnetic environment of the nuclear spin is affected by a variety of influences, including valence shell orbital hybridization and bonding of the spin-active atom, charge and bond polarity, electronegativity of nearby atoms and solvent effects. Changes in that environment will cause small changes in the resonant frequency of the nuclear spin, a phenomenon known as **chemical shift**, the origins of which we now consider.

An applied magnetic field $\vec{B}_0$ induces a magnetization $\vec{M}$ in bulk matter. This magnetization is proportional to the applied field $\vec{M} = \rho_0\chi\vec{B}_0$, where χ is the **magnetic**

susceptibility of the bulk matter. The product $\chi\vec{B}_0$ can be thought of as giving the magnitude and direction of an induced dipole, $\vec{\mu}_{ind}$. The concentration of induced dipoles per unit volume is ρ_0. If χ is negative, the induced dipoles that arise in response to the imposed field are opposite in direction to the applied field, and the bulk magnetization $\vec{M}$, the vector sum of the individual induced dipoles summed over unit volume, will be counter to that of $\vec{B}_0$. The result is that the induced magnetization partially counters the applied field, resulting in a magnetic field $\vec{B}_{eff} = \vec{B}_0 + \vec{B} < \vec{B}_0$ less in the bulk matter than would be detected with the same applied field in a vacuum. If χ is positive, the result is an increase in the field within the sample over that in a vacuum, $\vec{B}_{eff} > \vec{B}_0$. Substances with negative χ are **diamagnetic**, and those with positive χ are **paramagnetic**. For NMR, the bulk magnetic susceptibility is important only in that distortions of the magnetic field occur at boundaries between phases with different susceptibilities. This discontinuity can affect NMR line shapes. However, the molecular basis for magnetic susceptibility is another matter, as the same phenomenon also gives rise to chemical shift.

The path of a charged particle having velocity perpendicular to a homogeneous magnetic field $\vec{B}_0$ is a circle with a radius of curvature $r = mv/qB_0$ perpendicular to the imposed field, where m is the mass of the particle, v is its speed, and q is the particle's charge. This is the basis for devices such as the magnetic sector mass spectrometer. Electrons in atomic or molecular orbitals are also in motion, and respond to the presence of a magnetic field by moving in a circular path within the constraints determined by the orbitals in which the electrons reside. This motion represents a movement of charge (current), and produces a magnetic field. The direction of the electronic precession is clockwise when viewed along the direction of the imposed field $\vec{B}_0$, and the induced local field $\vec{B}_{ind}$ circulates clockwise around the path of electron motion. The result is an effective field in the plane of precession that is less than the applied field at the center of the orbit of precession, $|\vec{B}_{eff}| = |\vec{B}_0 + \vec{B}_{ind}| < |\vec{B}_0|$, while outside the orbit of precession $|\vec{B}_{eff}| > |\vec{B}_0|$ (see Figure 2.4). The effective field causing Zeeman splitting of nuclear spin states will be less than the applied field within the radius of the electron motion, while outside of the radius, the effective field will be greater than the applied field. The result is a small but detectable difference in the Zeeman splitting (and Larmor frequency) observed for a given nuclear spin depending on where the spin is placed relative to the moving electrons, with the energy difference depending on the effective field:

$$\Delta E_{\alpha\beta} = \hbar\gamma B_{eff} \qquad (2.11)$$

The dependence of the resonance frequency of a nucleus upon the local electronic environment provides one of the most important sources of information available from NMR, and is known as **chemical shift**. If a nucleus is located in a region where the local induced field $\vec{B}_{loc}$ is in the same direction as the applied field $\vec{B}_0$, the effective field detected by the nucleus is larger than the applied field, $|\vec{B}_{eff}| = |\vec{B}_0 + \vec{B}_{loc}| > |\vec{B}_0|$. The nucleus is said to be **deshielded**, and will resonate at a higher frequency than expected based only on the applied field and Equation 2.4. Conversely, a nucleus that is in a region in which $|\vec{B}_{eff}| = |\vec{B}_0 + \vec{B}_{loc}| < |\vec{B}_0|$, is **shielded**, and resonates at a lower frequency than expected from the applied field $\vec{B}_0$. In the early

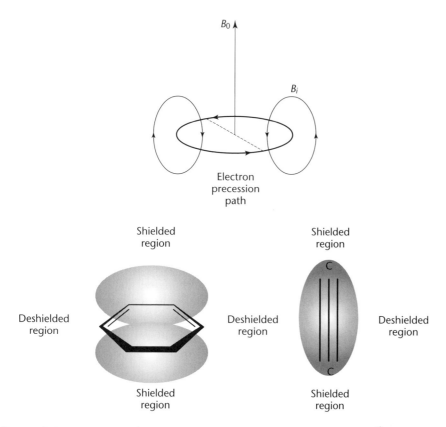

Figure 2.4 Precession of an electron in an applied magnetic field $\vec{B}_0$ generates an induced local field around the path of precession $\vec{B}_i$. This field can either subtract from the applied field such that the resulting local field $B_{eff} < B_0$, or add to it, so that $B_{eff} > B_0$. Regions where $B_{eff} < B_0$ are shielded, and regions where $B_{eff} > B_0$ are deshielded. Two typical situations are shown in the lower part of the figure. Lower left, the delocalized π electrons of an aromatic ring generate a "ring current" under the influence of the imposed field $\vec{B}_0$. The region above and below the ring will be shielded, and regions along the edge will be deshielded, as indicated. Electrons in a carbon–carbon triple bond (lower right) precess around the axis of symmetry generating a shielded region in which the atoms bonded to each carbon reside. Delocalized electron density in each case is represented as a diffuse cloud. In both cases, electronic precession is in the same direction as shown at the top of the figure, as is the direction of $\vec{B}_0$. Recall that current $\vec{I}$ and electronic motion are in opposite directions, so the "right-hand rule" still applies.

days of NMR spectroscopy, when electromagnets were commonly used to induce Zeeman splitting, it was simpler to vary the magnetic field strength B_0 and keep the EMR frequency constant when scanning a spectrum. In order to reach resonance for a deshielded nucleus, the magnetic field strength was reduced, and so deshielded resonances were said to be **downfield**. By increasing the applied field, resonances of shielded nuclei were reached, and these were said to be **upfield**. Although most NMR spectroscopy is now done with fixed magnetic fields and variable EMR frequency, the terminology is still often used.

The extent of shielding or deshielding experienced by a nucleus depends on the strength of the applied field, since the frequency of electron precession that generates the induced field is field dependent. The proportionality between the chemical shift and the applied field is given by the shielding constant σ_i for a nucleus in a particular environment. We can express a relationship between the observed resonance frequency of nucleus i and the applied field:

$$\omega_i = \gamma B_0 (1 - \sigma_i) \tag{2.12}$$

At present, we will ignore the important fact that σ_i for a particular spin depends upon the orientation of the molecule containing that spin with respect to the applied magnetic field. In other words, σ_i is a tensor (called the **shielding tensor**), and its value depends upon the orientation of the molecule relative to $\vec{B}_0$. This is because the precessional paths of electrons normal (perpendicular) to $\vec{B}_0$ are limited by the shapes of the orbitals where the electrons reside. The most important effects are those that result from electronic precession that follow the symmetry of the orbital. Hence, precession that occurs in a benzene π orbital when it is perpendicular to the applied field dominates the chemical shifts of nearby nuclei, while the dominant effects of electronic precession in an isolated carbon–carbon bond occur when the bond is parallel to the applied field (Figure 2.5). The dependence of the chemical shift upon molecular orientation is called **chemical shift anisotropy**, and is an important factor in solid-state NMR experiments. However, in **isotropic media**, no single orientation of a molecule relative to the applied field is preferred over any other and molecular tumbling is rapid (we'll find out how rapid later). In isotropic liquid samples that are most commonly used for high-resolution NMR, the orientation dependence of σ_i collapses and a scalar constant is sufficient to describe the shielding.

Chemical shift is usually measured relative to a standard reference. Chemical shift is obviously in units of energy or something proportional to energy, such as frequency. It turns out to be inconvenient to report chemical shifts in such units, since small changes in magnetic field will result in a change in the absolute values of the resonance frequencies of both the reference standard and the nucleus of interest. However, if one takes the ratio of the difference in resonance frequencies between the signal of interest and the reference to that of the reference as defined in Equation 2.13, the magnetic field term will cancel from each term. This leaves δ, a dimensionless ratio characteristic of the chemical shift of the signal of interest relative to a standard:

$$\delta_i = \frac{\omega_i - \omega_{ref}}{\omega_{ref}} = \frac{v_i - v_{ref}}{v_{ref}} \tag{2.13}$$

Because the chemical shift (in Hz or rad/s) is generally quite small relative to the absolute resonance frequency of the standard, this ratio tends to be in the range of **parts per million (ppm)**, and δ is usually reported as ppm relative to a standard (that needs to be specified). Figure 2.6 shows a mixture of several common organic solvents each containing only one type of proton environment so all the protons in

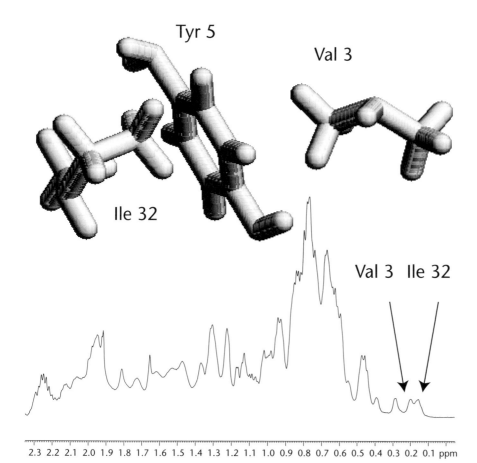

Figure 2.5 Electronic shielding of methyl groups by aromatic rings in protein structures. The top of the figure shows the positions of the methyl groups of isoleucine 32 (Ile 32) and valine 3 (Val 3) in the shielding region of the aromatic ring of tyrosine 5 (Tyr 5) in the NMR-derived solution structure of putidaredoxin. At the bottom is a portion of the ^{1}H NMR spectrum of putidaredoxin, showing the positions of the methyl resonances of Ile 32 and Val 3. These resonances are shifted by the "ring currents" of Tyr 5 into the shielded (upfield) region of the spectrum.

a given molecule resonate at the same frequency. The chemical shifts of these solvents are reported in ppm relative to the commonly used standard tetramethylsilane.

To a first approximation, the intensity of an NMR signal is proportional to the concentration of spins contributing to the signal. Thus, the protons of a methyl group (CH_3), which all contribute to the same signal, give rise to a resonance with three times the intensity of that of a methine (CH) proton, all other things being equal. The **integration** (that is, the area under the Lorentzian curve describing the NMR line) of a methyl resonance will be three times as large as that of a methine resonance in the same molecule at the same concentration. From the integration of

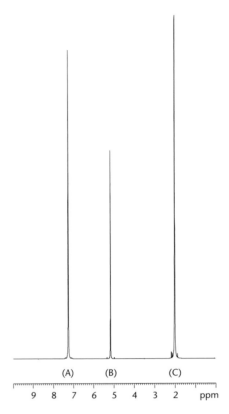

Figure 2.6 The ^{1}H NMR spectrum of a mixture of (A) benzene (C_6H_6), (B) methylene chloride (CH_2Cl_2) and (C) acetone (C_3H_6O), showing differences in chemical shift. All protons are equivalent in each molecule, so no coupling is observed. The spectrum was obtained using $CDCl_3$ as solvent in an 11.74 T magnet. The small, evenly spaced "satellite" peaks on either side of the major peaks are due to the 1.1% natural abundance ^{13}C, which couples to the attached protons resulting in the splitting of the corresponding proton signals.

individual signals in an NMR spectrum, one can determine the relative number of nuclei that contribute to a given resonance line.

Information available from NMR: 2. Scalar coupling

If chemical shift were the only information available from NMR, the technique would not have much more to offer than any other type of spectroscopy. In fact, in terms of raw information content, infrared spectroscopy is at least as information-rich as NMR concerning molecular structure and environment, since practically every chemical bond in a molecule is a potential chromophore for infrared absorption. The true value of NMR lies in the ability of the spectroscopist to detect coupling between nuclear spins, and to delineate coupling pathways. In this way, bonding patterns and stereochemistry can be determined, making NMR the only technique besides crystallography that can be used for detailed molecular structure determination.

There are two types of coupling observed in NMR, one that is mediated through the network of chemical bonds connecting the coupled nuclei, and the other that is transmitted directly through space. Through-bond coupling, often referred to as **scalar** or **J-coupling**, is propagated by the interactions of nuclear spin with the spins of bonding electrons. Consider a nucleus A with a spin $\mathbf{I} = 1/2$. Nucleus A can occupy either of two spin states, $\mathbf{m} = +1/2$ or $\mathbf{m} = -1/2$. Electrons that reside in bonding orbitals overlapping with nuclear spin A will be affected by the spin state of A, and the electron spin states (remember, electrons also have spin) will change slightly in energy in response to the spin of the nucleus. This perturbation of electronic spin states can be propagated to another nucleus (nucleus B) if nucleus B also overlaps the affected orbitals. The result is a slight change in the resonance frequency of nucleus B depending on whether nucleus A is in the $\mathbf{m} = +1/2$ or the $\mathbf{m} = -1/2$ state, and nuclei A and B are said to be J-coupled. Coupling is a two-way street, and if nucleus A affects B, nucleus B also affects A, and to the same extent. The difference between the resonance frequency of spin B when A is $\mathbf{m} = +1/2$ versus when A is $\mathbf{m} = -1/2$ (and vice versa) is called the **coupling constant** and is usually designated as J_{AB}. We will see later that the J-coupling term in the spin Hamiltonian operator does not depend on magnetic field strength, and is a property of the molecular structure. Because of this, unlike chemical shift, the magnitude of the coupling constant does not depend on the magnitude of B_0 (see Figure 2.7).

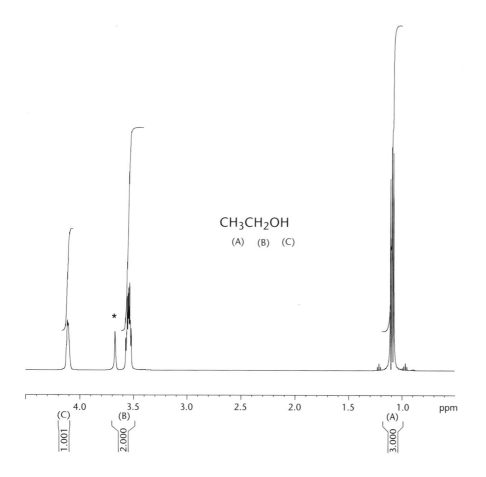

Figure 2.7 The integrated ^{1}H NMR spectrum of ethanol. Normalized signal integrations are shown below the chemical shift scale, and correspond to the number of protons contributing to the particular signal. The signal marked with an asterisk is due to trace H_2O.

The appearance of the spectrum of a compound containing coupled spins depends on the magnitude of the coupling constant(s) and to how many other spins a given spin is coupled. In the simplest case, two distinct nuclei, A and X, are coupled, and the coupling constant J_{AX} is much smaller than the difference between the resonance frequencies of A and X in hertz. This condition $J_{AX} << |v_A - v_X|$ is called the **weak coupling** case, and spectra in which only weak coupling is observed are called **first-order** spectra. In a first-order two-spin spectrum, each resonance is split into two lines (a **doublet**) that correspond to the transition frequencies of the resonant spin when the coupling partner is in either the lower or upper spin state, and the splitting between the two lines exactly equals the coupling constant. The chemical shift of a resonance in this case is taken as exactly at the midpoint of the doublet (Figure 2.8). As the coupling constant becomes larger relative to the difference in chemical shift ($J_{AX} \approx |v_A - v_X|$), the spectra become more complicated (non-first order), and the coupling is no longer considered weak. To indicate the loss of first-order character, such spin systems are normally called AB spin systems, with the closeness of the chemical shifts indicated by the use of alphabetically adjacent labels for the nuclei.

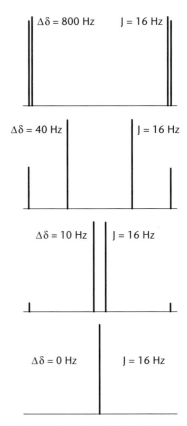

Figure 2.8 Spectral simulations showing field effects on line positions and intensities of coupled spins. The simulation is of a two-spin system with a constant coupling $J = 16$ Hz. In the top spectrum, the difference in chemical shift Δv of the two spins is 800 Hz, yielding a near first-order situation ($\Delta v >> J$). Also shown are the non-first-order cases where Δv is 40 Hz, 10 Hz and the limiting case of identical chemical shifts ($\Delta v = 0$).

Information available from NMR: 3. Dipolar coupling

Anybody who has held two bar magnets close to each other knows that magnetic dipoles interact, and the classical treatment of nuclear spins as magnetic dipoles would predict a similar interaction between neighboring nuclear spins. The direct through-space interaction between nuclear spins is called **dipolar coupling**, and although splittings due to dipolar coupling are not observed in NMR spectra of isotropic liquids, dipolar coupling is still important in a variety of phenomena, including nuclear spin relaxation and the **nuclear Overhauser effect** (**NOE**). In solids and oriented phases such as liquid crystals, splittings due to dipolar coupling are observed directly, and can be quite large (on the orders of hundreds or even thousands of hertz). In anisotropic media, dipolar splitting leads to extremely complicated spectra and many of the technical problems with which solid-state NMR spectroscopists must deal involve the decoupling and selective recoupling of dipolar-coupled spins in order to make the NMR spectrum more interpretable. Recently, the use of weakly orienting media and intrinsic magnetic anisotropies at high magnetic fields has allowed **residual dipolar couplings** (**RDC**) to be measured in solution. Because of the orientational dependence of these RDCs, they provide an important new source of structural and dynamic information from solution NMR techniques. These phenomena will be dealt with in more detail in Chapter 10.

The quantities necessary to describe the coupling of two nuclear dipoles $\vec{\mu}_1$ and $\vec{\mu}_2$, in a magnetic field $\vec{B}_0$ applied along the **z** axis in the Cartesian coordinate frame are illustrated in Figure 2.9. The **x** and **y** components of the spin dipoles vary with time as the spins precess and so average to zero with time. However, the **z** components $\vec{\mu}_z$ parallel to the applied field $\vec{B}_0$ do not average out, and the force experienced by $\vec{\mu}_2$ due to the magnetic field of $\vec{\mu}_1$ depends on the internuclear distance and angle formed by the internuclear vector with respect to $\vec{B}_0$. The observed dipolar splitting D_{AX} due to dipolar interactions between two nuclei A and X is summarized in the expression:

$$D_{AX} = \frac{\mu_0}{4\pi} \frac{\mu_1\mu_2}{r^3} (3\cos^2\theta - 1) \tag{2.14}$$

where D_{AX} is the dipolar coupling constant between nucleus A and nucleus X, θ is the angle between $\vec{B}_0$ and the internuclear vector $\vec{r}_{AX}$ and μ_0 is the permittivity of free space. In isotropic liquids, θ varies randomly (and rapidly) from 0 to π, and the average value of D_{AX} is 0. Hence, no dipolar splitting is observed in isotropic liquids. Note that when the term $3\cos^2\theta = 1$, ($\theta = 54.7°$) the coupling also vanishes. Dipolar couplings are usually eliminated in solid-state NMR spectra by spinning the sample rapidly in the applied magnetic field around an axis at an angle $\theta = 54.7°$ to the applied field, a technique known as **magic angle spinning** (see Figure 2.10). This results in all components of internuclear vectors that do not lie at this angle with respect to the field being averaged to zero. Those components that lie along the spinning axis give rise to zero dipolar coupling because of the relationship in Equation 2.14.

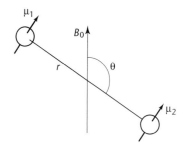

Figure 2.9 Dipolar coupling between two nuclear magnetic dipoles $\vec{\mu}_1$ and $\vec{\mu}_2$ is determined by Equation 2.14, with the angle made by the internuclear vector with respect to the applied field $\vec{B}_0$ determining the size of the interaction.

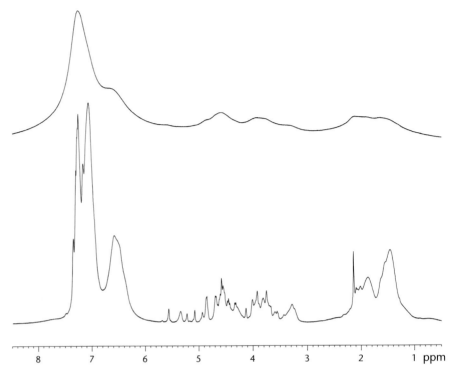

Figure 2.10 Elimination of dipolar couplings by magic angle spinning. The top trace shows the static ^{1}H spectrum of a solvent-swollen resin-bound trisaccharide. The bottom trace was acquired with 3.5 kHz spinning at the magic angle and demonstrates the line narrowing that can be achieved with removal of the dipolar couplings. Spectra were acquired in a 14 T magnet.

For solution NMR, dipolar coupling has two important consequences. The first is that most nuclear spin relaxation is mediated by dipolar coupling, at least for $\mathbf{I} = 1/2$ nuclei. As such, the extent of dipolar coupling will dictate a number of experimental variables (such as repetition rate) as well as affecting line width. The second consequence is the NOE, which is observed experimentally as a change in intensity in the signal of one nucleus when the signal of a nearby nucleus (to which the first is dipolar coupled) is perturbed. The NOE greatly increases the information content of NMR, since it identifies those nuclei that are close to each other in space. Note that since dipolar coupling is not mediated by bonding electrons, the two coupled spins do not have to be part of the same molecule in order for NOEs to be detected between them. This makes it possible to perform NMR experiments to investigate the structures of multimolecular complexes, solvent–solute interactions, and other interesting phenomena involving more than one molecule (see Figure 2.11). The effective range of the NOE is only a few angstroms, except in special cases, due to an r^{-6} distance dependence, where r is the internuclear distance. The NOE will be discussed more thoroughly in Chapter 4.

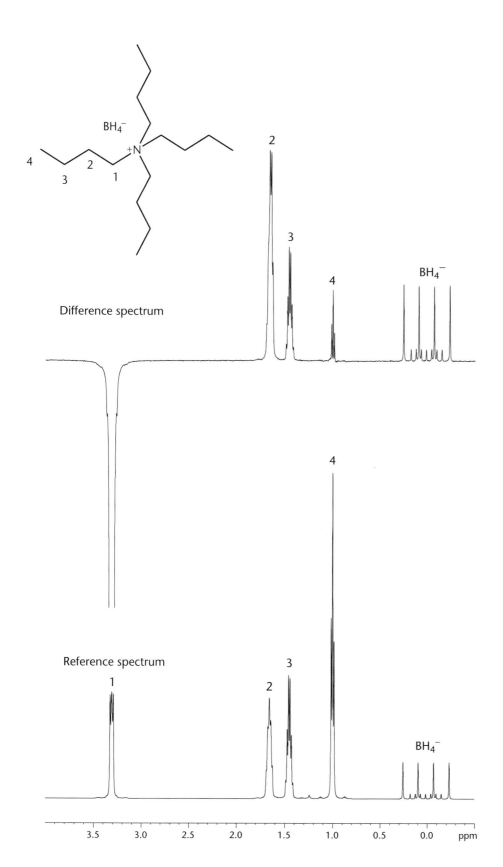

Figure 2.11 Through-space (dipolar) coupling effects in solution NMR spectra as demonstrated by the nuclear Overhauser effect. The ^{1}H spectrum of tetrabutylammonium borohydride (structure shown in inset) in deuterochloroform is shown at the bottom (reference spectrum). This ionic compound forms a tight ion pair in non-polar solution, placing the protons of the borohydride anion (BH_4^-) in close proximity to the 1-CH_2 group. The selective irradiation of the 1-CH_2 signal at 3.3 ppm perturbs the spin populations of nearby protons to which the 1-CH_2 protons are dipolar coupled. This results in changes in signal integration that are observed in the NOE difference spectrum (obtained by subtracting a reference spectrum from the spectrum containing the NOE perturbations). The irradiation of the 1-CH_2 signal (at $\delta = 3.32$ ppm) results in a large negative peak at the position of that signal in the NOE difference spectra, while positive peaks represent NOE enhancements of nearby signals. The largest NOEs are observed at the BH_4^- (centered at $\delta = 0.05$ ppm) and 2-CH_2 signals ($\delta = 1.68$ ppm), as these are closest in space to the irradiated protons. Smaller effects are seen at 3-CH_2 ($\delta = 1.46$ ppm) and 4-CH_3 ($\delta = 1.02$ ppm) signals.

Table 2.2 Approximate range of NMR techniques for investigation of dynamics.

Time scale (s) ≈ rate^{-1}	Types of processes	NMR technique
10^{-15}–10^{-10}	Local motions in molecules	Relaxation analysis
10^{-9}–10^{-4}	Large-scale motions in macromolecules	Solid-state NMR line-shape analysis, rotating frame relaxation measurements
10^{-3}	Chemical exchange, conformational averaging	Saturation transfer, exchange contributions to relaxation
10^{-2}–10^{-1}	Reactions and exchange	Liquid line-shape analysis
10^{0}–10^{3}	Slow reactions	Chemical shift and integration analysis

Information available from NMR: 4. Dynamics

NMR also provides information concerning molecular dynamics on a variety of time scales. Types of processes for which rates can be measured by NMR methods include conformational equilibria, molecular reorientation and correlation times, chemical exchange processes, tautomerism and reversible reactions, weak complex formation and large scale and correlated motions in macromolecules. The range of time scales is quite large, from 10^{-12} s (the time scale of motions that affect dipolar relaxation) to chemical exchange processes with time constants on the order of hours or days. Table 2.2 lists orders of magnitude of time scales accessible by NMR methods.

The most straightforward application of NMR to the study of dynamics is the analysis of chemical shifts in a situation where chemical exchange is taking place. Consider an NMR-active nucleus that samples two distinct environments, A and B, each of which gives rise to resolved resonances. If exchange is slow (leaving aside for a moment how "slow" may be defined), two different resonances will be observed, one at frequency v_A and the other at v_B, with relative integrations proportional to the fraction of time spent by the nucleus in each environment. As far as the NMR experiment is concerned, two different nuclei are being observed. In such a case, exchange is **slow** on the **chemical-shift time scale**.

Now imagine that exchange is speeded up. One consequence of faster exchange is that the lifetime of a spin state in a given environment is shortened, and so lines get broader, as expected based on the uncertainty principle. (Recall that the line width at half-height is given by $\Delta v_{1/2} = (\pi\tau)^{-1}$, where τ is the state lifetime.) This phenomenon is called, obviously enough, **exchange broadening**. If the rate increases even more, we reach a point where the rate of exchange is similar to that of the difference in Larmor frequencies, and the nuclei switch between sites at random time intervals that are, on average, comparable to the time for one period of Larmor rotation. The result is an observed frequency that is distributed around a maximum at:

$$v_{obs} = \chi_A \, v_A + (1 - \chi_A) \, v_B \tag{2.15}$$

where χ_A is the fraction of exchanging nuclei occupying site A, and $(1 - \chi_A)$ the fraction occupying site B.

Since exchange rates increase with increasing temperature, one can often move between the slow and fast exchange regimes for a given process by changing temperature. The temperature at which the two slow-exchange resonances at v_A and v_B collapse into a single resonance at v_{obs} is called the **coalescence temperature**, and the exchange rate $k_{c(AB)}$ at the coalescence temperature is given by:

$$k_{c(AB)} = \frac{\pi \Delta v_{AB}}{\sqrt{2}} \qquad (2.16)$$

Equation 2.16 allows NMR to be used to measure simple one-point kinetics for exchange processes with energies of activation in the range of 40–100 kJ/mol. Since Δv_{AB} is field dependent, a range of rate constants at different coalescence temperatures can be measured for a given reaction using several NMR spectrometers with different field strengths.

As the exchange rate increases still further, so that $k_{AB} >> \Delta v_{AB}$, the **fast exchange** regime is reached. Now, line broadening due to exchange is given by:

$$\Delta v_{1/2} = 1/2\pi(v_A - v_B)^2 k_{AB}^{-1} \qquad (2.17)$$

Interestingly, the line width at fast exchange is inversely proportional to the exchange rate, rather than directly proportional, as in the slow exchange regime. That is, the faster the exchange occurs, the narrower the line becomes. This will be dealt with in more detail later, but for now it suffices to think of it in the following way. At very slow exchange rates, chances are good that if a nuclear spin is at site A, it will stay there throughout the experiment, and the line width will be determined by the excited state lifetimes at the individual sites. As exchange rate increases, the chance that exchange between sites will occur during the observation also increases, and exchange becomes fast enough to compete with relaxation for limiting the lifetime of excited states at the individual sites. As a consequence, exchange broadening is observed. However, near the coalescence point, a new set of states (ground and excited states) begins to form. These represent a weighted time-average of the original two sets of ground and excited states from sites A and B, with a transition frequency given by Equation 2.15. As the exchange rate increases further, uncertainty concerning the exact state of the system decreases, and so does the line width of the resonance at frequency v_{obs}. The transition between the slow and fast exchange regimes is illustrated in Figure 2.12.

A graphical way of thinking about the transition from the slow-exchange to fast-exchange regimes is shown in Figure 2.13. Populations of spins in exchanging sites A and B are shown as sets of circles, with the darker circles representing Larmor precession in the rotating frame at the frequency of site A, and the lighter circles representing precession at the frequency of site B. When exchange is slow, there are two distinct populations present that will be seen as two lines in the NMR spectrum. As the exchange rate increases to the intermediate regime, the populations become indistinct, with each spin spending some portion of a precession in one site, but with significant likelihood of a switch to the other site before a complete pre-

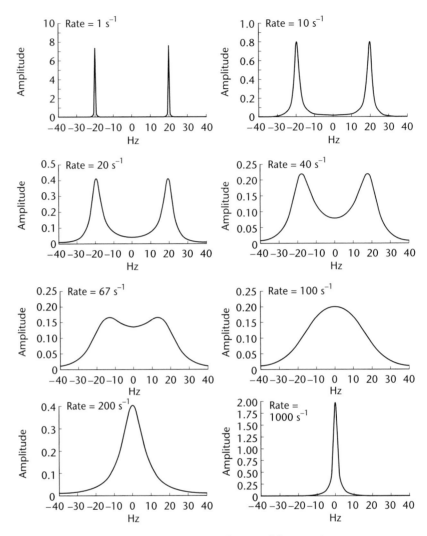

Figure 2.12 Effect of cross-over between slow and fast exchange on appearance of an NMR spectrum. Exchange of spins between two equally populated sites, one site giving rise to a chemical shift of –20 Hz relative to the center of the spectrum and the other with a shift of +20 Hz. The exchange rate of nuclei between the two sites is given in exchange events per second. Note that the amplitude varies in each graph, although the total integration (area under the curves) is constant.

cession in the rotating frame. In the fast-exchange regime, the spins switch sites many times in the course of a precession, thus giving rise to an average precessional frequency that is observed as a single line.

In the example shown in Figure 2.12, there is presumed to be no *J*-coupling between the exchanging sites. If nuclei at sites A and B are coupled to each other (and the exchanging nuclei are both $\mathbf{I} = 1/2$), the dependence of the observed spec-

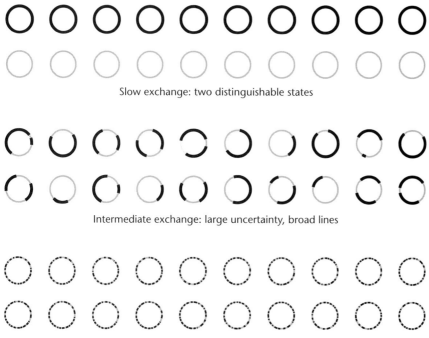

Slow exchange: two distinguishable states

Intermediate exchange: large uncertainty, broad lines

Fast exchange: average of both states, single line

Figure 2.13 Graphical representation of slow, intermediate and fast two-site exchange on the chemical-shift time scale. Each circle represents the Larmor precession of a single nuclear spin, with a complete precession represented by a circle of a single shade, black or gray. Precession at the frequency of one site is represented by a black curve, precession at the other site by a gray curve. A "population" of 20 spins is represented. In the slow exchange case, precession occurs many times at each site before exchange, giving rise to two distinct populations that are observed as two discrete lines in the NMR spectrum. As the exchange rate increases into the intermediate exchange case, the possibility of site exchange during a single precession becomes more likely, increasing the uncertainty of the frequency measurement and broadening lines. In the fast exchange case, intersite exchange occurs many times during a single cycle, giving rise to a single average line.

trum on k_{AB} is more complex, and the rate constant at coalescence is given by Equation 2.18:

$$k_{c(AB)} = \sqrt{\frac{\left(\Delta v_{AB}^2 - 6J_{AB}^2\right)}{2}} \qquad (2.18)$$

J-coupling time scale, decoupling experiments and exchange decoupling

Splitting due to *J*-coupling between spins at two sites A and B (even if the two sites are not related by exchange) will be observed if the lifetime of the spin states is long enough. One can think of nucleus A as a little spectrometer detecting what is going on at spin B, and reporting on that observation via the observed splitting (and vice versa). As with chemical exchange, distinguishability of spin states depends on the rate at which transitions between those states occurs. As noted earlier, the energies of the lower (α) and upper (β) spin states are nearly equal, even for nuclei with large gyromagnetic ratios in very strong magnetic fields. Therefore, the populations of the α and β states of nucleus B are nearly identical (which is why the two peaks of the doublet due to nucleus A split by nucleus B are equal in intensity), and the transition probability $W_{\alpha\beta}(B)$ is roughly the same in either direction. If $W_{\alpha\beta}(B)$ is small, the rate of transition $k_{\alpha\beta}(B)$ between states will also be slow, and nucleus A will detect both stationary states of nucleus B. As the transition rate $k_{\alpha\beta}(B)$ increases, the two states will become less distinguishable, in the same way that two environments connected by chemical exchange will become indistinguishable as the exchange rate between them increases. When the transition rate becomes larger than the coupling constant J_{AB}, the A doublet will collapse into a singlet, a phenomenon called **decoupling**.

Decoupling can occur in a number of ways. Chemical exchange will cause decoupling if the rate of exchange is fast relative to the coupling constant under observation. This is because there is only a 50:50 probability that the nucleus entering the coupled site is in the same spin state as the nucleus that is leaving, so that fast chemical exchange gives rise to an average spin state being detected by the coupling partner. A common example of exchange decoupling is found with primary alcohols, such as ethanol, dissolved in non-polar solvents. The hydroxyl proton in ethanol is coupled to the protons on the adjacent methylene group. In the absence of an exchange catalyst, this coupling is readily observed as splitting of both the –OH and 1-CH_2 proton resonances. But upon addition of a trace of acid or an increase in temperature, the exchange rate increases and the splitting is no longer observed (Figure 2.14).

Interaction between nuclear spins and radio-frequency (RF) EMR: 1. RF decoupling

The α and β states of an **I** = 1/2 nucleus are stationary states, and their wave functions (referred to earlier as α and β) are eigenfunctions of the Hamilitonian operator in the time-independent Schrödinger equation, with eigenvalues that are the energies of those states, E_α and E_β. However, the concept of stationary states is really only applicable when no time-dependent perturbation of the system is occurring (and we have avoided worrying about the time-dependent Schrödinger equation, which is really more important for the hard-core spectroscopist than the time-independent version). When a time-dependent perturbation of a quantum oscillator takes place, the stationary states of the oscillators are mixed, and the correct wave description of the system is that of a **virtual state**, which contains elements of the wave functions of the time-independent stationary states. In NMR, virtual states can be generated by the application of EMR at the frequency of the transition between the α and β states. Since practically all of the nuclear spin transitions with which we are concerned occur in the radio frequency range, we will dispense with the EMR abbreviation and substitute **RF (radio frequency)** in its place. In the

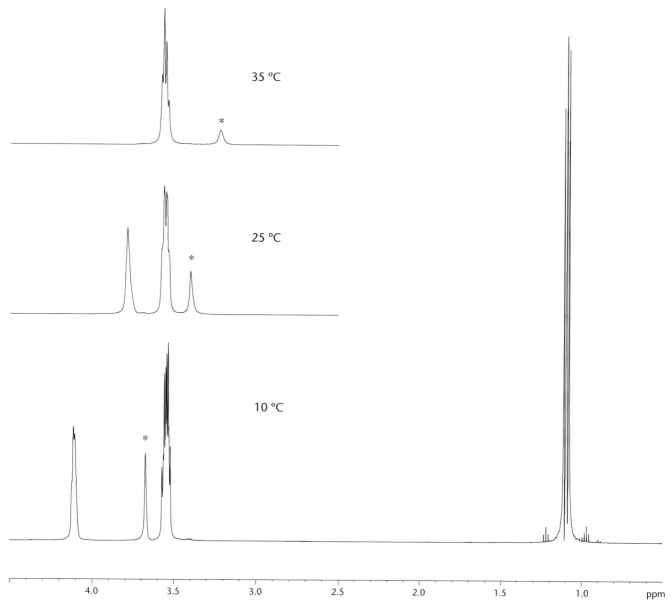

Figure 2.14 Coupling as a function of chemical exchange in ethanol, CH_3CH_2OH, in deacidified chloroform. The bottom spectrum was obtained at 10°C. At this temperature, chemical exchange of the OH proton (~4.1 ppm) is slow enough so that coupling to the CH_2 protons (3.53 ppm) is still observed, although some exchange broadening is evident. At the higher temperatures (inserts), the exchange rate is sufficiently fast to decouple the OH and CH_2 protons, as evidenced by the loss of splitting of the OH resonance and the change in shape of the CH_2 multiplet. Trace amounts of water give rise to the peak marked with an asterisk, which is in exchange with the OH proton. Note that the line widths of the OH and water peaks broaden in parallel as the temperature rises and the exchange rate increases. All spectra were obtained at 11.74 T (500 MHz ^{1}H).

presence of such a perturbing RF the two stationary states no longer exist as such, but are replaced by a virtual state that contains elements of both stationary states. The result for a nucleus A which is normally coupled to the perturbed spin B is that A no longer detects both spin states of B, but only an average state, and so the normal splitting of A due to B collapses. This phenomenon is known as **RF decoupling**. Upon removal of the perturbing RF field at nucleus B, the virtual state no longer exists, and the original stationary states immediately reappear, so the collapse of multiplets due to RF decoupling is only observed if RF is actively applied at the perturbed spin while the signal of the coupling partner is observed (see Figure 2.15).

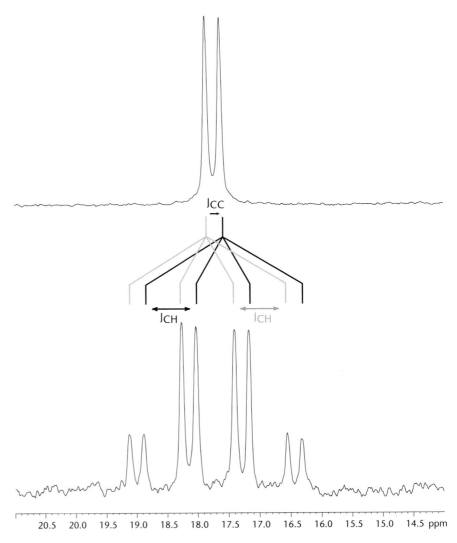

Figure 2.15 ¹H, ¹³C *J*-coupling in the 14 T (150 MHz) ¹³C NMR spectrum of ¹³C-labeled alanine. The multiplet shown at the bottom is the signal observed in the ¹³C spectrum for the methyl group of alanine. The larger splitting corresponds to a 1-bond coupling between the methyl protons and the methyl ¹³C ($^1J_{CH}$ = 130 Hz). The smaller splitting is the one-bond coupling due to the attached ¹³C-carbon at the α-position of alanine ($^1J_{CC}$ = 35 Hz). Irradiation of the ¹H resonances during acquisition of the ¹³C spectrum (top) results in the collapse of the $^1J_{CH}$ splittings (decoupling of the ¹H and ¹³C signals).

Problems

*2.1 High-field NMR magnets often have "fringe fields" that extend a significant distance from the center of the magnet, into spaces (including nearby rooms) next to, above and below the magnet. An unshielded 11.74 T magnet has a fringe field that extends upward such that in a room situated above the NMR room with a floor 7 feet above the top of the magnet there is a fringe field of about 5 G at floor level. If you are sitting in that room, what is the resonant frequency of protons in your big toe at this field? How about the ^{31}P in your anklebone?

*2.2 (a) The borohydride anion, BH_4^-, has a distinctive ^{1}H NMR spectrum (see Figure 2.11). Boron has two common isotopes, ^{10}B ($\mathbf{I}$ = 3, 18% natural abundance), and ^{11}B ($\mathbf{I}$ = 3/2, 82% natural abundance). Although both are quadrupolar nuclei, the symmetry of the borohydride anion minimizes the contribution of quadrupolar relaxation in either case, and an 81 Hz coupling between ^{11}B and attached protons is observed. The ^{1}H–^{10}B coupling is ~30 Hz. Describe what you expect the ^{1}H, ^{10}B and ^{11}B spectra of BH_4^- to look like (assuming that there is no isotope effect on the chemical shift of the protons). Describe both the number of lines expected and their relative intensities. What happens to the ^{11}B spectrum if one of the ^{1}H spins is replaced with ^{2}H ($\mathbf{I}$= 1)? Note that the ^{2}H–^{11}B coupling is much smaller than the ^{1}H–^{11}B coupling (~10 Hz).

(b) In the ^{1}H NMR spectrum of an organoborane (R-BH_2, for example, where R is an alkyl group) the boron-attached protons tend to be very broad, featureless lumps without any of the splitting observed for BH_4^-. Why?

*2.3 Using Boltzmann's equation, calculate the equilibrium ratios of populations in the two spin states of ^{1}H, ^{15}N and ^{29}Si in a 5.87 T field (250 MHz ^{1}H) at 298 K. What difference does the sign of the gyromagnetic ratio (positive for ^{1}H, negative for ^{15}N and ^{29}Si) make to your calculation?

*2.4 Given the equation for a Lorentzian line shape (Equation 1.16), what is the minimum relaxation time required for an NMR signal in order for a coupling of 12 Hz to be resolved at half-height?

*2.5 Consider the following system: proton A is coupled to proton B, which exchanges with a site C to which A is not coupled. All chemical shifts are distinct, with an equilibrium constant K_{BC} = 1 for site exchange. Sketch the spectrum you expect for each of the following situations:
(a) Slow exchange on the chemical shift and *J*-coupling time scale.
(b) Fast exchange on the coupling scale, but slow on the chemical-shift time scale.
(c) Fast exchange on both time scales.
(d) Slow exchange on both time scales, but K_{BC} = 0.5.
(e) Fast exchange on both time scales, but K_{BC} = 0.5.

2.6 Derive Equation 2.8, $\cos \theta = \mathbf{m}[\mathbf{I}(\mathbf{I} + 1)]^{-1/2}$ for the angle that the magnetic dipole μ makes with respect to the applied field in the classical picture using Equations 2.5–2.7.

*2.7 The amplitude of the magnetic field component (B_1) of the applied RF field in NMR is often described in terms of the frequency around which the observed nuclei will precess in that field. For example, a 10 kHz B_1 field is one in which a proton will precess at 10 kHz.

(a) How long will it take a proton to precess once in a 10 kHz B_1 field?

(b) NMR spectrometers are often run in "locked" mode. This means that concurrent with whatever experiment you are running, another NMR signal is being acquired from a "lock nucleus", usually ^2H in the solvent. The lock signal is used as a reference by the spectrometer hardware to keep signals from "drifting" during the course of an experiment in case of small changes in sample homogeneity or magnetic field drift. It is important that the lock signal be strong enough to observe, that is, enough RF power is put into the sample at the lock frequency that a response is detected, but not so much RF power that the lock signal becomes saturated. The more intense the RF field applied (that is, the higher the RF power), the faster a signal will saturate in that field. The optimum field for avoiding saturation is given by the equation:

$$1 >> \gamma^2 B_1{}^2 T_1 T_2$$

where γ is the gyromagnetic ratio of the observed nucleus and T_1 and T_2 are characteristic relaxation times of the observed nucleus. For a deuterium nucleus in $CDCl_3$, with a relaxation time of 0.1 s for both T_1 and T_2, suggest a reasonable amplitude for the RF (the "lock power" on your NMR spectrometer) to be used in avoiding saturation of the lock signal of the spectrometer.

(c) The T_1 and T_2 of deuterons in D_2O are usually significantly shorter than those in $CDCl_3$. What does this imply about the lock power that one can use for samples in D_2O relative to those in deuterochloroform?

2.8   In older NMR spectrometers, which used electromagnets, it was generally easier to hold the RF frequency constant and change the magnetic field in order to scan the NMR spectrum. The 90 MHz NMR was a fairly standard "high-field" NMR in the 1970s. Determine by what amount the magnetic field would need to be changed in a 90 MHz NMR in order to detect two resonances that are separated by a distance of 10 ppm, with one at 0 ppm and the other at 10 ppm. Provide your answer both in gauss and in tesla units. Which of the two signals would be detected at higher magnetic field?

2.9   Two protons exchange between sites A and B that are not coupled to each other and differ in chemical shift by 400 Hz on a 400 MHz NMR spectrometer, The two signals exchange in a reaction that is known to have a free energy of activation of 60 kJ/mol, and exhibit a coalescence temperature T_c of 250 K. Using the following expression:

$$\Delta G_{activation} = RT_c[23.74 + \ln(T_c/k_{AB})]$$

calculate the rate of exchange k_{AB} at the coalescence temperature. Based on this information, would you be able to calculate the expected coalescence temperature for the same reaction on a 500 MHz NMR?

3

ELEMENTARY ASPECTS OF NMR: II. FOURIER TRANSFORM NMR

Interaction between nuclear spins and RF: 2. A single spin in the rotating frame of reference

The introduction of RF decoupling at the end of the last chapter was our first description of a real NMR experiment, leading us into territory where we need to consider how RF interacts with nuclear spins in more detail. The quantum approach to this interaction is quite complicated, involving the mixing of stationary states to generate time-dependent virtual states. Although that bridge must eventually be crossed, the semiclassical approach (treating the RF as a wave) is sufficient for the present.

The classical interaction between a nuclear magnetic dipole and a fixed magnetic field was discussed in Chapter 2. A torque (which is the rate of change in angular momentum) $\vec{N}$ is exerted on a nuclear magnetic moment $\vec{\mu}$ by a magnetic field $\vec{B}_0$ that is placed by convention along the **z** axis in the **laboratory frame of reference**. The laboratory frame of reference is a Cartesian system of coordinates that defines spatial orientation in the room housing the NMR magnet. The torque is determined by the cross-product $\vec{N} = \vec{\mu} \times \vec{B}_0$ and is normal to the plane defined by $\vec{\mu}$ and $\vec{B}_0$ (see Figure 3.1). As $\vec{B}_0$ is aligned along the **+z** axis, the torque is exerted in the **x**, **y** plane perpendicular to both the **z** axis and the component of $\vec{\mu}$ in the **x**, **y** plane. The torque $\vec{N}$ forces Larmor precession of $\vec{\mu}$ around the **z** axis (see Figure 3.1).

Now consider the interaction of the nuclear dipole moment $\vec{\mu}$ with an applied RF. EMR of all types, including RF, has associated with it an **electric field** vector

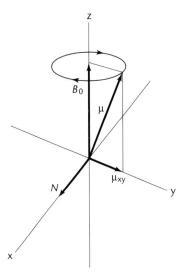

Figure 3.1 The torque $\vec{N} = \vec{\mu} \times \vec{B}_0$ induced at a magnetic dipole $\vec{\mu}$ in the presence of a magnetic field $\vec{B}_0$ results in rotation of the dipole in a circular path around the **z** axis.

(usually designated $\vec{E}_1$) that oscillates between maximum positive and negative amplitude at the frequency determined by $E = h\nu$ (where E is the energy, not the electric field). The electric field vector is perpendicular (**orthogonal**) to the wave's velocity or **propagation vector** $\vec{k}$. Orthogonal to both the electric field vector and the propagation vector is a magnetic field vector that oscillates in phase with the electric field, designated $\vec{B}_1$ (Figure 3.2). The subscript 1 is usually appended to the magnetic component of the RF in order to distinguish this oscillating magnetic field from the static magnetic field $\vec{B}_0$. In optical spectroscopy, where oscillating electric dipoles provide the chromophores, it is the interaction between the electric field vector and the chromophore that is important. However, in NMR we are concerned with the interaction between the magnetic component of the RF, $\vec{B}_1$, and the nuclear magnetic dipole.

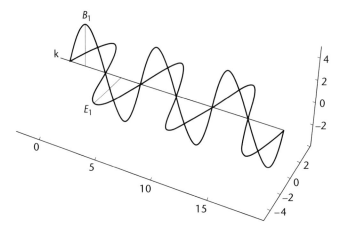

Figure 3.2 Directions of oscillating electric and magnetic field vectors, $\vec{E}_1$ and $\vec{B}_1$, associated with an electromagnetic wave propagating along the direction $\vec{k}$.

Before discussing the interaction between the $\vec{B}_1$ field of the applied RF and a nuclear spin dipole, it is worth reviewing some vector algebra. We will repeatedly use two types of products between two vectors $\vec{a}$ and $\vec{b}$ that define an angle θ. This angle is positive for a rotation of vector $\vec{a}$ that aligns it with vector $\vec{b}$. The first product is the dot product, $c = \vec{a} \cdot \vec{b}$. The result of the dot product c is a scalar quantity whose magnitude is given by $c = |a||b|\cos\theta$, where $|a|$ and $|b|$ represent the magnitude of the $\vec{a}$ and $\vec{b}$ vectors, respectively. The dot product can be visualized as the projection of one vector onto the other. Note that $\vec{a} \cdot \vec{b} = \vec{b} \cdot \vec{a}$, and the product vanishes when the two vectors are perpendicular to each other. The second product is the cross-product, $\vec{a} \times \vec{b} = \vec{p}$, where $\vec{p}$ is a vector with a magnitude of $|a||b|\sin\theta$ and direction perpendicular to the plane determined by $\vec{a}$ and $\vec{b}$ according to the right-hand rule. Note that $\vec{a} \times \vec{b} = -\vec{b} \times \vec{a}$ and that this product vanishes when the $\vec{a}$ and $\vec{b}$ are parallel to each other. The cross-product has already been seen in the expression $\vec{N} = \vec{\mu} \times \vec{B}_0$ above.

Now we are ready to consider the interaction of RF with a nuclear spin dipole. Imagine that the magnetic field vector $\vec{B}_1$ of the applied RF lies along the **x** axis of the laboratory frame of reference, with its origin (0 amplitude for $\vec{B}_1$) at the origin of the same frame. $\vec{B}_1$ will also exert a torque on $\vec{\mu}$; that torque will be normal to the plane defined by $\vec{\mu}$ and $\vec{B}_1$, that is, to the **x** axis and the **y**, **z** component of $\vec{\mu}$ (see Figure 3.3). The magnitude and direction of the torque at any given instant will depend on the angle between $\vec{\mu}$ and $\vec{B}_1$, as well as the magnitude of $\vec{B}_1$. It turns out that the torque on $\vec{\mu}$ due to $\vec{B}_1$ will average to zero unless the oscillating frequency of the RF is the same as the Larmor frequency of the precessing nuclear dipole, and this situation can be visualized as follows.

The $\vec{B}_1$ vector is shown as **plane-polarized**, that is, fixed in the plane defined by the **x** axis of the laboratory frame of reference and the propagation vector of the RF

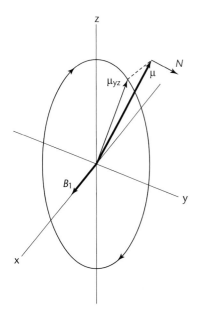

Figure 3.3 The application of an RF field such that the $\vec{B}_1$ component lies along the **x** axis and exerts a torque $\vec{N}$ at $\vec{\mu}$ normal to the plane defined by $\vec{\mu}$ and $\vec{B}_1$, that is, to the **x** axis and the **y**, **z** component of $\vec{\mu}$. The magnitude and direction of the torque at any given instant will depend on the angle between $\vec{\mu}$ and $\vec{B}_1$, as well as the magnitude of $\vec{B}_1$.

wave $\vec{k}$. However, one can always treat a plane-polarized wave as the resultant of two in-phase circularly polarized waves with equal and opposite senses of rotation for their field vectors $\vec{B}_{clockwise}$ and $\vec{B}_{counter}$ (Figure 3.4). As the circularly polarized waves progress, $\vec{B}_{clockwise}$ and $\vec{B}_{counter}$ do not change in magnitude, only in direction, and they describe clockwise and counterclockwise corkscrews in space. Now consider what happens to the nuclear dipole $\vec{\mu}$ precessing at its characteristic Larmor frequency around the static field $\vec{B}_0$ in the presence of the RF field $\vec{B}_1$, that is also oscillating near the Larmor frequency of the nuclear dipole. Following the two circularly polarized components of plane-polarized field $\vec{B}_1$, we see that one of the two components moves in the same direction as the precessing nuclear dipole, while the other moves in the opposite direction. The component that moves in the direction opposite to that of the nuclear dipole $\vec{\mu}$ will exert an instantaneous torque on $\vec{\mu}$, but as the angle between the vectors changes, that torque will average to zero over time. However, the component that moves in the same direction as the dipole exerts a torque that does not cancel with time. (Assuming a nuclear spin with a positive gyromagnetic ratio, this is the component $\vec{B}_{clockwise}$.) The net torque exerted by $\vec{B}_{clockwise}$ results in precession of the nuclear dipole $\vec{\mu}$ around $\vec{B}_{clockwise}$. This precession of the nuclear dipole around $\vec{B}_{clockwise}$ is exactly analogous to Larmor precession, and the frequency of the precession is determined the same way, that is $\omega_{precession} = \gamma B_{clockwise}$. The result of this precession is to tip the nuclear dipole further away from the $\mathbf{z}$ axis. Since Larmor precession around $\vec{B}_0$ is typically much higher in frequency than any precession induced by $\vec{B}_1$, the nuclear dipole will precess many times around $\vec{B}_0$ before it proceeds even part of the way around $\vec{B}_1$, and the nuclear dipole moment describes a trajectory on a sphere in the laboratory frame of reference as shown in Figure 3.5.

Visualizing two $\vec{B}_1$-derived vectors rotating in opposite directions and the trajectories of the nuclear dipole as it precesses around both $\vec{B}_0$ and $\vec{B}_1$ can be very complicated. But many of these complications can be resolved simply by changing the frame of reference in which one thinks about the precession. The motion of the planets in the sky was extremely complicated to think about until the Sun, instead of Earth, was chosen as the center of the frame of reference. The apparently complicated planetary behavior could then be resolved into simpler motions that were

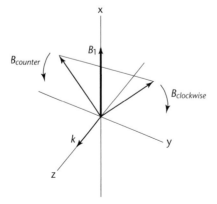

Figure 3.4 The plane-polarized oscillating magnetic field component $\vec{B}_1$ of the electromagnetic wave shown in Figure 3.2 viewed along the direction of the propagation vector $\vec{k}$ can be thought of as the vector sum of two components rotating in opposite directions, the circularly polarized $\vec{B}_{clockwise}$ and $\vec{B}_{counter}$ vectors. For clarity, the positions of the $\mathbf{x}$ and $\mathbf{z}$ axes have been rotated.

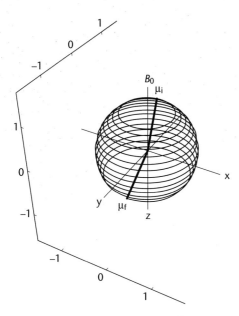

Figure 3.5 Effect of an oscillating magnetic field $\vec{B}_1$ lying in the **x**, **y** plane of the laboratory frame of reference on the motion of a magnetic dipole $\vec{\mu}$ that is already precessing around an imposed static field $\vec{B}_0$. $\vec{B}_1$ is assumed to oscillate at the same frequency as the Larmor precession of the dipole, and so has one component that rotates in the **x**, **y** plane around the **z** axis at the same frequency as the nuclear dipole. The torque $\vec{B}_1$ exerts on the dipole does not average to zero with time, and forces a slower precession of $\vec{\mu}$ around itself. The dark vectors connecting the beginning and end of the spiral to the origin represent the position of the nuclear dipole at the beginning ($\vec{\mu}_i$) and end of the evolution ($\vec{\mu}_f$) described by the spiral line. Axes are labeled in arbitrary displacement units. In this figure, precession of $\vec{\mu}$ around $\vec{B}_0$ is 50 times as fast as that around $\vec{B}_1$. However, this is only for clarity: in real NMR situations the ratio is usually much higher.

readily amenable to calculation and prediction. So it is with nuclear magnetism. If a frame of reference is chosen so that the **x** and **y** axes rotate along with the nuclear dipole at the Larmor frequency (the **rotating frame of reference**), describing the interaction between the oscillating field $\vec{B}_1$ and the nuclear magnetic dipole is greatly simplified. Why is this? Once again separating $\vec{B}_1$ into oppositely rotating circularly polarized vectors $\vec{B}_{clockwise}$ and $\vec{B}_{counter}$, we see that $\vec{B}_{clockwise}$ is fixed in the rotating frame, since we have stipulated that the RF be applied at the Larmor frequency. By further stipulating that the **y** axis of the rotating frame is parallel to $\vec{B}_{clockwise}$ (now labeled as $\vec{B}_1$ in Figure 3.6), the rotating field vector observed in the laboratory frame is replaced by a fixed field vector $\vec{B}_1$ along the **y** axis in the rotating frame. Now the precessional motion of the nuclear spin dipole $\vec{\mu}$ around $\vec{B}_{clockwise}$ is much easier to picture, since $\vec{\mu}$ is now fixed relative to the **x** and **y** axes in the rotating frame, since it also precesses at the Larmor frequency. Only the **x**, **z** component of $\vec{\mu}$ is affected by the torque $\vec{\mu} \times \vec{B}_{clockwise}$, and the trajectory is described by rotation of the spin dipole around $\vec{B}_1$ (Figure 3.6).

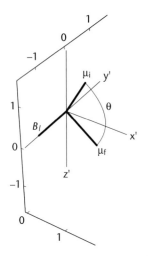

Figure 3.6 Same effective $\vec{B}_1$ as in Figure 3.5, but in a frame of reference rotating at the same frequency as the Larmor precession of the dipole $\vec{\mu}$. $\vec{B}_1$ is now fixed along the **y** axis in the rotating frame, and precession of $\vec{\mu}$ around $\vec{B}_1$ through the angle θ is in the **x**, **z** plane. The same precession of $\vec{\mu}$ around $\vec{B}_1$ is taking place in both Figures 3.5 and 3.6.

Note that in Figure 3.6, $\vec{\mu}$ and $\vec{B}_{clockwise}$ are arbitrarily placed 90° from each other in the **x**, **y** plane, so the entire $\vec{\mu}$ is affected by the rotation, not just the **x**, **z** component. The other component of $\vec{B}_1$, $\vec{B}_{counter}$, moves at twice the Larmor frequency in the rotating frame of reference, and can be safely ignored, since any torque it exerts on the nuclear dipole will be rapidly averaged to zero. If RF is applied that is not at the Larmor frequency, both components of $\vec{B}_1$ will have a non-zero frequency relative to $\vec{\mu}$ and any torque exerted will be averaged to zero over time. Although off-resonance RF can give rise to observable effects in the NMR spectrum (including small changes in resonance positions known as **Bloch–Siegert shifts** that occur in the presence of an off-resonance RF field), only a field at the Larmor frequency will force precession away from the **z** axis.

Interaction between nuclear spins and RF: 3. An ensemble of spins in the rotating frame of reference

So far, only the effect of an applied RF on a single spin has been discussed. However, the NMR experiment usually deals with large ensembles of spins, and we need to consider what happens to such an ensemble when an RF field is applied. In all further discussions, rotating frame of reference will be assumed unless otherwise noted. For the moment, also assume that all spins in the ensemble precess at the same Larmor frequency. All of the nuclear dipoles in the ensemble have a component along the **z** axis, with that component being positive, or lower energy, for α (**m** = +1/2) and negative (higher energy) for β (**m** = –1/2). (This assumes a positive gyromagnetic ratio; for a negative γ, the opposite situation exists, that is, **m** = –1/2 gives rise to the positive component.) If the system is at thermal equilibrium, there are slightly more spins in the ensemble in the α state than in the β state, meaning that the resultant of the **z** components of the ensemble is a net magnetization (often labeled $\vec{M}_0$) along the +**z** axis. However, there is no preference for one orientation of the nuclear dipoles relative to the **x** or **y** axes in the rotating frame; for every nuclear dipole that points along +**x**, +**y**, there is another pointing –**x**, –**y** at any given moment. This is because, although the nuclei are precessing around the imposed magnetic field at the same Larmor frequency, their phases of precession are completely random at equilibrium. As a result, there is no net magnetization in the **x**, **y** plane at thermal equilibrium.

Consider an RF field applied at the Larmor frequency such that the circularly polarized component of $\vec{B}_1$ rotating in the same direction as the rotating frame is fixed along the +**x** axis of the rotating frame. This is the only component that is important, so we dispense with calling it $\vec{B}_{clockwise}$ and simply refer to it as $\vec{B}_1$. $\vec{B}_1$ exerts a torque on $\vec{M}_0$, rotating it in the **y**, **z** plane as determined by the right-hand rule (see Figure 3.7). The angular frequency ω of this rotation is determined by $\omega = -\gamma B_1$. As a result of this rotation, a non-zero component of the net magnetization $\vec{M}$ is now generated along the **y** axis.

Microscopically, we can interpret this motion of the macroscopic magnetization $\vec{M}$ due to $\vec{B}_1$ as a decrease in the population difference between the α and β states (i.e. a smaller **z** component to the magnetization). Furthermore, *coherence has been generated* between the individual nuclear spins. There is now a non-zero component of the magnetization along the −**y** axis, indicating that spin dipoles are precessing around $\vec{B}_0$ in phase. When this happens, the in-phase oscillations of the coherent nuclear spins will induce a current in a correctly configured receiver coil, and **magnetic resonance** is detected. One important generalization we can draw from this discussion is that in order to detect an NMR signal, some time-dependent (oscillatory or **evolving**) coherence must be generated within the ensemble. In the classical description, this means that an excess of spins in the ensemble are now in phase, with their **x**, **y** components pointing in the same direction at the same time, giving rise to a nonzero resultant of the net magnetization in the **x**, **y** plane precessing at the Larmor frequency in the laboratory frame.

The angle θ between the magnetization vector $\vec{M}$ and the **z** axis after precession around $\vec{B}_1$ is called the **tip angle**. Clearly, if a component of the net magnetization in the **x**, **y** plane is required to generate an NMR signal, then the maximum signal is obtained when $\theta = \pi/2$ rad (90°). This means that the RF is applied long enough to rotate the magnetization completely into the **x**, **y** plane *and then turned off*. There is now no **z**-component of $\vec{M}$, and we can interpret this as meaning that in the ensemble, the populations of the α and β states are equal after tipping $\vec{M}$ completely into the **x**, **y** plane. This is not equivalent to what happens when a transition is saturated by RF, as described in the previous chapter. In RF saturation, the RF is applied continuously, resulting in multiple complete rotations around the RF axis. As such, any coherence generated is quickly rotated out of the **x**, **y** plane, resulting in no detectable signal.

What happens if, instead of stopping at a 90° tip angle, we apply the RF long enough for precession to continue until $\vec{M}$ lies along the −**z** axis ($\theta = \pi$ rad, or 180°)? As there is no longer any component of $\vec{M}$ in the **x**, **y** plane, no signal will be observed in this circumstance. However, now the equilibrium populations of the α and β states are reversed (inverted); there are more spins in the β than in the α state, a situation called **population inversion**. Rotation by another $\pi/2$ rad will return $\vec{M}$ to the **x**, **y** plane, but oriented along +**y**. This would also generate a maximum signal in the receiver coil of the spectrometer, but the signal would be opposite in sign from that seen after a $\theta = \pi/2$ tip (see Figure 3.8).

Detection of an NMR signal

Typically, multiple nuclei of the same type but with different chemical shifts are observed in an NMR experiment. For spins with the same γ but different chemical shifts, a common rotating frame of reference is used. This frame of reference rotates at one frequency, usually chosen to be the frequency of the applied RF, and the precession of the magnetization due to nuclei in various environments is observed in terms of the difference between the frequency of the rotating frame, v_r, and the Larmor frequency of a nucleus in a given environment, v_i. The apparent frequency of precession of nucleus in the rotating frame is then given by $v_{apparent} = (v_i - v_r)$. The frequency $v_{apparent}$ is a "beat" frequency generated by interference between the Larmor frequency of a given nucleus and the reference frequency represented by v_r. The beauty of the rotating frame description is that it nicely illustrates how an NMR spectrum is actually acquired. Because NMR

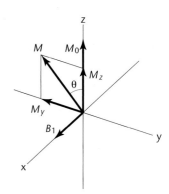

Figure 3.7 A component of an oscillating field $\vec{B}_1$ is fixed along the **x** axis in the rotating frame of reference, and exerts a torque on the net magnetization of the ensemble of nuclear spins, $\vec{M}_0$. The magnetization rotates through an angle θ that is equal to $-\gamma B_1 t$, where t is the time for which the $\vec{B}_1$ field is present. The result is the magnetization vector $\vec{M}$, which has the same magnitude as $\vec{M}_0$, but is tipped away from the **z** axis, giving a nonzero component along the **y** axis. This component precesses at the Larmor frequency in the laboratory frame and generates an oscillating signal in the receiver coil of the NMR. When $\theta = \pi/2$ rad (90°), the **z** component of $\vec{M}$ will be equal to zero, and the **y** component will be equal in magnitude to $\vec{M}_0$.

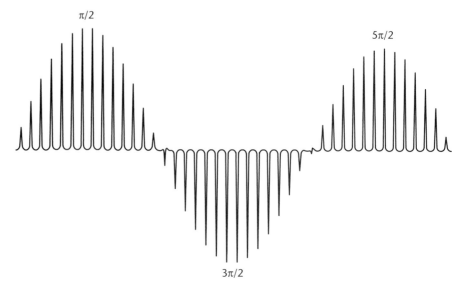

Figure 3.8 Pulse width calibration plot. The pulse width starts at 1 μs and is incremented in 1 μs steps. Maximum positive signal is seen when the tip angle is 90° ($\pi/2$ rad) and 450° ($5\pi/2$ rad) and maximum negative signal is at 270° ($3\pi/2$ rad). Nulls occur at 180° (π rad), 360° (2π rad), and 540° (3π rad).

frequencies range from tens to hundreds of megahertz, it is more convenient to measure an NMR signal as the beat, or difference frequency, between a reference frequency and the resonant frequency of the observed nucleus, a process known as **heterodyning**. The beat frequency generated by heterodyning the reference and signal frequencies is usually in the range of hundreds or thousands of hertz, and is easier to handle electronically than the high frequency signal of the nuclear spin resonance. The reference frequency used (often referred to as the **carrier frequency**) is usually (although not always) at the center of the observed spectrum.

Imagine an uncoupled ensemble of spins of type A with Larmor frequency ν_A in a rotating frame of reference that is also rotating at frequency ν_A. The net magnetization due to the spins A, $\vec{M}_A$, starts along the +**z** axis. If RF at frequency ν_A is applied for a sufficient time, $\vec{M}_A$ will tip completely into the **x**, **y** plane (tip angle of $\pi/2$ radians), generating coherence among the ensemble of spins A. The spins A process coherently at their characteristic Larmor frequency, and, in the laboratory frame of reference, so too would $\vec{M}_A$. In the rotating frame, however, $\vec{M}_A$ will appear stationary, since $\nu_{apparent} = (\nu_A - \nu_r) = 0$.

Now assume a slightly different situation. Spins A still precess at ν_A, as does the rotating frame of reference. But now assume that spin A is coupled to a spin X, with a coupling constant J_{AX}. About half of the coherent spins A will see X in the α state, while the other half will see X in the β state. Their resonance frequencies will differ by J_{AX}, and the vector $\vec{M}_A$ will therefore split into two components that exhibit this difference in precessional frequency. In the rotating frame, the two vectors will appear to precess in opposite directions, one with a frequency of $+ J_{AX}/2$ and the other a frequency of $-J_{AX}/2$ relative to the axes of the $\mathbf{x}$, $\mathbf{y}$ plane. Both the uncoupled and coupled cases are shown in Figure 3.9.

Next, consider what happens to spin X (assumed to have the same gyromagnetic ratio as A, and so described using the same frame of reference). Both A and X spins will have net magnetization along the $+\mathbf{z}$ axis at thermal equilibrium, and we will assume that they can be simultaneously tipped by the application of RF into the $\mathbf{x}$, $\mathbf{y}$ plane. (This simultaneous tipping of magnetizations with different resonant frequencies is performed by an **RF pulse**, which will be described in detail later in this chapter). To further complicate the situation, assume that the carrier frequency v_r (that we will also take to be the frequency at which the rotating frame precesses) lies halfway between the Larmor frequencies of spins A and X. Both magnetization vectors $\vec{M}_A$ and $\vec{M}_X$ will now have a nonzero apparent precession in the rotating frame. If spins A are coupled to spins X with a coupling constant J_{AX}, four vectors are observed, with precessional frequencies given by $v_{apparent}(A) = (v_A - v_r \pm J_{AX}/2)$ and $v_{apparent}(X) = (v_X - v_r \pm J_{AX}/2)$. This situation is shown in Figure 3.10.

How does this spin motion ("evolution of coherences") translate into an NMR spectrum? The first step is to detect the signal, and avoiding the complications of heterodyning and coupling for a moment, we will use some simple analogies to describe what the NMR detector coil "sees". Imagine that you are sitting on the $\mathbf{x}$ axis of the rotating frame of reference in Figure 3.11, and that you are observing what was happening on the $\mathbf{y}$ axis, and on that axis only. As the vector corresponding to spin A coherence precesses, you see the $\mathbf{y}$ component of $\vec{M}_A$ oscillate between a maximum and a minimum of $+M_A$ and $-M_A$ with a frequency of $v_{apparent} = (v_A - v_r)$. If precession begins on the $\mathbf{y}$ axis, the amplitude along $\mathbf{y}$ will be given by

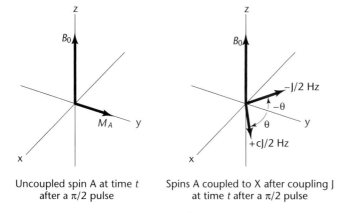

Uncoupled spin A at time t
after a $\pi/2$ pulse

Spins A coupled to X after coupling J
at time t after a $\pi/2$ pulse

Figure 3.9 Motion of the magnetization $\vec{M}_A$ due to an ensemble of spins A in the rotating frame after a $\pi/2$ pulse is applied. It is assumed that the frame of reference rotates at the Larmor frequency of A, so in the uncoupled case (left), no motion of the $\vec{M}_A$ vector is apparent. If coupling is present (right), the $\vec{M}_A$ will split into two equal components, one moving with apparent frequency $+J/2$ in the rotating frame, and the other with an apparent frequency $-J/2$. Each vector corresponds to the half of the ensemble which detects the coupled spin (that we will call X) in a particular spin state, α or β. The angular distance from the $\mathbf{y}$ axis of each component after time t is given by $\theta = \pi J t$.

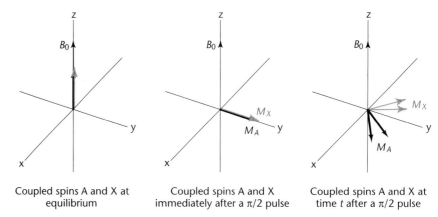

| Coupled spins A and X at equilibrium | Coupled spins A and X immediately after a π/2 pulse | Coupled spins A and X at time t after a π/2 pulse |

Figure 3.10 Motion of magnetization due to an ensemble of coupled spins A (black arrows) and X (gray arrows) in the rotating frame after a $\pi/2$ RF pulse is applied. The frame of reference rotates at a frequency v_r halfway between the Larmor frequencies of A and X. After free precession in the **x**, **y** plane for a time t, each vector is split into two components separated by an angle $2\pi Jt$, with J being the coupling constant for A and X. The vector sum of the two components of magnetization A will also have moved through an angle $2\pi(v_A - v_r)t$, while the vector sum of the two components of magnetization X will have moved through an angle $-2\pi(v_X - v_r)t$. Therefore, the frequency of each component of magnetization A in the rotating frame is $v_{apparent}(A) = (v_A - v_r \pm J_{AX}/2)$, and $v_{apparent}(X) = (v_X - v_r \pm J_{AX}/2)$ for each component of X.

$M = M_A(\cos 2\pi(v_{apparent})t)$, and will look the same whether $v_{apparent}$ is positive or negative, since the cosine is an "even" function, that is, symmetric around zero. But the direction of precession of $\vec{M}_A$ in the rotating frame depends on the sign of $v_{apparent}$, so by looking down only one axis, we cannot tell whether an observed frequency is positive or negative relative to the reference frequency. Thus, the true Larmor frequency of $\vec{M}_A$ could be either $v_r + v_{apparent}$ or $v_r - v_{apparent}$. So a second **orthogonal** point of view, 90° out of phase from the first and looking at the **x** axis, is required to completely characterize the frequency and direction of rotation of $\vec{M}_A$ as it precesses. From the second point of view, the amplitude would start at zero (since precession begins with $\vec{M}_A$ completely along **y**) and the amplitude along **x** would be given by $M = M_A(\sin 2\pi(v_{apparent})t)$. Since the sine function is "odd" (antisymmetric around zero), the amplitude would depend on the sign of $v_{apparent}$, as shown in Figure 3.11. Of course, either sign of $v_{apparent}$ could give rise to the same sine-dependent amplitude for M along **x**, depending on whether the cosine function observed along **y** starts with a negative or positive amplitude at $t = 0$ (the $\vec{M}_A$ vector along either the negative or positive **y** axis at $t = 0$). Thus, both **x** and **y** components of $\vec{M}_A$ must be detected in order to determine the direction of precession in the rotating frame and thereby determine the true Larmor frequency of $\vec{M}_A$.

Time-domain detection in the NMR experiment: the free induction decay and quadrature detection

The considerations described above would be of little more than academic interest if NMR experiments were still done using sequential excitation. In a sequential

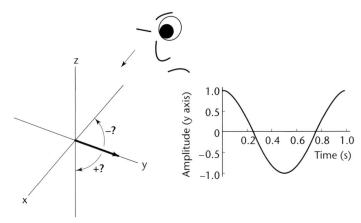

Looking down the **x** axis at the **y** component, one sees the same
evolution no matter the direction of precession

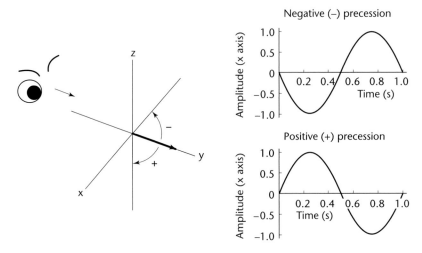

Looking down the **y** axis at the **x** component, what one sees
depends upon the direction of precession

Figure 3.11 Orthogonal detection of a magnetization vector precessing at a
frequency of 1 Hz. One eye does not see stereo!

excitation experiment, the Larmor frequency of a given nucleus would be determined from the frequency of the RF that was applied at resonance, and as the RF
spectrum was scanned, each spin would resonate at the appropriate frequency.
The resonance would be noted as a response on the frequency spectrum and that
would be that. However, modern NMR is almost exclusively done using simultaneous excitation methods. This means that all of the spins to be detected are
brought into coherence simultaneously regardless of their chemical shift, and the
response is recorded as a time domain signal to be deconvoluted into a frequency
domain spectrum after acquisition. Signals are generated as a function of time by
the coherent precession of nuclei, hereafter referred to simply as **coherence**.
Coherence gives rise to an oscillating macroscopic magnetization that in turn generates a current in the receiver coil. These signals are heterodyned with a carrier

frequency generated by the spectrometer, and the interference pattern (caused by differences between the carrier frequency and the Larmor frequencies of the various coherences) so generated is what we see on the display screen of the NMR spectrometer as raw data. The precession of coherences is "free" around $\vec{B}_0$ (they are not being forced in any fashion after the initial excitation of the nuclei by RF) and induces the oscillating current in the receiver coil. With time, coherence among the precessing spins is lost because of relaxation, and the induced signal eventually dies away, or **decays**. Based on these features, the name **free induction decay**, or FID, is applied to the time-domain signal that is the heterodyned combination of the signal induced at the detector and the carrier frequency. The process of separating the two orthogonal "points of view" required to characterize completely the resonant frequencies of the various spins is called **quadrature detection**.

The signal generated by coherence A in the receiver coil of the NMR spectrometer is at the true Larmor frequency v_A. As mentioned earlier, it is common practice to resort to heterodyning in order to analyze frequencies in the megahertz range. Heterodyning amounts to mixing the signal with the carrier frequency to generate the beat frequency $v_{apparent} = (v_A - v_r)$. This low frequency resultant is used to generate the NMR spectrum. Heterodyning, or frequency conversion, takes place in an electronic device known as a **mixer**. Mixers are based on the trigonometric relationship:

$$2(\cos x)(\cos y) = \cos(x + y) + \cos(x - y) \tag{3.1}$$

One input into the mixer is at the carrier frequency, with amplitude proportional to $\cos(2\pi v_r t)$. The other input comes from the receiver, and is proportional to $\cos(2\pi v_A t + \theta)$, where θ is an arbitrary phase factor. After combining the two inputs as per Equation 3.1, two outputs are obtained, one that is a function of the difference between the two frequencies and the other a function of the sum. The sum term will be very high frequency (double the resonant frequency of the nucleus of interest) and can be ignored. The difference signal is proportional to $\cos[2\pi(v_A - v_r)t + \theta]$, and is usually in the audio frequency range (i.e. if put through an amplifier and a speaker, it can be heard). It is in this signal that we are interested. However, we recall that one signal is insufficient for quadrature detection, and a second mixer is needed, for which one of the inputs is the receiver signal, and the other proportional to $\sin(2\pi v_r t)$. The desired output of this mixer is the difference signal $\sin[2\pi(v_A - v_r)t + \theta]$. The two signals are then stored and processed separately.

The time domain signal on the display screen of a modern NMR spectrometer is one of these two signals. Generally, only one component is displayed, but it is important to remember that the other component is there. The FID is not displayed for decoration. An experienced spectroscopist can tell at a glance a good deal about what the frequency domain spectrum will look like from the time-domain signal, and it is worth spending some time with the examples provided later on in this chapter to understand what the FID can tell you.

Digitization of the free induction decay

The modern NMR spectrometer was made possible by advances in digital electronics. Virtually every aspect of spectrometer function is controlled by digital logic, from signal acquisition to data storage and processing. Even the RF frequencies are generated digitally. The heterodyned output of the mixers, however, is **analog** (that is, described by a smooth function) and must be converted into a series of discrete values (computer words), or **digitized**, in order to be processed and stored. The digitization process consists of sampling the analog signal at discrete intervals using an ***analog-to-digital converter*** (**ADC**). The signal is detected as an oscillating voltage, and the value of the voltage at the point of sampling is stored in a digital register. Almost any modern measuring device of any type contains ADCs. The **dynamic range** of an ADC determines the minimum and maximum voltages that can be compared. A defined small voltage increment ΔV_i is required to "flip" a bit from 0 to 1 (the smallest possible increment in digital information). The minimum signal that can be measured is thus ΔV_i. The maximum signal depends on the size of the computer word that can be generated by the ADC. A 12-bit digitizer can count from 0 to 2^{11} (2048) with one bit reserved for sign. With a 12-bit ADC, a signal as small as ΔV_i or as large as $2048 \times -V_i$ can be measured accurately.

How often does the ADC need to sample the analog signal? The analog signal needs to be sampled at a rate that insures one can accurately determine the highest frequency in the signal $\nu_{max} = |\nu_A - \nu_r|$, where ν_r is the carrier frequency (at the center of the spectrum when quadrature detection is in use) and ν_A is the frequency of the resonance furthest from ν_r (closest to the edge of the spectrum). The required sampling rate turns out to be $2\nu_{max}$, referred to as the **Nyquist frequency**. The basis for this requirement is easy to visualize, as shown in Figure 3.12. If one samples below the Nyquist frequency, the highest frequencies in the signal are detected, but incorrectly. They are shifted to a lower frequency, or **aliased**, and will show up in the NMR spectrum at the wrong place. It is usually easy to detect aliasing; aliased signals are often observed as a distorted peak shape, "out-of-phase" relative to other signals in the spectrum, and their spectral positions will shift relative to other signals when the carrier frequency is moved. The different phase characteristics of the aliased signal arise from the fact that their phase advance at the beginning of the sampling time is different to those signals that are not aliased (see Figure 3.12).

Experimentally, one avoids aliasing by setting a spectral width for acquisition that is wide enough to contain all the frequencies of interest. Although the spectral width is usually a readily accessed parameter, what one is actually setting when setting the spectral width is a parameter called the **dwell time**, that is, the duration of a single sampling event. The dwell time determines the sampling rate $\nu_{sampling}$, as the dwell time is equal to $(\nu_{sampling})^{-1}$. The correct sampling rate is one that does not alias any of the signals in the spectrum. The correct dwell time DW for a spectrum in which ν_A is the signal furthest from ν_r is given by DW $\leq 1/2(\nu_A - \nu_r)$, the inverse of the Nyquist frequency. From this discussion, it should be clear that the spectral width is determined by the dwell time, and is equal to 1/DW.

A second question one needs to ask when setting up an experiment is: how long does one need to sample the FID? This experimental parameter (called the **acquisition**

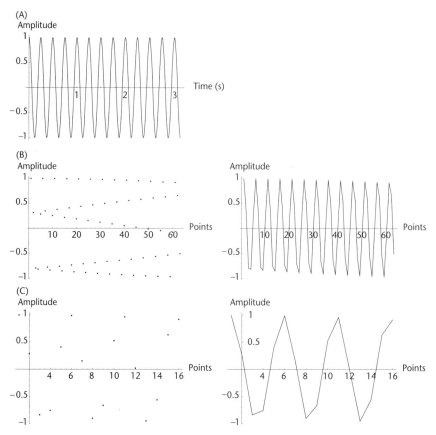

Figure 3.12 (A) Plot of a continuous cosine function with a frequency of 4 Hz (Nyquist frequency of 8 Hz). (B) On the left, discrete sampling of the function plotted in (A) at a rate of 100 Hz, well over the Nyquist frequency. Note that for all plots, the same time interval is sampled. On the right is the same plot, but points are joined by a curve. (C) On the left, discrete sampling of the same function as in (A), but at a rate of 5 Hz, below the Nyquist frequency. The aliased frequency obtained by joining the points, shown on the right, is about 1 Hz. Note that the phase angle at time $t = 0$ also differs between (B) and (C), which would lead to different phase characteristics after Fourier transformation. Aliased peaks usually phase differently than correctly sampled peaks.

time) is the dwell time multiplied by the number of data points, n, collected. The acquisition time is an important experimental parameter, since it is one of the variables that determine how often one can repeat an NMR experiment, and also determines how much space the experiment will take in the computer's memory storage device. In fact, one should sample the signal as long as it persists (if possible), but no longer. Three cases are illustrated in Figure 3.13, the first in which the signal remains strong throughout the entire acquisition time, the second in which the signal dies out well before acquisition ends, and the third in which the signal has nearly decayed to zero by the time the acquisition ends. The acquisition time is too short in the first case, too long in the second and appropriate in the third. Most of the points collected in the second case contain only noise, and are wasted. In fact, these points are devoid of coherent information and they will reduce the signal-to-

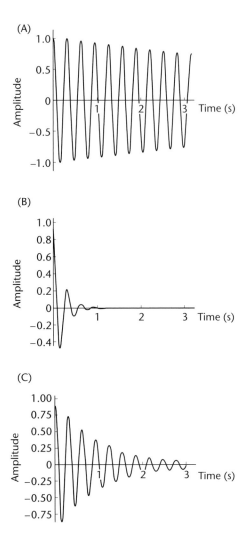

Figure 3.13 (A) Acquisition of FID for too short a time. (B) Acquisition of FID for too long a time. (C) Appropriate acquisition time.

noise ratio (S/N) after processing; one would be better off not collecting these points at all, or replacing them with zeros (an operation called zero-filling, that will be discussed later in this chapter). This is illustrated in Figure 3.14.

The first case, where too short an acquisition time is used, results in truncation of the time-domain signal and gives rise to artifacts in the processed spectrum (Figure 3.15). Truncation errors are easily recognized in the processed spectrum as oscillations in the baseline at the bottom of resonances, and can be removed after the fact by appropriate manipulations of the FID. These will be dealt with in the next section.

Fourier transformation: time-domain FID to frequency-domain spectrum

We have thus far been mysterious about how one goes from the time-domain FID to a frequency-domain spectrum, which is after all the goal of this effort. The time

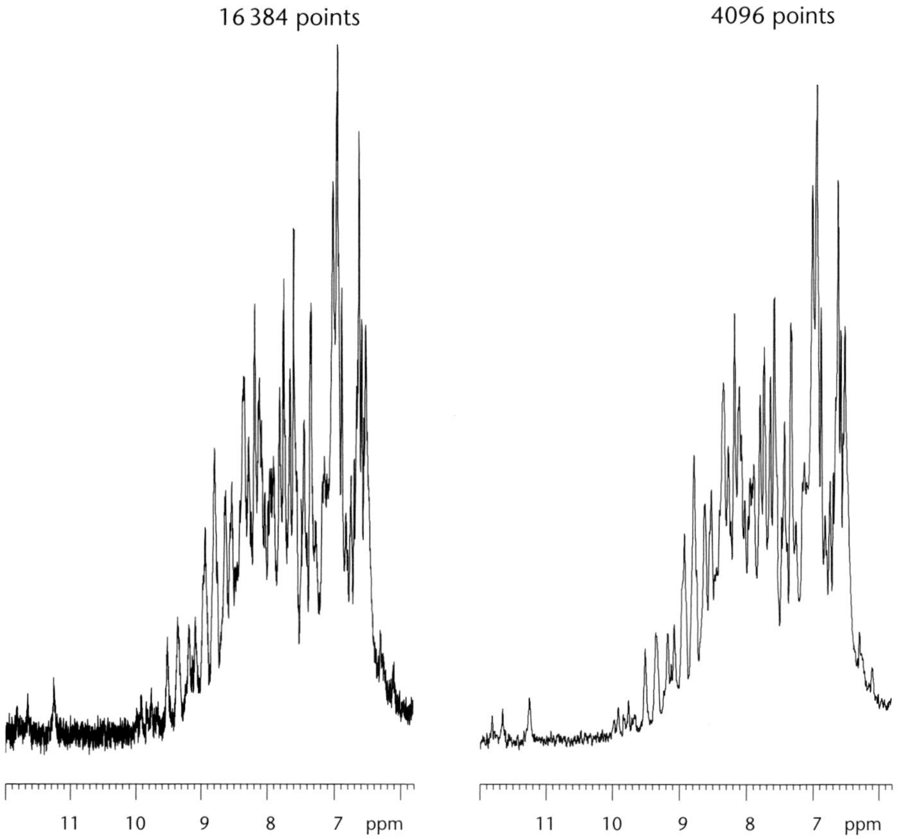

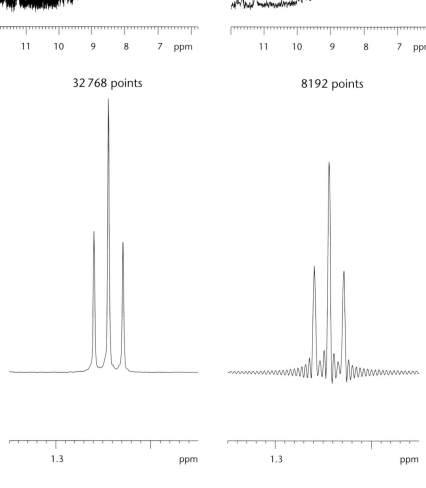

16 384 points

4096 points

11 10 9 8 7 ppm   11 10 9 8 7 ppm

Figure 3.14 Amide region of the ^1H spectrum of a 1 mM protein sample (adrenodoxin) in 90% H_2O/D_2O acquired on a 14 T magnet. The left spectrum was acquired with too many points, introducing additional noise.

32 768 points

8192 points

1.3   ppm   1.3   ppm

Figure 3.15 Methyl signal of the ^1H spectrum of ethylbenzene in $CDCl_3$ acquired on an 11.74 T magnet. The right spectrum was acquired with too few points in the FID, introducing truncation artifacts.

and frequency domains can be related to each other by a mathematical function called the **Fourier transform**. Fourier was a French mathematician who traveled with Napoleon's army to Egypt. The story goes that he went to sleep on the Great Pyramid and when he awoke, he was inspired to write the equations that bear his name. However that may be, the relationships he discovered can be used to transform any smooth time-domain function $s(t)$ into a frequency-domain function $S(\omega)$ and vice versa:

$$S(\omega) = \int_{-\infty}^{\infty} s(t) e^{i\omega t} dt \tag{3.2}$$

$$s(t) = \frac{1}{2\pi} \int_{-\infty}^{\infty} S(\omega) e^{i\omega t} d\omega \tag{3.3}$$

where ω is frequency in angular units and i is the square root of -1. Applied to the time-domain FID, which represents system response as a function of time, a *Fourier* transformation (FT) will generate a frequency-domain spectrum, that is, the system response as a function of frequency.

The concept of FT is simple to grasp intuitively. Imagine a pure frequency wave that starts in the time domain at negative infinity and goes on to positive infinity. Since the wave has a single fixed frequency, a plot of wave amplitude versus frequency (as opposed to time) would have a single (infinitely narrow) line at that frequency, and would be zero at every other frequency. This is essentially what happens when one Fourier transforms a time-domain FID into a frequency-domain spectrum, except that since the FID is a damped harmonic function, finite line widths are observed for each line in the frequency spectrum.

Fourier transforms have some interesting properties that are worth discussing in more detail. First, the Fourier transform is a linear operation, that is, the Fourier transform of the sum of two functions is the sum of the Fourier transforms of the individual functions. Spectroscopists sometimes take this for granted, but this means that the frequency domain spectrum one obtains upon FT of a series of co-added (or subtracted) FIDs is identical to the result one would obtain by FT of each individual FID prior to co-addition or subtraction. Each FID itself can also be considered the sum of the contributions from each signal present, and linearity implies that the presence of one signal does not change the appearance or intensity of another signal.

Secondly, the Fourier transform has some important symmetry properties. Any time-dependent function $f(t)$ can be expressed as a linear combination of **even** and **odd** functions as follows:

$$f_{even}(t) = \frac{1}{2}\left[f(t) + f(-t)\right] \tag{3.4}$$

$$f_{odd}(t) = \frac{1}{2}\left[f(t) - f(-t)\right] \tag{3.5}$$

such that $f(t) = f_{even}(t) + f_{odd}(t)$. Even functions are symmetric around zero [e.g. $f(t) = f(-t)$] whereas odd functions are antisymmetric [$f(t) = -f(-t)$]. The Fourier transform of the even portion of the linear combination, represented by the symbol $\Im(f_{even}(t))$, is also a real and even function of the frequency ω. In turn, the FT of the odd portion gives rise to an imaginary and odd function of frequency (see Problem 3.2). Note that Equations 3.4 and 3.5 imply a definite relationship between the even and odd functions that combine to give $f(t)$. Given values for $f_{even}(t)$, $f_{odd}(t)$ must be uniquely defined for a **causal** function (one that starts at zero time). Because of this, it is possible to generate $f_{odd}(t)$ given a complete map of $f_{even}(t)$ using a mathematical operation called a **Hilbert transform**. For a more complete description of the properties of the Fourier transform and Hilbert transform, the reader should consult Reference 1.

Now let us consider the function representing the FID. Recall that the outputs of the mixers are two signals, one cosine-modulated $\{\cos[2\pi(v_A - v_r)t + \theta]\}$ and an even function and the other sine-modulated $\{\sin[2\pi(v_A - v_r)t + \theta]\}$ and odd. These expressions are stated in terms of the difference between the resonance frequency v and the carrier frequency v_r (both in hertz). For simplicity, we will replace the term $2\pi(v_A - v_r)$ in these expressions with the more general radial frequency ω. (It is important to keep in mind that we are still dealing with the difference frequency $2\pi(v_A - v_r)$, and all the data processing is performed using these audio range frequencies.) These signals can be transformed separately to give real (even) and imaginary (odd) functions of frequency and then recombined to obtain quadrature detection. The two mixer outputs are clearly related to the Fourier transform expression by the relationship:

$$e^{-i\omega t} = \cos\omega t - i\,\sin\omega t \tag{3.6}$$

A plot of the function on the left-hand side of Equation 3.6 in complex space is shown in Figure 3.16. In this form, the expressions are readily incorporated into an algorithm for fast Fourier transformation called the Cooley–Tukey algorithm (see below). The output of the algorithm is a series of complex numbers, which can be divided into two frequency domain functions, called the **real** and **imaginary** components of the frequency-domain spectrum, respectively (see Figure 3.17).

The frequency domain spectra produced by the Fourier transformation contain components of two different types of line shape. One, the **absorptive** line shape, is one sign at all values, and is essentially the Lorentzian line described in Chapter 1. The absorptive line shape is given by:

$$A(\omega) = \frac{2T_2}{1 + T_2^2(\omega_A - \omega)^2} \tag{3.7}$$

where ω_A is the transition frequency for nucleus A, and T_2 is the characteristic relaxation time of nucleus A governing the line width of the resonance. The other line shape is **dispersive**, and is given by:

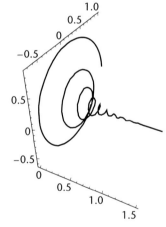

Figure 3.16 A plot of a damped harmonic function of the form $e^{-i\omega t} = \cos\omega t - i\,\sin\omega t$, plotted in complex variable space. This is the "true" form of the FID acquired in the NMR experiment. However, usually only a projection is shown on the computer screen, from one of the two mixer outputs.

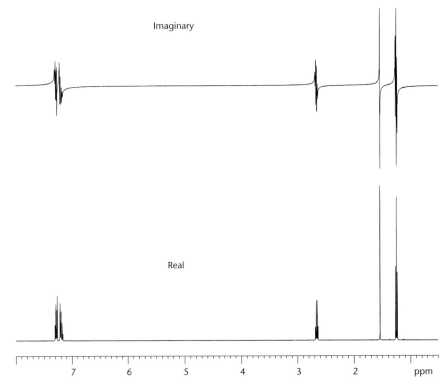

Figure 3.17 Real and imaginary frequency domain spectra obtained after FT showing pure absorptive line shape for the real spectrum and pure dispersive line shape for the imaginary spectrum.

$$D(\omega) = \frac{2T_2^2 (\omega_A - \omega)}{1 + T_2^2 (\omega_A - \omega)^2} \qquad (3.8)$$

For the dispersive line shape, as the sign of $\omega_A - \omega$ changes, so does the sign of the amplitude. It is generally desirable to view spectra in the absorptive mode, because the amplitude drops off more rapidly as one moves from the center frequency in the absorptive mode than in the dispersive mode, giving better resolution. However, it is easy to establish the center frequency of a resonance in the dispersive mode, since the amplitude is zero at that point. When one deals with very broad lines, the dispersive line shape is often informative, and electron spin resonance spectra are often viewed in dispersive mode.

Discrete Fourier transformation

Two problems resulting from the very nature of Equations 3.2 and 3.3 must be dealt with if FT is to be used efficiently in NMR spectroscopy. Firstly, both equations involve integration of continuous functions, an operation not easily handled directly by digital computers. Secondly, they are both single-variable functions, and if one wants to generate a response spectrum as a function of frequency from the time-domain FID, one must reintegrate Equation 3.2 at each discrete frequency ω in order to generate a complete frequency-domain spectrum. In other words, Fourier transformation acts as a tunable filter that looks for the presence of a particular frequency in a time-domain signal. If that frequency is not represented in the signal, Equation 3.2 will integrate to a value of zero. Unless these calculations can be done very quickly, simultaneous acquisition NMR spectroscopy would not represent much of a time-saving over the sequential acquisition method, since one would spend a long time doing the calculations. Obviously this problem has been solved or you wouldn't be reading this. A fast FT algorithm called the Cooley–Tukey (CT) algorithm allows a digital approximation of Equation 3.2 to be solved rapidly on a digital computer (for a useful starting reference on the CT algorithm, see Reference 2). The CT algorithm is widely used for all sorts of digital signal processing, including NMR spectroscopy.

The CT algorithm works in the following way. After digitization, a time-domain signal is stored as a series of two sets of n points, one set from each of the two mixer circuits, where n is the total FID acquisition time divided by the dwell time. For every point in the time domain, we will eventually generate a point in the frequency domain, so a smooth continuous frequency spectrum $F(\omega)$ will be divided up into n discrete points if n time points were collected. (One can also artificially increase n after acquisition, a process called zero-filling, which is discussed below.) A digital approximation to the FT function is obtained by evaluation of the series expansion:

$$f_k = \sum_{j=0}^{n-1} \left[\exp\left(i\omega_k t_j\right) \right] F_j \tag{3.9}$$

where f_k is the frequency-domain spectrum point corresponding to frequency ω_k, t_j is the jth time-domain point, with F_j the intensity of the time-domain signal at time point t_j. Since there are n points total, in order to evaluate for each of n frequencies, the calculation would need to be performed n^2 times in order to complete the transformation from the time domain to the frequency domain. For a spectrum with a fairly standard digital resolution of 16 384 points (2^{19}), this means that about 3×10^8 calculations would be required for the transformation. This is very large, even for a fast computer. But the CT algorithm breaks the problem down into smaller units. Equation 3.9 can be expressed as the sum of two lower-order polynomials as in Equations 3.4 and 3.5:

$$f_{even,k} = \sum_{j=0}^{n/2-1} \left[\exp\left(i\omega_k t_j\right) \right] F_j$$

$$f_{odd,k} = \sum_{j=n/2}^{n-1} \left[\exp\left(i\omega_k t_j\right) \right] F_j \tag{3.10}$$

Note that these first two divisions in fact correspond to the outputs from the two mixer circuits, where one set corresponds to the even cosine-modulated (real) data and the other to the odd sine-modulated (imaginary) data. Only $(n/2)^2 = n^2/4$ calculations are needed in order to evaluate each of these expressions, for a total of $n^2/2$ calculations. However, the same subdivision can be repeated until one obtains a set of $n/2$ polynomials, each with only two terms. To solve this set of equations would require a total of only 2^n calculations. However, calculation of the phase relationship between the resulting solutions requires some additional calculations, so a total of $2n \log n$ calculations will finally be required to complete the transform. Still, this is well within the realm of what even a modest computer can handle, making **fast Fourier transform** (FFT) a standard tool for the spectroscopist. Note that in order for the CT algorithm to work correctly, there must be 2^n points in the time-domain signal, a consideration when setting this parameter for data acquisition or subsequent zero-filling of the time-domain data.

The frequency spectrum resulting from FFT analysis is made up of the same number of points n that were present in the digitized time-domain signal. The **digital resolution** of the spectrum is the width (in frequency units) of a single point in the frequency domain, and is the total spectral width $2(v_{edge} - v_r)/n$. The spectral width is, as pointed out earlier, the inverse of the dwell time DW. Note that increasing the digital resolution of an experiment by increasing the acquisition time (i.e. increasing n) does not improve the actual spectral resolution past that determined by the natural line widths of signals in the spectrum. However, the apparent digital resolution can be improved by **zero-filling**, that is, adding a set of time-domain points with zero amplitude to the end of the FID. Zero-filling increases the number of points to be transformed (usually to some power of 2) without increasing the noise content of the spectrum. The resulting frequency-domain spectrum consists not only of the expected "real" amplitudes at the frequency intervals determined by $2(v_{edge} - v_r)n$, but a series of interpolated amplitudes between those points. Thus, there is one interpolated point between each pair of "real" amplitude points for each time the size of the time-domain data set is doubled by zero-filling. Zero-filling prior to FT often improves the appearance of the spectrum (Figure 3.18) and does not require an increase in the acquisition time or the data storage requirements of the raw data from the experiment. Therefore, zero-filling is commonly applied in processing of NMR data, especially in multidimensional NMR experiments, where additional digital resolution requires large increases in experiment time.

In most cases, collection of insufficient points (poor digital resolution) is an avoidable way of losing resolution. On most modern spectrometers, the availability of in-line processing options known as **digital filters** makes **oversampling** a viable alternative for data acquisition. Oversampling digitizes data at a frequency ($\sim10^5$ Hz) much higher than that required by the Nyquist theorem. As a result, the Fourier transform of the oversampled FID covers a large frequency range. However, if the oversampled data are treated prior to Fourier transform with an appropriate digital filter, it is possible to select the region of the spectrum that is of interest for processing. After digital filtering, the data can be treated as data acquired using standard Nyquist-restricted sampling.

One of the benefits of oversampling is an improvement in dynamic range available from the ADC for a given sample over that normally available using the Nyquist digitization cutoff. The theoretical improvement in dynamic range is proportional to the square root of the ratio of the spectral width of the oversampled spectrum to that of the desired region (the region that would normally be selected by the Nyquist limit) (see Reference 3).

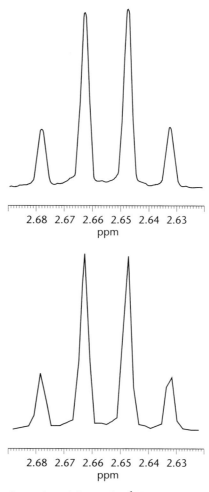

Figure 3.18 The methylene signal from the ^{1}H spectrum of ethylbenzene in CDCl$_3$ acquired on an 11.74 T magnet. The top spectrum results from acquiring the FID with 16k points and zero-filling to 32k points before FT. FT of the same FID without any zero-filling resulted in the bottom spectrum.

Digital filtering in combination with oversampling allows the selection of a spectral window of interest, while preventing aliasing of any signals resonating outside this region (Figure 3.19). Furthermore, significant random noise reduction can be obtained. We have already discussed how signals are aliased by incorrect Nyquist sampling, but noise that occurs at frequencies higher than the Nyquist frequency is also folded over (aliased) into the spectrum. In order to prevent noise foldover, the analog NMR signal has traditionally been passed through an electronic filter that removes frequencies higher or lower than the desired spectral range. Two of the most common such filters are the **Bessel filter** and **Butterworth filter**. The filter names refer to the mathematical function each filter approximates. Neither filter has a very sharp cutoff. That is, the **bandpass** region (the frequency range that passes through the filter) drops off as one moves to the edge of the filtered region, but with a distinct slope, so that signal response is not uniform over the width of the spectrum. Digital filters can be designed to give much sharper cutoffs at the

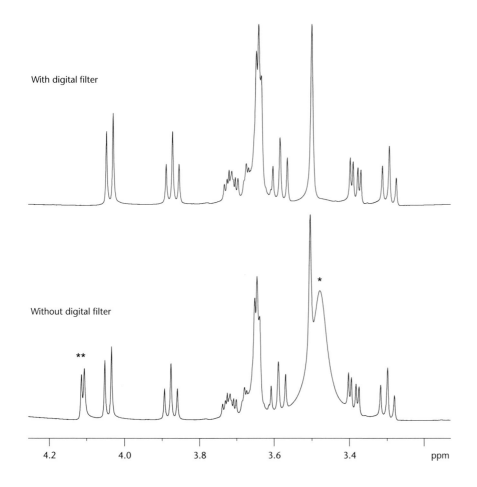

With digital filter

Without digital filter

Figure 3.19 Application of digital filtering to a sample of sucrose in D_2O, acquired on an 11.74 T magnet. The bottom spectrum shows the entire spectrum. The boxed region indicates the sweep width that is used in the middle and top spectra (565 Hz). The middle spectrum is acquired by simply reducing the sweep width. The signals highlighted (* and **) are aliased. The top spectrum was acquired with the reduced sweep width and digital filtering. The aliased signals are clearly absent.

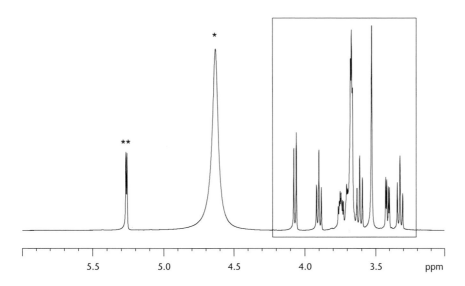

bandwidth edges, limiting noise foldover, and providing flat (uniform) responses over the desired spectral regions. This also results in flatter baselines over the spectral region of interest. (For a more detailed description of the use of digital filtering in NMR, see Reference 4.)

Off-resonance detection, that is, the placement of the carrier frequency outside of the spectral envelope of interest, is facilitated by digital filtering and oversampling. This method of detection removes artifacts associated with imperfect quadrature detection ("quad images"), low frequency noise such as 60 Hz noise from AC electronic equipment and carrier "spikes" (Figure 3.20).

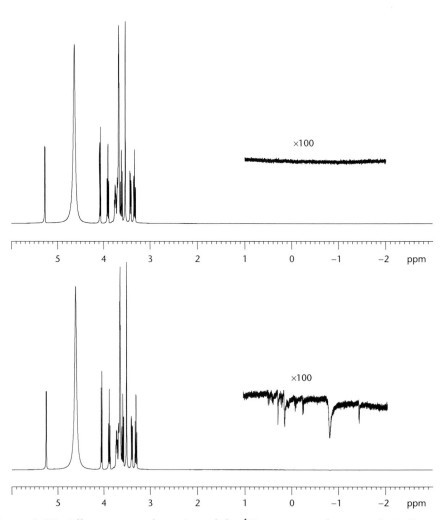

Figure 3.20 Off-resonance detection of the ^{1}H spectrum of sucrose in D_2O acquired on an 11.74 T magnet. The bottom spectrum was acquired with one scan and the usual method of detection. The inset shows the quad images. The top spectrum shows the removal of quadrature artifacts by using off-resonance detection.

Spectral phasing

Figure 3.17, which shows the transforms of the real and imaginary components of a FID, is rather simplistic in that it shows the output of the one transform of each of the mixer outputs as being absorptive, and the other as dispersive. In reality, the detector output that enters the frequency mixer will have an arbitrary phase factor ϕ. This makes it unlikely that either of the two Fourier transforms of the mixer output will contain pure absorptive or pure dispersive line shapes, unless special pre-conditions are met. Normally, the output spectra will instead be a mix of the two lineshapes:

$$\mathrm{Re}[s(\omega)] = A(\omega)\cos\phi + D(\omega)\sin\phi$$
$$\mathrm{Im}[s(\omega)] = A(\omega)\sin\phi - D(\omega)\cos\phi \tag{3.11}$$

where the $\mathrm{Re}[s(\omega)]$ and $\mathrm{Im}[s(\omega)]$ are the real and imaginary frequency-domain spectra, and the $A(\omega)$ and $D(\omega)$ are the absorptive and dispersive components of those spectra, respectively. Rearranging Equations 3.11, one gets:

$$A(\omega) = \mathrm{Re}[s(\omega)]\cos\phi + \mathrm{Im}[s(\omega)]\sin\phi$$
$$D(\omega) = \mathrm{Re}[s(\omega)]\sin\phi - \mathrm{Im}[s(\omega)]\cos\phi \tag{3.12}$$

Appropriate linear combinations of $\mathrm{Re}[s(\omega)]$ and $\mathrm{Im}[s(\omega)]$ result in spectra that are pure phase absorptive and dispersive. The phase correction factor ϕ itself consists of two parts, a frequency-independent component θ, often called the **zero-order phase correction**, and a frequency-dependent factor $\chi(\omega)$, referred to as **first-order phase correction**. That is, $\phi = \theta + \chi(\omega)$. The frequency-independent portion θ is required because the output of the detector channels will be out of phase with the carrier frequency by some arbitrary amount. The first-order correction factor $\chi(\omega)$ can be minimized by several considerations. One cause of first-order phase shift is the delay of sampling of the FID after generation of observable coherences; coherences with different precessional frequencies will advance in phase at different rates, and so the first sampling point will catch different coherences at different points in their evolution (Figure 3.21).

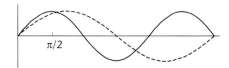

Figure 3.21 Origin of first-order phase shifts. A plot of two sine functions with angular frequencies 2 rad/s (dashed line) and 3 rad/s (solid line). Horizontal axis is time in seconds. If first data point is sampled at $\pi/6$, the lower frequency has advanced in phase by $\pi/3$, while the higher frequency has advanced by $\pi/2$. After FT, in order to make the frequency domain spectra have similar line shapes, these phase differences must be accounted for.

RF pulses and pulse phase

You may have noticed that all of the previous discussions require that all coherences be generated simultaneously. This is the case if, for example, one set of phase adjustments is to be used for the entire spectrum. Implicit in these discussions is that there is some way of exciting all of the resonances simultaneously, regardless of their chemical shift. This is accomplished by the use of **RF pulses** that are applied to the sample for a well-defined amount of time with a fixed amount of power (energy per unit time). As noted in the first chapter, a damped harmonic oscillation cannot be described by only a single frequency, and the more rapidly the oscillation is damped, the greater the range of frequencies represented in that description. A modern NMR spectrometer is capable of generating very monochromatic RF, but we cannot (and do not want to) escape the fact that any finite oscillation having a beginning and end in real time is not described by a single frequency. The shorter the amount of time the RF field is applied to a sample (the shorter the pulse), the larger the number of frequencies that are present in addition to the frequency from which the pulse is generated (the carrier frequency, ν_r). The range of frequencies extends symmetrically around ν_r, and is related to the duration of the RF pulse by the expression:

$$\Delta\nu = (\nu_{null} - \nu_r) = \tau_P^{-1} \tag{3.13}$$

where τ_P is the duration of the RF pulse and ν_{null} is the frequency (in hertz) at which the first null in the bandwidth of the pulse occurs.

The relationship between bandwidth and pulse length is analogous to the line-width phenomenon; the more rapidly a signal decays the broader the line that results from FT of that signal. In fact, the **power spectrum** of a pulse (the square of the amplitude versus frequency) is the square of the FT of the pulse (Figure 3.22). The amplitude of the pulse (and the corresponding power spectrum) is measured in terms of the magnitude of the magnetic field component of the RF, $|B_1|$. The units of $|B_1|$ are those of magnetic field intensity (e.g. gauss, tesla) but it is often more convenient to consider $|B_1|$ in terms of the effect of the field on an observed nuclear spin ensemble. This is described by the precessional frequency $\omega_P = \gamma B_1$ expected for that spin around $\vec{B}_1$, where γ is the gyromagnetic ratio of the affected nucleus. (This concept was introduced in Problem 2.7 in the last chapter.) The energy available to the ensemble of spins per unit time is thus proportional to $|B_1|$, and an RF field of sufficient magnitude to induce 2500 cycles of precession of a given nucleus around $\vec{B}_1$ per second is described as a 2.5 kHz RF field. Conversely, the extent of precession around $\vec{B}_1$ is determined by the length of time the RF is applied. If an RF pulse of magnitude $|B_1|$ is applied for a time τ_P, the magnetization due to the ensemble of spins is rotated by an angle $\theta = \gamma B_1 \tau_P$. Assuming that the carrier frequency of the RF is exactly at the resonance frequency of the nuclear spins (i.e. the RF is **on resonance**) and the net magnetization $\vec{M}$ starts from the **z** axis, the angle θ corresponds to the tip angle that $\vec{M}$ makes with respect to the **z** axis after the pulse. (We will see shortly why we require the RF to be on resonance.) If the RF is applied long enough for $\vec{M}$ to reach the **x**, **y** plane, the pulse is called a $\pi/2$ (or 90°) pulse. If the pulse is applied for twice as long as for a $\pi/2$ pulse, the magnetization

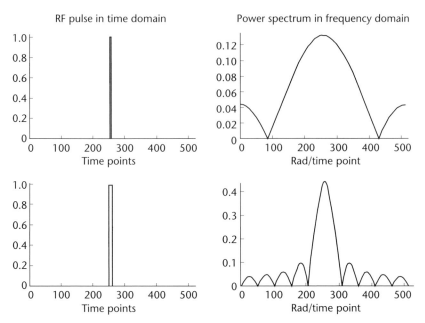

Figure 3.22 Relationship between radio frequency (RF) pulse length (time domain, shown on left) and power spectrum (frequency domain, shown on right). Pulses are simple square pulses, step functions in the time domain. The top left pulse is shorter (two time points wide in 512 time points). The bottom left pulse is nine time points wide. The power spectrum (plotted as the absolute value of the real part of the Fourier transform of the pulses) is shown in a frequency domain with units of rad/time point. At each null, the sign of the amplitude changes, but only the absolute value of the amplitude is shown. Note that the shorter the pulse the wider the frequency range before the first null.

vector reaches the −**z** axis (a π, or 180° pulse). A pulse three times as long as a $\pi/2$ pulse returns the magnetization back to the **x**, **y** plane, but with the signs of both directions opposite to that resulting from a $\pi/2$ pulse. A pulse four times as long as a $\pi/2$ pulse (2π or 360° pulse) will return the magnetization back to where it started, along the +**z** axis.

Another important consideration is the **phase** of the RF pulse. The magnitude of the torque experienced by $\vec{M}$ is determined not only by the magnitude of $\vec{B}_1$ but also by the angle ϕ between $\vec{B}_1$ and $\vec{M}$. The torque is the cross-product of $\vec{M}$ and $\vec{B}_1$, and is given by $|B_1 \cdot M|\sin\phi$. If ϕ is zero, there is no torque. Therefore, the phase of $\vec{B}_1$ (which is determined by where in the cycle the RF oscillation begins) is critical in determining how it affects a particular magnetization vector (Figure 3.23).

It is easy to see how the phase of a pulse is determined by understanding how a frequency is generated electronically. The amplitude of the RF at a given point in time is determined by a series of digital addresses that are accessed sequentially. The rate at which the addresses are accessed determines the frequency of the RF, and the first address accessed determines the phase. This series of digital inputs is then converted into an analog (smooth) RF output signal by a device called (logically) a *d*igital-to-*a*nalog *c*onverter (DAC; see Figure 3.24). Note that, in general, modern

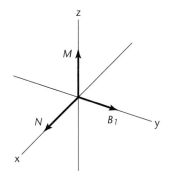

Figure 3.23 An RF field $\vec{B}_1$ lying along the **x** axis in the rotating frame of reference will exert a torque $\vec{N}$ on the magnetization $\vec{M}$ which lies along the +**z** axis. This torque will cause $\vec{M}$ to rotate around the **x** axis towards the −**y** axis (counterclockwise).

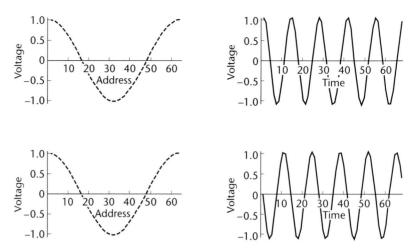

Figure 3.24 Digital generation of frequencies and phase shifts. Each address in a memory register contains a number corresponding to a voltage, as shown at upper left, and is sampled sequentially in order to generate an oscillating voltage. In the top case, every address in the 64-address register is sampled once every 0.16 s (frequency of 0.16 Hz, or an angular frequency of 1 rad/s), starting at register 1 to generate a cosine wave. If the same sampling rate is used, but sampling is begun at register 16 (amplitude of 0), as in the bottom case, a negative sine wave is generated at the same frequency, which corresponds to a phase shift of 90° in the rotating frame (from **x** pulse to a **–y** pulse).

spectrometers do not generate the very high frequencies used for spin excitation directly. Typically, a very low frequency is generated digitally, and then a series of mixer circuits is used to generate the frequency actually required for the NMR experiment.

Pulse power and off-resonance effects from RF pulses

An assumption of the previous discussions is that the excitation profile of the pulse is even for all resonances; i.e. a 90° pulse is a 90° pulse for all of the spins of interest. Another way of stating this is to say that the pulse power (energy/time) supplied is the same to all spins in the ensemble, regardless of chemical shift. However, from Figure 3.22, it is clear that the power distribution within the pulse is not even. It is greatest for a spin exactly at the carrier frequency of the pulse, and the power drops off as one gets further in frequency from the carrier in either direction. If the pulse is short enough and the frequency range of the spectrum narrow enough, the approximation that a pulse has an even excitation profile for all the resonances of interest is usually a good one, especially for ^{1}H spins, which usually have a narrow chemical shift range (~0–10 ppm) relative to their resonant frequency. However, as chemical shift ranges increase, the approximation breaks down, and one often finds that for ^{13}C, which has a lower resonant frequency than ^{1}H and a larger chemical shift range (~0–200 ppm), what is a 90° pulse for one part of the chemical shift range is not nearly as efficient at other parts of the spectrum. In fact, if the pulse is long enough, the null points in the excitation profile of the pulse may lie within the spectral envelope of interest, that is, some spins may not be excited at all by the pulse and not be detected in the FID, or they can be out of phase with other signals even if they are not aliased. Problems 3.11–3.14 at the end of this chapter deal with this important concept in detail.

For practical purposes, it is important to realize that pulse power is a critical variable in the NMR experiment. The wider the spectral range to be excited by the pulse, the shorter the pulse and the greater the power that must be supplied in order to get even excitation. Conversely, for longer, more selective pulses, the power must be lower in order to prevent damaging the NMR probe or cooking the sample. While spectroscopists discuss pulse power in terms of the magnitude of $|B_1|$, the power supplied by the spectrometer for pulse generation is usually measured in decibels (dB). The decibel is a logarithmic measure of amplifier power output in comparison to available power, $dB = 10 \log (P_1/P_2)^2$, where P_1 is the output pulse power and P_2 is a reference power level. A general rule of thumb is that a decrease in 6 dB of pulse power requires doubling the pulse length in order to obtain a similar degree of rotation of magnetization around $\vec{B}_1$ on resonance.

Phase cycling: improved quadrature detection using CYCLOPS

The heterodyned outputs of the two mixers used to determine the absolute frequency of the output signal from the NMR spectrometer (see Equation 3.1) are physically handled by two different detector circuits, and small differences between those circuits necessarily exist. Thus, the resulting spectra will be affected by small errors in cancellation of the "image" signal (that is, a signal at $v_r - v_A$ when the true frequency of the resonance is $v_r + v_A$). This residual image signal is called a **quadrature image**, an artifact that can be minimized by a hardware adjustment (Figure 3.25). However, in the presence of mixed strong and weak signals, quadrature images (or "quad images", as the cognoscenti call them) can be troublesome even on a properly adjusted spectrometer. Quad images can be largely removed by switching the jobs of the two detector circuits on alternating scans and co-adding the resulting FIDs. This is accomplished by a process called **phase cycling**: alternating the starting phase of the RF pulse so that the direction of the magnetization vector $\vec{M}$ in the $\mathbf{x}$, $\mathbf{y}$ plane of the rotating frame after the pulse is different in each step of the phase cycle. This is readily visualized by assuming that the rotating frame of reference rotates at the same frequency as the applied RF. If so, one of the circularly polarized vectors of $\vec{B}_1$ is always in a fixed position in the rotating frame. The direction along which the $\vec{B}_1$ vector points is completely arbitrary, and we can choose any direction to be "$\mathbf{x}$" in the rotating frame. What is important is that another pulse can be applied with the phase advanced relative to the first pulse by a fixed amount. If a second pulse is applied with a phase advance of +90° (clockwise) with respect to the first pulse (defined as lying along the $\mathbf{x}$ axis, and called an "$\mathbf{x}$ pulse"), its $\vec{B}_1$ vector will lie along the $-\mathbf{y}$ axis (a "$-\mathbf{y}$ pulse"). If the second pulse is advanced in phase by −90° relative to the first, it will lie along $\mathbf{y}$, and so forth.

The second consideration in phase cycling is that not only can pulse phase be changed, but the phase of the receiver output can be digitally manipulated. This means that phase cycling of the transmitter (RF pulses) can be combined with the

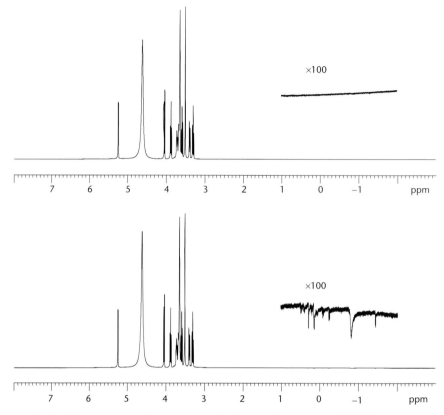

Figure 3.25 ¹H spectrum of sucrose in D₂O acquired on an 11.74 T magnet. The bottom spectrum was acquired with one scan and clearly shows quad images. The top spectrum was acquired by using four scans and applying CYCLOPS phase cycling.

receiver phase cycling to accomplish the task of removing quadrature images from a spectrum. The phase cycling scheme that is used to accomplish quad image suppression is called CYCLOPS. Like all phase cycles, CYCLOPS consists of a series of experiments in which pulse and receiver phases are incremented for each experiment. The cycle ends when the next step returns both pulse and receiver phase to the initial values. In CYCLOPS, both the pulse phase and receiver phase are incremented in +90° steps. The result is shown in Figure 3.26. In the first step, after a +**x** pulse rotates the magnetization from the +**z** axis to the −**y** axis, the output of the first mixer (arbitrarily chosen as representing modulation along **y** in the rotating frame) is a negative cosine-modulated signal, while the output of the mixer showing modulation along the **x** axis is a negative sine-modulated signal. The digitized outputs from each are placed into separate registers in the computer memory (arbitrarily marked A and B). In the second step, the pulse phase is advanced by 90° (a −**y** pulse) and the **x** and **y** receiver outputs are now negative cosine and positive sine signals, respectively. By switching which digitized mixer output goes to which computer register, the jobs of the two receiver circuits are switched, resulting in what is essentially a phase shift for both receiver circuits (−90° for **y**, +90° for **x**). In order to insure that both receivers follow the coherence faithfully, both must experience

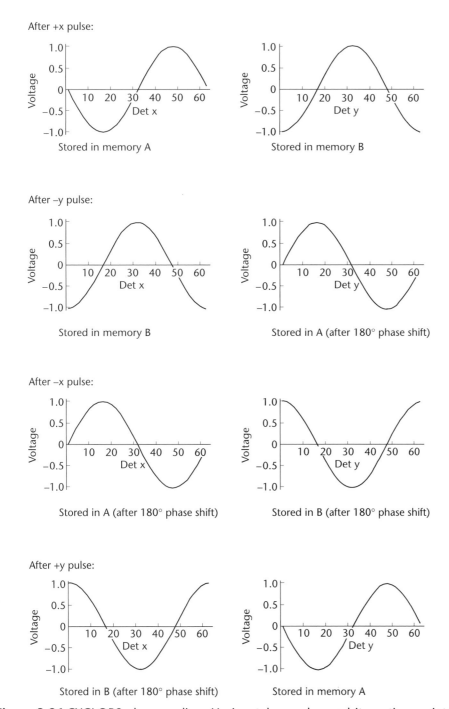

Figure 3.26 CYCLOPS phase cycling. Horizontal axes show arbitrary time points and vertical axes are mixer output voltages. Note that the phase shifts needed for non-cancellation of signals in the computer memory have not yet been applied to the voltages in the figure.

the same phase shift as the pulse. A further +180° shift on the output of the **y** detector (resulting in a change in sign) is required to give an output of the correct phase (−90° + 180° = 90°). In the third step, the pulse is now along −**x** and again the jobs of the two receiver circuits are exchanged (now a 90° shift for **y**, −90° for **x**). Since both channels are now back to where they started, an additional 180° shift must be applied to each output in order to keep up with the phase of the pulse, which has advanced by 180°, so the signs of both channel outputs are changed before storage. By the fourth step, the pulse phase has advanced by 270°, and the receivers are exchanged once again (−90° shift for **y**, 90° for **x**). Since a phase advance of −90° is the same as one of +270°, the output from **y** is stored directly into register A as is, while once again the sign of output from **x** is changed (an additional +180° shift), so both outputs are advanced in phase the same as the pulse phase. The net result of this cycling is that, although the output that is added to each computer register A and B is at the same phase after each step, the two receiver channels will each detect all four different modulations (+sine, +cosine, −sine, −cosine) at some step in the phase cycle. After a phase cycle is completed, Fourier transformation of the combined outputs of the two detector circuits (after heterodyning) will give rise to a spectrum in which any errors resulting from physical differences between the two channels are cancelled.

Factors affecting spectral quality and appearance: shimming, window functions and apodization

Many factors affect spectral quality, and experience is by far the best teacher regarding how to optimize the NMR experiment in terms of desired information from a particular sample. Nevertheless, some considerations are generally applicable and we will discuss those now.

One of the most basic considerations in optimizing spectrometer performance is the homogeneity of the magnetic field over the volume occupied by the sample. The resonance frequency of a given nucleus depends on the magnetic field strength, and if the field changes depending on where in the sample cell one looks, the result will be broad lines because not all nuclei of the same type are resonating at exactly the same frequency. (Spatial dependence of chemical shift has important practical applications in magnetic resonance imaging, in which temporary field inhomogeneities, called **pulsed field gradients**, are purposefully introduced, and this will be discussed in Chapter 11. However, for now, assume that one wishes to avoid *static* field inhomogeneities as much as possible.) One of the biggest factors controlling field homogeneity can only be controlled during the magnet installation, and that is **cryoshimming**. Most modern NMR spectrometers use superconducting cryomagnets to provide their magnetic fields. A large current is contained within a coil of superconducting material, generating a magnetic field that is designed to be relatively homogeneous and of a defined strength at the point at which the sample is placed. The coil is submerged in liquid helium at ~3 K in order to maintain superconductivity. Unless something catastrophic happens (an earthquake, attachment of large paramagnetic objects to the magnet, or someone forgetting to refill the insulated magnet container, or **dewar**, with liquid cryogens) the field can be maintained with minimal loss for very long periods of time. However, the field generated by the coil is not sufficiently homogeneous to allow high-resolution NMR to be performed without adjustment. In order to improve field homogeneity, a series of **shim coils** is also present, and by varying the amounts of current that pass through each, the homogeneity of the field at the sample point is improved. Two sets of shims are present on most magnets, the superconducting shims that are set upon installation and the user-adjustable room temperature shims. The superconducting shims are adjusted upon charging of the magnet (called "bringing the magnet to field"), and so must be properly set at that time. Because magnet installation is often performed by harried vendor engineers, it is important for the customer to pay close attention to this portion of the installation of any new magnet, since the operation of the spectrometer can be severely compromised by poor cryoshimming.

The room temperature shims (which are adjusted by the operator, and are optimized for each sample) are divided into two classes, the spinning (axial) and non-spinning (radial) shims. The **spinning shims** (Z, Z^2, Z^3, Z^4 ...) deal with inhomogeneities of the field along the normal to the magnet coils (usually also along the length of the NMR sample tube for solution state experiments) and are readjusted for each sample. They are called spinning shims because sample spinning along the normal of the field coil (usually with an air-driven turbine) averages inhomogeneities in the **x** and **y** directions of the field (with Z assumed to be along the sample length). Inhomogeneities that are not canceled by spinning, i.e. those shims that affect the Z axis homogeneity of the field must be adjusted even for spinning samples. The lower order shims (Z^1 and Z^2) generally affect line width of the whole peak, with Z^1 affecting the symmetric line shape and Z^2 affecting peak asymmetry. Z^3 and Z^4 similarly affect symmetric and asymmetric line shape, but usually lower on the resonance line. The effects of Z^5 and Z^6 are observed at the base of the resonance. It is important to remember that sample spinning *is not recommended* for most experiments involving multiple dimensions and long acquisition times, where spectrometer stability is at a premium. In these cases, the non-spinning shim sets (X, Y, X^2–Y^2, XY, XZ, YZ, etc.) should also be checked if adequate homogeneity is not obtained using the spinning shims. However, non-spinning shim adjustments are generally small if they have already been optimized for a given sample tube size and probe.

On most modern spectrometers, options for automated shimming are available. **Gradient shimming**, that takes advantage of controlled perturbations in the homogeneity of the magnetic field around the sample called **field gradients**, is a particularly useful method for automated shimming. Field gradients will be discussed more thoroughly in Chapter 7.

After the fact: window functions and zero filling

The FT of the FID as acquired is rarely the best presentation of the data for every purpose. In some cases, a slightly higher *S/N* might be desired, or an increase in resolution. It is usually possible to emphasize one or another of these spectral features by multiplication of each point in the FID by a **window function** prior to FT. A window function is a function that modulates the amplitude of the FID as a function of time in order to emphasize a desired feature or remove an unwanted one.

One of the most commonly used window functions is an exponential multiplier (EM), a function that decays in value asymptotically to zero with increasing time t. The EM is used to improve the S/N of the transformed spectrum relative to what would be observed without application of the window function. The principle behind the EM is quite simple. When one examines an FID, it is clear that most of the signal is contained in the earlier part of the decay, and most of the noise in the later part. If one were to emphasize the first part of the FID at the expense of the second, the result would be an increase in apparent S/N of the transformed spectrum. The EM function has the form:

$$\exp(-t/a) \tag{3.14}$$

where a is an adjustable multiplier corresponding to a relaxation time that in turn can be related to an observed line width. In fact, the value of a is often input in the form of a desired half-height line broadening in Hz, equal to $1/\pi a$. The application of Equation 3.14 to an FID is shown in Figure 3.27.

The improvement of S/N by the EM function is not without cost. Remember that the faster a signal decays, the broader the line into which it transforms. Any function that emphasizes the earlier parts of the FID at the expense of the later parts will broaden lines in the transformed spectrum, because the overall effect is to bring the signal to zero more quickly than was observed experimentally. Thus, the improvement in S/N comes at the expense of decreased resolution. Hence, multiplication of an FID by Equation 3.14 is often called **exponential line broadening**.

A very different effect is obtained by multiplication of the FID by a **sine bell**, a function of the form:

$$\sin(\pi t/t_{max}) \tag{3.15}$$

Signal amplitude at each time point in the FID is multiplied by Equation 3.15, where t is the time at which the point was acquired and t_{max} is the last time point multiplied by the window function (not necessarily the last point of the FID). Multiplication by a sine bell results in the de-emphasis of the signal from the earliest part of the FID (where the broadest components of the lines are found), and emphasizes those parts of the signal that are still oscillating late in the acquisition and so have the narrowest line widths. The transformed spectrum will have narrowed lines, but at the expense of a decrease in S/N (Figure 3.28). The sine bell is a harsh window function, as it essentially guts the initial signal-rich part of the spectrum. This can be ameliorated by shifting the phase of the multiplier so that the function does not start at zero. If the sine bell is shifted by 90°, a cosine bell is generated, which will broaden lines by emphasizing the beginning of the FID, much as an EM does.

Another commonly used window function is the Lorentz–Gauss transformation, often called simply a Gaussian multiplier (GM):

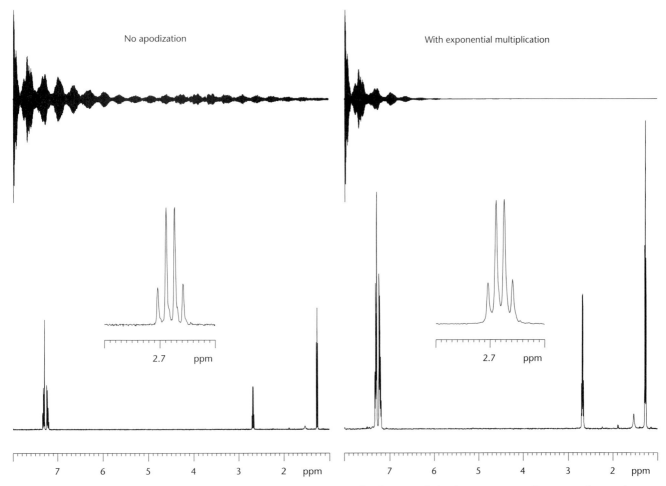

Figure 3.27 The FID as acquired is shown at the top left. At the bottom left, the corresponding transformed spectrum is shown. On the right, the FID has been multiplied by a 1 Hz exponential line-broadening function and then transformed. The sample is ethylbenzene in $CDCl_3$ and the data were acquired on an 11.74 T magnet. The expansions of the methylene proton resonance show the improved S/N and decreased resolution obtained with the exponential multiplication.

$$\exp[t/T_2 - \sigma^2 t^2/2] \qquad (3.16)$$

where T_2 is the relaxation time governing the natural line width of the resonances of interest, and σ is an adjustable factor that determines where the GM reaches a maximum value. This function converts Lorentzian line shapes to Gaussian lines, an improvement in resolution because for a given line width, the Gaussian shape has a narrower width at the base. Since the function decreases with time, a gain in S/N is obtained, but some resolution enhancement is also obtained because the function reaches a maximum at an adjustable point rather than at the beginning of the FID as with EM (Figure 3.29).

A variety of other window functions is available, each with their own advantages and disadvantages. One of the more commonly used is **convolution difference** (CD).

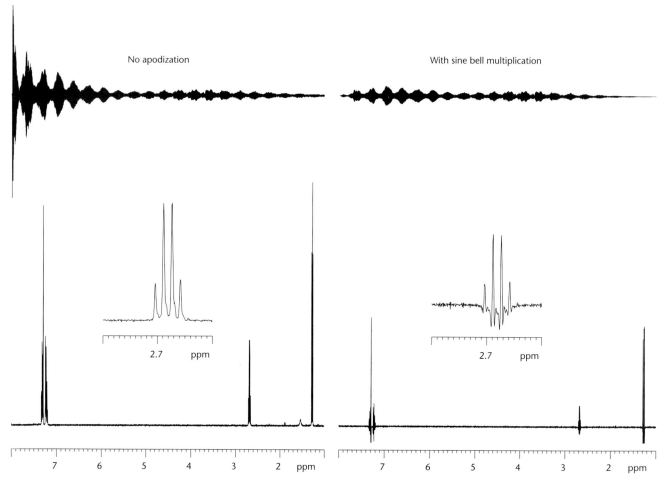

Figure 3.28 Top left shows the FID of ethylbenzene in CDCl$_3$ acquired on an 11.74 T magnet. The bottom left shows the transformed spectrum. On the right, the FID has been multiplied by a sine bell function and then transformed. The expansions of the methylene proton resonance show the improved resolution and reduced S/N obtained with sine bell multiplication.

If one applies a very harsh exponential line broadening to an FID, this eliminates virtually all of the signal except the first few data points, giving a signal that transforms into very broad lines. If one then subtracts the FID processed in this fashion from the original FID that is not modified, the result is subtraction of the rapidly decaying components of the FID, which will also be subtracted from the transformed spectrum, as is shown in Figure 3.30. Because the CD acts only near the beginning of the FID, the tailing part of the FID near the end of acquisition is unmodified. If acquisition was terminated before all of the signals were completely damped, the resulting truncation of the FID is essentially the same as would be obtained by multiplication of the FID by a step function. The Fourier transform of a step function is a sinc function (*sin x/x*), and truncation of the FID results in artifacts in the transformed spectrum known as "sinc wiggles" around the base of a transformed resonance (Figure 3.31). Generally, some other window function (such as a cosine bell) must be applied along with a convolution difference in order to avoid truncation artifacts.

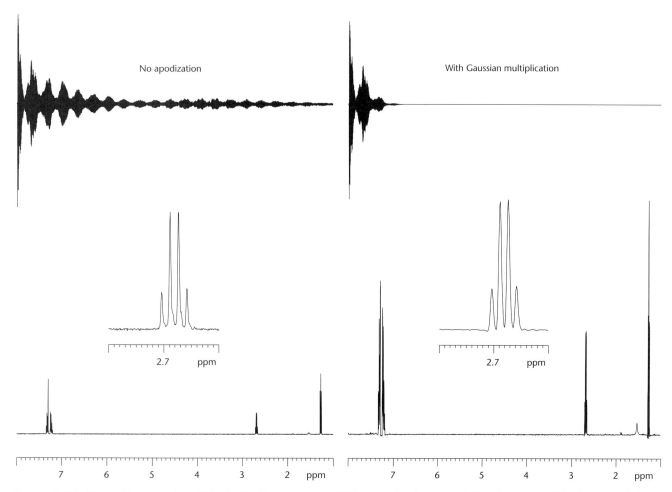

Figure 3.29 Top left shows the FID of ethylbenzene in CDCl$_3$ acquired on an 11.74 T magnet. The bottom left shows the transformed spectrum. On the right, the FID has been multiplied by a Lorentz–Gauss function and then transformed. The expansions of the methylene proton resonance show that S/N and resolution can be increased by Lorentz–Gauss multiplication.

The best window function is one that optimizes resolution at the least cost in *S/N*. This is accomplished by using a window function that fits the decay envelope of the FID as closely as possible, a so-called **matched filter**. It is usually worth spending some time to choose an appropriate window function and then optimizing the parameters so as to obtain the best combination of S/N and resolution.

Linear prediction

The use of zero-filling to improve the apparent digital resolution of a spectrum without decreasing the S/N has already been discussed. More sophisticated methods can be used to increase the number of time points (and the apparent digital resolution) of the frequency domain spectrum obtained therefrom. The most commonly used technique for improving the apparent resolution of a spectrum is called **linear prediction** (Figure 3.32). In multidimensional NMR experiments,

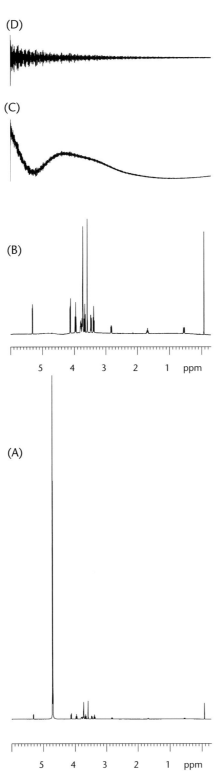

Figure 3.30 Convolution difference applied to a spectrum of 2 mM sucrose in 90% H_2O/10% D_2O acquired on an 11.74 T magnet. The FID of the sucrose spectrum as acquired is shown in (C) and the transformed spectrum is shown in (A). The intense signal at 4.7 ppm is residual H_2O (most of the water signal has already been suppressed by other means). In (D), the FID is shown after the convolution difference has been applied and (B) shows the Fourier transform of this FID. The intense H_2O signal has been removed, while the narrower sucrose signals remain largely unaffected.

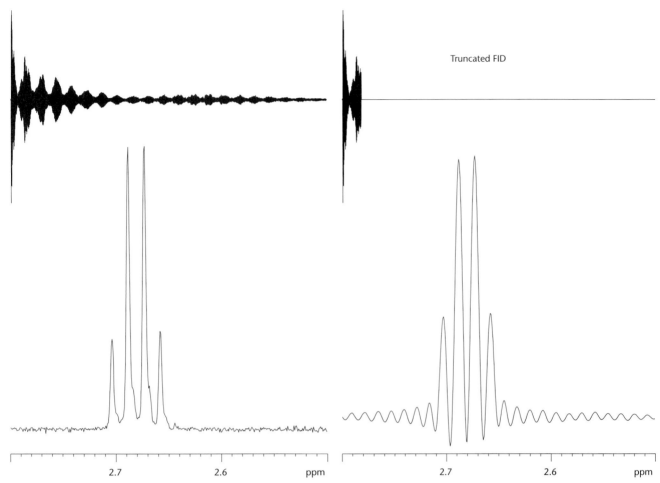

Figure 3.31 At the top left, an FID of ethylbenzene in CDCl$_3$ has been acquired for sufficient time in an 11.74 T magnet. The bottom left shows an expansion of the methylene proton resonance after FT. On the right, that same FID has been truncated and then transformed. The truncated FID can be treated as the multiplication of the complete FID by a step function. The transform of a step function is a sinc function, which gives rise to the "sinc wiggles" around the peaks in the transformed spectrum.

acquisition time is often at a premium, and it may be necessary to cut acquisition times so that the completely decayed signals are not acquired. Instead, acquisition is truncated before the signal is completely decayed. As noted above, this would lead to truncation errors if the data were not treated with a window function prior to transformation in order to bring the signal smoothly to zero. However, any window function that brings the signal smoothly to zero at the end of the acquisition time would do so by removing valuable signal intensity at the end of the signal. In cases where signal acquisition ends before the signal is completely decayed, it is often desirable to perform linear prediction on the FID. This operation extends the FID beyond the end of the acquisition time.

The concept of linear prediction is straightforward. If one knows the frequency and phase of a given signal as a function of time, it should be possible to predict the

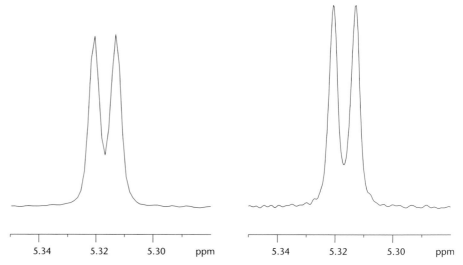

Figure 3.32 The anomeric proton resonance from the spectrum of 2 mM sucrose in 90% H_2O/D_2O acquired on an 11.74 T magnet. The spectrum on the left was obtained by acquiring an FID with 8k points and zero-filling to 32k points before FT. The spectrum on the right resulted from linear predicting to 32k points before FT.

amplitude of that signal at any point in time. Linear prediction does just this, predicting the amplitude of the FID based on the previous time-dependent behavior. Of course, using this logic *ad absurdum*, one might expect that only a few points need actually ever be really sampled, and then the rest of the points would be predicted. In actuality, improper use of linear prediction can produce serious artifacts, including false signals, and should be used judiciously. Still, most commercially

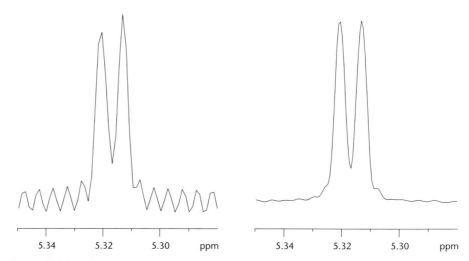

Figure 3.33 The anomeric proton resonance from the spectrum of 2 mM sucrose in 90% H_2O/D_2O acquired on an 11.74 T magnet. The spectrum on the left was obtained by FT of an FID acquired with 4k points. The spectrum on the right resulted from linear predicting to 16k points before FT.

available NMR data processing software allows one to do linear prediction. Linear prediction is also commonly used to back calculate the first point of an FID (which is incorrectly sampled on some instruments for technical reasons). After linear prediction, a window function such as a sine bell or Gaussian multiplier is applied to the data in order to get rid of truncation errors (Figure 3.33). If the window function is constructed so that only the predicted points are attenuated, not hard-won signal, the removal of truncation error is accomplished without a loss of real signal.

Problems

3.1 Electronic filters are described in terms of their transfer function A. The transfer function is the output gain (in decibel units) as a function of input frequency where $|A| = [(1 - \omega'^2)^2 + \omega'^2 d_0^2]^{-1/2}$ and $\omega' = \omega/\omega_0$ with ω_0 being the center (resonant) frequency of the filter, and d_0 being the damping factor. The Butterworth and Bessel filters are both "critically damped" filters [which means that they do not ring like a tuning fork (underdamped) or die with a thud near the edges (overdamped)]. The Butterworth filter has a d_0 of 1.414, while the Bessel filter has a d_0 of 1.732. Find the values of ω' at which each filter reaches an absolute value of A of 0.1, 0.01 and 0.001. What does this suggest about the shapes of the edges of the filter and how signals near the edge of the filter will be affected?

3.2 (a) Show that the Fourier transform of $f_{even}(t)$, the even part of $f(t)$, is given by a real even function:

$$\Im\left(f_{even}\left(t\right)\right) = 2\int_0^\infty f_{even}\left(t\right)\cos\omega t\, dt$$

(b) Show a similar relationship for the Fourier transform of $f_{odd}(t)$ demonstrating that the result is an odd imaginary function.

*3.3 A single square pulse of length τ can be represented mathematically by defining a function $f(t)$ given by $f(t) = \cos(v_0 t)$ from $t = -\tau/2$ to $t = +\tau/2$, and $f(t) = 0$ for $t \le -\tau/2$ or $t \ge +\tau/2$. The cosine Fourier transform of any function $f(t)$ is given as:

$$F\left(v\right) = \int_{-\infty}^{+\infty} \cos\left(vt\right) f\left(t\right) dt$$

In this case,

$$F\left(v\right) = \int_{-\tau/2}^{+\tau/2} \cos\left(v_0 t\right) \cos\left(vt\right) dt$$

(a) Find a suitable relationship for $(\cos x)(\cos y)$ from your high-school geometry text and substitute this formula into the one above for $F(v)$ to get this function as a sum of two simple integrals. (The math involved is the basis for the electronic mixer device described in Chapter 3.)

(b) If the carrier frequency v_0 is on the order of 500 MHz and the pulse length τ is 5×10^{-6} s, how many periods of the function $f(t)$ will occur during that pulse?

(c) We are usually interested only in frequencies that are near to ν_0, the center frequency from which the pulse is generated. Sketch the integrands of the two terms you got as answers for part (a) between $t = -\tau/2$ and $t = +\tau/2$ for the two cases where $(\nu - \nu_0) = 0$ and $|(\nu - \nu_0)| = 1/\tau$. Two of the integrals are easy to evaluate and two are very small. Show these terms and evaluate them.

*3.4 Spin populations can be thought of in a very chemical way in terms of "concentrations" of spins (as long as you are not dealing with a coherence). The usual diagram for an $\mathbf{I} = 1/2$ system using a chemist's formalism is:

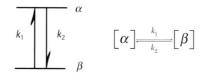

where k_1 and k_2 are rate constants for interconversion of spin populations with concentrations given by $[\alpha]$ and $[\beta]$. Calculate ΔG for this reaction for ^{1}H in an 11.74 T magnetic field. Also, calculate the equilibrium constant for this reaction (as defined by $K = [\alpha_0]/[\beta_0]$). Note that $\Delta S = 0$ for this reaction, since there is no disorder change associated with a single spin flip.

3.5 (a) The magnetization $\vec{M}_{z0}$ along the applied field $\vec{B}_0$ is proportional to the difference in the populations $M_z = n\mu([\alpha]-[\beta])$. The kinetic equation for the rate of change of α to β is $\dfrac{d[\alpha]}{dt} = -k_1[\alpha] + k_2[\beta]$, with a similar equation for $\dfrac{d[\beta]}{dt}$. Derive an equation dM_z/dt and calculate the rate constant T_1^{-1} in terms of the rate constants k_1 and k_2. (Hint: immediately convert the equations for the rate of change of $[\alpha]$ and $[\beta]$ to one involving dM_z/dt).

     (b) The rate constants k_1 and k_2 are nearly equal for ordinary NMR purposes. Why? Use this fact to derive an approximate relationship between k_1 and k_2 and T_1^{-1}. (Hint: what is the relationship between the equilibrium constant K and the k_i?)

*3.6 If an RF field is applied "on resonance" (at the same resonance as a transition), the spins become saturated with time. This can be approximated under some circumstances by the following kinetic expression:

$$\left[\alpha\right]\underset{k_2+k_{RF}}{\overset{k_1+k_{RF}}{\rightleftharpoons}}\left[\beta\right]$$

where k_{RF} is the rate of flips due to the RF field. Note that k_{RF} is the same in both directions. The thermal rates are not equal (although they are close) and reflect the temperature of the solution that produces the random motions that in turn cause relaxation (this will be treated in detail in the next chapter). Likewise, the k_{RF} is the result of interaction with the RF field. Now physicists will say that the applied RF field has an "infinite temperature". Why would anybody (even a physicist) say such a thing?

3.7   Repeat Problem 3.5 for the following case, using the chemical shorthand derived in that problem. Find the steady state $M_{z(ss)}$ by asking what happens when $dM_z/dt = 0$. Calculate the ratios $M_{z(ss)}/M_z^0$ where M_z^0 is the equilibrium magnetization you calculated in Problem 3.5.

3.8   The Bloch equations (see Chapter 4) predict $\dfrac{M_{z(ss)}}{M_z^0} = (1 + \gamma^2 B_1 T_1 T_2)^{-1}$ for an RF field on resonance, where T_1 and T_2 are relaxation times for return to the +**z** axis and overall loss of coherence, respectively. Compare this with your result from Problem 3.7 to deduce k_{RF}.

3.9   NMR is often done on mixtures of species that are chemically interconverting according to:

$$\left[A\right] \underset{k_{BA}}{\overset{k_{AB}}{\rightleftharpoons}} \left[B\right]$$

where this is an ordinary chemical equation. Generally, one might see a resonance of a proton that is not exchanged during the reaction, and if k_{AB} and k_{BA} are small enough, you would see signals from this proton due to both A and B. If the two signals are at two different frequencies ω_A and ω_B, their relative intensities will reflect the equilibrium constant K for this reaction (assuming equilibrium is reached on the time scale of the NMR experiment). If k_{AB} and k_{BA} are in the range of T_1^{-1} for this proton in either species, then saturation transfer is observed by irradiating at spin A for a relatively long time with a weak RF field, and immediately observing the ^{1}H spectrum. The result would be a lowering of signal intensity at *both* peaks A and B due to saturation at spin A only.

Write a set of kinetic equations that describe this system, and calculate the ratio of the magnetization of the spins on B with irradiation at ω_A to that without saturation. Assume for simplicity that $k_{AB} = k_{BA}$ and that the k_{RF} is big enough to equalize the populations of both spin states of A, and that there is no k_{RF} term for species B. Hint: see Problem 3.5.

*3.10   Two ^{1}H spins differ in chemical shift by 10 ppm, and are coupled to each other by a $J = 8$ Hz. Assuming a frame of reference that rotates at a frequency exactly half-way between the two spins, calculate the apparent frequency of precession of all of the vectors corresponding to the two spins in that frame of reference at 11.74 T.

*3.11   The effective field experienced by a nucleus due to an RF pulse applied along the **x** axis of the rotating frame of reference is given by:

$$\vec{B}_{effective} = B_1 \vec{x} + \Delta B \vec{z}$$

where B_1 is the applied RF field amplitude (pulse strength) and $\vec{x}$ and $\vec{z}$ are unit vectors along the **x** and **z** axes, respectively. $\Delta \vec{B}$ is the distance between the frequency of the nucleus being observed and the carrier frequency of the applied pulse. The effective field makes an angle θ with respect to the **z** axis, where $\tan \theta = B_1/\Delta B$ (see Figure 3-H11). A spin population starting exactly on the +**z** axis can never be flipped by an angle greater than ϕ, no matter how long the pulse is. What is the relationship between ϕ and θ?

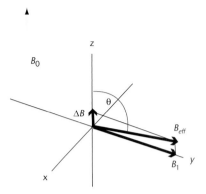

A note on units: The most convenient way to describe the magnitude of B_1 and B_0 (applied and fixed fields) is in units of hertz. Much as we say that an 11.74 T magnet is a 500 MHz magnet (meaning that protons precess at 500 MHz in such a field), so too can we describe the magnetic field component of RF B_1 in terms of the induced precession. The larger the B_1 field, the faster the precession. Therefore, a 2.4 kHz applied RF field will cause the magnetization to precess around it 2400 times per second. The total rotation induced by a pulse is given by the frequency of the pulse multiplied by the duration of the pulse in seconds. Thus, a 10 kHz pulse with a 200 ms duration will cause the nuclear spins to precess around B_1 2000 times.

*3.12 A spectrometer operating at 500 MHz for protons can rotate ^{13}C spins through a flip angle of 90° ($\pi/2$) in 30 μs. Certain multipulse NMR sequences require 180° (π) pulses on ^{13}C. The operator sets the ^{13}C pulse length for 60 μs, but finds that this does not work very well. In fact, it turns out that if ΔB is not too large, the 60 μs pulse is not even really a 90° pulse. What is the ΔB in ppm for ^{13}C at which a spin cannot be flipped by more than 90°? Compare this result with the chemical shift range for ^{13}C, which is about 200 ppm.

*3.13 Unlike a 180° pulse, a 360° pulse (that returns the magnetization to its starting position) can be generated regardless of the value of ΔB. Because of the large water signal in biological samples (protons have a concentration of ~110 M in biological samples), it is difficult to detect the low concentration (often <1 mM) of interesting signals for such samples by FT NMR. It is convenient to generate pulses that do not flip the water magnetization (or return it to the +**z** axis, by rotating it through 360°) but are 90° pulses for the signals of interest. Remember that a 90° pulse can be made as long as you want it by controlling the RF power (B_1). Design a pulse that is a 90° pulse on resonance (at the carrier frequency) in the middle of the NH region for proteins (about 8 ppm ^1H), but is a 360° pulse at the water line (4.8 ppm). Assume a 500 MHz ^1H frequency. Hint: think about the 360° pulse first. The answer should give both the pulse length and amplitude in hertz.

*3.14 Estimate how good the pulse you calculated in Problem 3.13 is as a 90° pulse (i.e. what is the angle ϕ that a spin makes starting from the +**z** axis with that axis after the pulse?) for spins that resonate at 7 and 9 ppm (i.e. 1 ppm away from the carrier frequency at which the 90° pulse was generated). Do the same calculation for the 30 μs pulse in Problem 3.12 using the ΔB calculated there. Note that in both cases, the magnetization does not end up precisely parallel to the **y** axis; some phase correction will be required.

*3.15 Justify the rule of thumb that a decrease in 6 dB of pulse power requires doubling of the pulse length in order to obtain a similar degree of rotation of magnetization around $\vec{B}_1$ for a spin on resonance, that is, a spin that has a Larmor frequency at the carrier frequency of the pulse.

References

1. F. G. Stremler, *Introduction to Communication Systems*, 2nd edition. Addison-Wesley Publishing (1982) Harlow, Essex.
2. J. Stoer and R. Bulirsh, *Introduction to Numerical Analysis*, 2nd edition. Springer-Verlag (1993) London.
3. M. A. Delsuc and J. Y. Lallemand, *J. Magn. Reson. B* 69, 504 (1986).
4. M. E. Rosen, *J. Magn. Reson. A* 107, 119 (1994).

4

NUCLEAR SPIN RELAXATION AND THE NUCLEAR OVERHAUSER EFFECT

Longitudinal (T_1) relaxation and the sensitivity of the NMR experiment

We have already alluded in general terms to the concept of nuclear spin relaxation. We now explore the basis and practical consequences of relaxation for the NMR experiment. Nuclear spin relaxation can be described by analogy to chemical reactions (Reference 1). If spins in an ensemble are able to move between spin states freely and are in contact with a heat bath, they will eventually reach thermal equilibrium, with the relative populations of each spin state determined by a Boltzmann distribution (see Equation 4.1). For nuclei with $\mathbf{I} = 1/2$, and two spin states α and β, the relative populations are given by the equilibrium constant K_{eq}:

$$K_{eq} = \frac{N_\beta}{N_\alpha} = \exp\left(\frac{-\Delta E_{\alpha\beta}}{kT}\right) \tag{4.1}$$

K_{eq} is equal to the ratio of the rate constant for the "forward" change in spin state (or a **spin flip** from α to β), $k_{\alpha\beta}$, to that of the reverse spin flip, $k_{\beta\alpha}$:

$$[\alpha] \underset{k_{\beta\alpha}}{\overset{k_{\alpha\beta}}{\rightleftarrows}} [\beta] \text{ and } K_{eq} = \frac{k_{\alpha\beta}}{k_{\beta\alpha}} \tag{4.2}$$

Because the energies of the α and β states differ by only a small amount, K_{eq} is very close to 1, so that the rate constants are essentially the same for either flip. This has important consequences for practical NMR. One is that any process that encourages spin flips in one direction will also encourage them in the other. This is precisely analogous to the concept of catalysis in chemistry and biology. Chemical catalysts affect the rate at which a reaction reaches equilibrium (i.e. they lower the activation barrier between products and reactants), but do not affect the relative concentrations of products and reactants at equilibrium. For this reason, one can never build up a very large population difference between two connected spin states, except under special circumstances. Restated, $\Delta N = N_\alpha - N_\beta$ will never be larger than is permitted by Equation 4.1 for two isolated connected states. (This is phrased carefully because many of the tricks employed in multinuclear NMR are designed to get around this limitation). As signal intensity is proportional to ΔN, the sensitivity of the NMR experiment is severely limited. A consequence of Equation 4.2 is that the return to thermal equilibrium will be slow relative to the rates of the forward and reverse reactions (as determined by $k_{\alpha\beta}N_\alpha$ and $k_{\beta\alpha}N_\beta$, respectively), as these are nearly equal. Again, since the observation of an NMR signal depends on there being an excess of nuclear spins in one state or the other ($\Delta N \neq 0$), no signal will be observed if ΔN is exactly zero. If the relative state populations are perturbed (as happens when an RF pulse is applied prior to detecting an FID) one needs to wait for the system to move a significant fraction of the way back towards equilibrium before repeating the experiment. This **relaxation delay** is necessary because otherwise, each successive RF pulse would further reduce ΔN until finally no signal at all is observed.

How does one measure the rate of return of an ensemble of spins to thermal equilibrium experimentally? Turning once again to the classical model for bulk sample magnetization, we recall that the small excess of spins aligned with the imposed static field $\vec{B}_0$ results in a magnetization vector $\vec{M}_0$ along the +**z** axis with a magnitude proportional to the differences in the populations between the upper and lower energy spin states. Application of an RF pulse with a magnetic field component $\vec{B}_1$ along the one axis of the **x**, **y** plane in the rotating frame results in precession of +**z** around $\vec{B}_1$ with an angular frequency given by γB_1, so that a pulse of length τ_P will result in a tip angle $\theta = \gamma B_1 \tau_P$ between $\vec{M}_0$ and the +**z** axis. Assume that the RF pulse is applied long enough for a tip angle of 45° ($\theta = \pi/4$) to be reached relative to the +**z** axis. There is now a component of $\vec{M}_0$ along the +**z** axis with a magnitude $M_z = M \cos\theta$ and a component in the **x**, **y** plane of magnitude $M_{x,y} = M_0 \sin\theta$. Since the magnetization vector along +**z** is 0.71 times the size it is at equilibrium, we can interpret this as the pulse having reduced the population difference between the two spin states ΔN by a factor of 0.71. As there is now a nonzero component in the **x**, **y** plane, the pulse has also generated detectable coherence among the spins (i.e. there is a nonrandom distribution of precessional phases among the nuclear dipole vectors). Increasing the tip angle of the pulse to 90° ($\theta = \pi/2$) results in no component of the magnetization along **z**, that is, a population difference of zero exists between the two spin states. Increasing the tip angle to 180° ($\theta = \pi$) results in the precession of the magnetization to the −**z** axis and no component in the **x**, **y** plane. This situation corresponds to a **population inversion**; the population excess that started out in the lower energy α state now resides in the higher energy β state. For this reason, a 180° pulse is often referred to as an **inversion pulse**.

Now consider what happens to the ensemble after an inversion pulse is applied. Immediately after the pulse, there is a vector of magnitude M_0 along the $-\mathbf{z}$ axis. The ensemble will begin to move back towards equilibrium, losing energy to the surroundings via processes that will be discussed in detail later. In the rotating frame, we can imagine the magnetization vector along the $-\mathbf{z}$ axis shrinking as the population excess in the higher energy state begins to drain back to the lower energy state. At some point there would be no net magnetization (that is, no population difference between the α and β states, and no vector along $\mathbf{z}$ in either direction. Finally, at some time after the experiment began, the situation would return to the original state, with all of the original magnetization lying along the $+\mathbf{z}$ axis.

If one wishes to measure the rate of return to thermal equilibrium of the ensemble, a 90° ($\theta = \pi/2$) RF pulse can be applied at a time τ following the inversion pulse. This pulse returns the magnetization to the $\mathbf{x}$, $\mathbf{y}$ plane, and the resulting signal is detected (Figure 4.1). This second pulse is often called a **read pulse**, since it permits one to "read" the state of the system after it evolves for a time τ. The intensity of the signal observed after the read pulse is proportional to the population difference between the two spin states at the time the read pulse is applied. If the net magnetization vector still lies along $-\mathbf{z}$ at the time of the read pulse (that is, the population of the β state still exceeds that of the α state) the phase of the observed signal will be inverted relative to that observed if no 180° pulse had been applied to $\vec{M}_0$ prior to the read pulse. However, if the delay time between the pulses τ is long

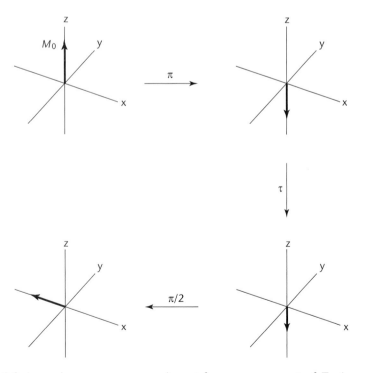

Figure 4.1 Inversion recovery experiment for measurement of T_1. A π pulse is applied to the equilibrium magnetization $\vec{M}_0$, inverting the equilibrium spin state populations. After a time τ, the spin distributions have moved back towards equilibrium, giving a small net magnetization vector in the $-\mathbf{z}$ direction.

enough, the net magnetization has time to return to the $+\mathbf{z}$ axis (the lower energy state is once again the more populated), and the phase of the detected signal after the second pulse will be the same as if no 180° pulse had been applied. If τ is sufficiently long, the intensity of the signal will return to that observed starting from equilibrium (i.e. as if no 180° pulse had been applied).

The 180–τ–90–detect series is called an **inversion recovery sequence**, and is the simplest example of a **multiple pulse NMR experiment** (Figures 4.1 and 4.2). The signal intensity S detected in the inversion recovery sequence is proportional to the magnitude of the magnetization M_z remaining along the $\mathbf{z}$ axis after time τ. The rate of return of S to an equilibrium value S_0 is given by the expression:

$$\frac{d}{dt}S = \frac{-(S - S_0)}{T_1} \tag{4.3}$$

where T_1 is a constant with units of time. Formally, T_1 is the inverse of the first-order decay rate constant R_1 that has units of inverse time. Integration of Equation 4.3 with respect to time, and evaluation of the constant of integration from the condition that at $\tau = 0$, $S = -S_0$ (that is, the observed signal is proportional to $-M_0$ when there is no delay between the inversion pulse and the read pulse) gives:

$$S = S_0 - 2S_0 \exp(-t/T_1) = S_0 - 2S_0 \exp(-R_1+) \tag{4.4}$$

By fitting the observed signal intensity (Figure 4.2) as a function of time to this expression, the rate constant for the exponential decay R_1 is obtained (see Problem 4.1).

Expressions 4.3 and 4.4 deal with the return of spin state populations in an ensemble to equilibrium after those populations have been perturbed in some way. Since these population differences are reflected in the rotating frame by the magnitude of M_z (the longitudinal component of the net magnetization), the characteristic time T_1 is often called the **longitudinal relaxation time**. The longer T_1 is, the longer it takes for thermal equilibrium to be re-established after a perturbation.

Now consider what T_1 relaxation means. Since the processes that cause T_1 relaxation are those that tend to restore the system to thermal equilibrium, they involve transfer of energy between the spin ensemble and the surroundings. In solids (where many of the first NMR experiments were performed), this implies a transfer of energy from the spins to the lattice, so another commonly used name for T_1 relaxation is **spin-lattice relaxation.** The ensemble at equilibrium does not differ significantly in entropy (order or information content) from the ensemble after a population inversion, so the return to thermal equilibrium as reflected by T_1 can be thought of as an enthalpy-driven process.

Transverse (T_2) relaxation and the spin-echo experiment

Let us now reconsider the one-pulse NMR experiment. Starting with the equilibrium magnetization $\bar{M}_0$ along $+\mathbf{z}$, a 90° pulse is applied, tipping the magnetization completely into the $\mathbf{x}$, $\mathbf{y}$ plane. This generates the maximum observable coherence

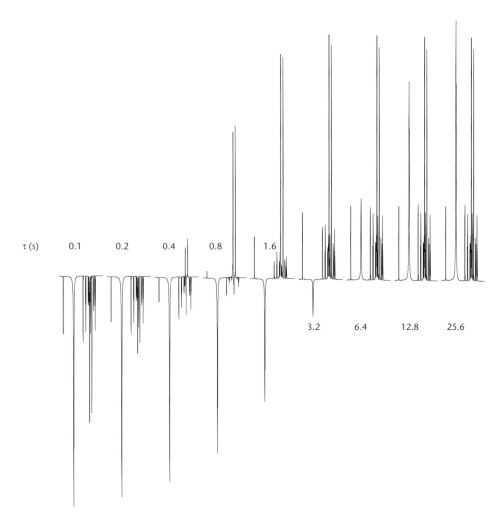

Figure 4.2 A series of transformed spectra corresponding to a ^{1}H T_1 inversion-recovery measurement on a 1 mM sample of sucrose in D_2O. The experiment was performed at 14 T (600 MHz ^{1}H). The recovery time is listed (in seconds) for each spectrum. The intensity as a function of time for two protons, the anomeric proton and the water signal, are provided in Problem 4.1.

in the ensemble, and detection of the FID immediately after the pulse ended would yield the largest possible signal. With time, the signal decays. There are several reasons for this decay. One reason is, of course, that T_1 relaxation returns some magnetization to the +**z** axis. But this is not the only reason for loss of coherence. Another process can result in a transfer of spin energy between two nuclei that are in opposite states. This process, called **spin–spin interaction**, results in a loss of phase memory (coherence) that in turn reduces the magnitude of the detectable **x**, **y** component of the magnetization. Spin–spin interactions will be discussed in detail later in this chapter.

Experimental factors can also speed up the loss of coherence. If the magnetic field is inhomogeneous (perhaps because of poor shimming), nuclei that should resonate at the same frequency exhibit slight differences in resonance frequency depending on their physical location in the sample tube. Pictorially, one can think of this process as a "fanning out" of the **x**, **y** component of the magnetization with time into a series of smaller vectors, some of which move more quickly than the rotating frame, some of which move more slowly. The different rates of precession correspond to different Larmor frequencies.

The rate at which coherence is lost is characterized by the **transverse relaxation time**, T_2. If we do not need to differentiate between the various factors contributing to loss of coherence, we can get an estimate for the apparent transverse relaxation time T_2^* from the FID envelope. A tracing connecting the tops of the maxima of the FID can usually be approximated by an exponential decay, and the rate equation dealing with this decay is simply:

$$\frac{d}{dt}S = \frac{-S}{T_2^*}$$
(4.5)

which, after integration with respect to time and evaluation of constants of integration, yields:

$$S(t) = S_0 \exp(-t/T_2^*)$$
(4.6)

where $S(t)$ is the amplitude of the decay envelope as a function of time, and S_0 is the intensity at $t = 0$. The rate constant for the decay is T_2^{*-1}. An even easier way to get an estimate of T_2^* for a given resonance is from the line width of that resonance in the transformed spectrum, recalling that the line width at half-height is given by $\Delta v_{1/2} = (\pi\tau)^{-1}$, where τ is the characteristic relaxation time of the resonance. We now can formally identify this relaxation time as T_2^* (see Figure 4.3).

If we want to separate the experimental contributions to T_2^* (field inhomogeneity) from those that are due to inherent behavior of the ensemble of nuclei (spin–spin and spin–lattice interactions) we can use a multiple pulse sequence called the **spin-echo experiment** (Figures 4.4 and 4.5). As with other experiments discussed so far, we start with the equilibrium magnetization $\vec{M}_0$ along the +**z** axis. A 90° pulse is applied along the **x** axis, tipping the magnetization completely into the **x**, **y** plane so that it lies along +**y**. Assume for convenience that the Larmor frequency of the spins is the same as that of the rotating frame, so no Larmor precession is expected in the rotating frame. As time passes, coherence is lost both to irreversible processes (T_1 relaxation and spin–spin interactions) and inhomogeneities in the field, resulting in the fanning out of individual components of the coherence (called **isochromats**) as described above. If, after a time τ, a 180° pulse is applied along the **x** axis, the **y** components of the isochromats are rotated by 180° from the +**y** to the −**y** axis (Figure 4.4). The **x** components parallel to $\vec{B}_1$ are unaffected by the pulse. After the pulse, those components that were moving faster than the rotating frame and were ahead in phase are still moving with the same angular velocity, but are now behind in phase, while those that were behind in phase prior to the 180° pulse are still moving more slowly than the rotating frame, but are ahead in phase. The result is that

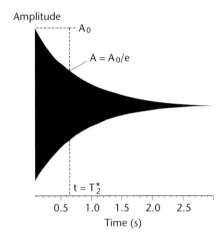

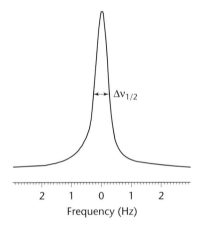

Figure 4.3 Relationship between T_2^* and the observed line width for an NMR resonance. T_2^* is the time required for the FID to decay from A_0 to A_0/e. In turn, the line width at half-height is given by $v_{1/2} = (\pi T_2^*)^{-1}$. T_2^* calculated from the line width (0.494 Hz at half-height) is 0.644 s. Calculated from the envelope, a value of 0.67 s is obtained. Errors may arise either from the imprecision of the first point in the FID or from imperfect line shape.

the faster isochromats catch up with the slower ones, resulting in a **refocusing** of the individual isochromats into a single vector at a time 2τ. This would be reflected in the FID as a new maximum, or **echo**, in signal intensity. Of course, the isochromats will continue to precess at their different angular velocities, and so will once again begin to dephase after the echo. However, application of another 180° pulse at time 3τ results in a new echo at 4τ, and so on. The echo each time would be lower in amplitude than the previous echo due to irreversible relaxation processes, but the decay envelope generated by the train of echos would show loss of coherence due only to irreversible processes, not field inhomogeneities (see Figure 4.5). The relaxation time characteristic of the irreversible loss of coherence obtained from the echo decay envelope in the spin-echo experiment is the "true" T_2, and is calculated using the same type of equation as T_1, except that the initial signal intensity (at

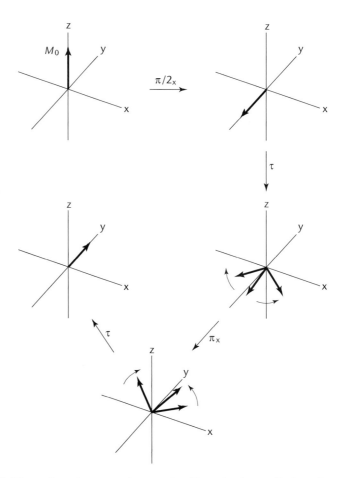

Figure 4.4 The spin-echo experiment. A $\pi/2_x$ pulse is applied to the equilibrium magnetization, tipping the magnetization to the $-y$ axis. The coherence dephases (spreads out) into several isochromats (represented by the separated arrows) during the time τ, after which a π_x pulse is applied. As the directions of precession of the individual isochromats (represented by the shaded curved arrows) are not affected by the π_x pulse, the individual isochromats refocus into the single coherence after a second time τ along the $+y$ axis. The spin echo is at a maximum after the second τ period (when all isochromats are refocused), after which the coherences once again begin to separate.

$\tau = 0$) is given by M_0, not $-M_0$. Processes contributing to T_2 result in the loss of coherence in the **x**, **y** (transverse) plane, therefore T_2 is often called the **transverse relaxation time**. Because spin–spin interactions are important in the loss of coherence, but do not necessarily contribute to T_1 relaxation (that is, they occur without energy loss to the surroundings), relaxation characterized by T_2 is also called **spin–spin relaxation**.

Obviously, processes that cause T_1 relaxation also result in T_2 relaxation. However, the reverse is not necessarily true: spin–spin interactions do not necessarily result in T_1 relaxation, but contribute to T_2. Unlike T_1 processes, which result in changes in state *populations*, T_2 processes result in the loss of *coherence*. Any process that results

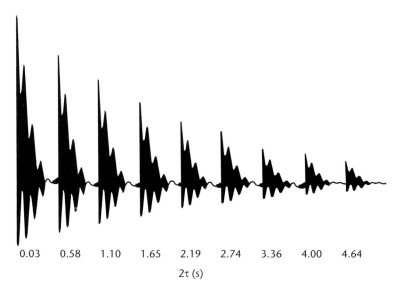

Figure 4.5 FIDs resulting from a series of spin-echo experiments for measuring the T_2 of ^{19}F in a 0.1 M sample of tetrabutylphosphonium tetrafluoroborate in $CDCl_3$. The maximum of each decay envelope provides the data that are used to calculate T_2 (see Problem 4.2).

in a loss of coherence in the ensemble contributes to T_2 relaxation. Using a thermo-dynamic analogy once again, T_2 relaxation contributes to a loss of order (and information) in the ensemble without necessarily reflecting a change in the energy of the system. As such, T_2 relaxation can be thought of as entropy-driven (see Figure 4.6).

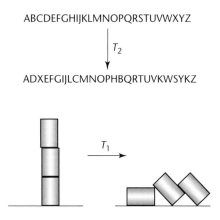

Figure 4.6 Entropically driven versus enthalpically driven relaxation processes. The letters of the alphabet, arranged in order, represent a highly ordered (coherent) ensemble. If one "detects" the first few letters, one can easily predict the order of the remaining letters (i.e. there is a high information content). "T_2" processes that randomly interchange letters ("letter–letter relaxation?") reduce the information content and increase the entropy of the system. After such processes, it is more difficult to predict what the next letter will be in the alphabetical progression. "T_1 processes" take a nonequilibrium system (cans stacked on top of each other) and lower the net energy of the system by releasing energy to the surroundings (the cans are knocked over).

Chemical shift and *J*-coupling evolution during the spin echo

At the echo maxima $(2\tau, 4\tau, 6\tau \ldots)$ in the spin-echo experiment, the effects of different rates of precession of individual isochromats due to chemical shift differences are removed (i.e. spins with different chemical shifts all end up on the same axis in the rotating frame at the echo maximum). This is because the torque exerted on the different isochromats by the laboratory magnetic field remain unchanged after the π pulses (the directions of precessions are unchanged). But what about other types of coherent evolution (e.g. *J*-coupling)? To visualize what happens during a spin echo to evolution resulting from *J*-coupling, imagine the evolution of coherence on spin *A* due to a coupling with another spin *X*. Half of the *A* spins see an *X* spin in the α state [with $\vec{\mu}_z(X)$ pointing along +z], while the other half see *X* in the β state ($\vec{\mu}_z(X)$ along –z). If we imagine that the coherent magnetization due to *A* will precess clockwise around the $\vec{\mu}_z(X)$ along +z and counterclockwise around $\vec{\mu}_z(X)$ along –z, we would observe two isochromats separating with time, one moving in a clockwise direction in the **x**, **y** plane and the other precessing counterclockwise, with a relative difference in their precessional frequencies of J_{AX}, the coupling constant between spins *A* and *X*. But what happens when a 180° pulse is applied, as in the spin-echo experiment? Besides flipping the coherences as shown in Figure 4.4, the two vectors corresponding to $\vec{\mu}_z(X)$ along +z and –z also interchange. (Since this is a homonuclear experiment using a nonselective pulse, both the *A* and *X* spins experience the 180° pulse.) Thus, the *A* spins that were seeing *X* spins in the α state prior to the 180° pulse are seeing *X* spins in the β state afterwards, and vice versa. Hence, the direction of their precession changes, and unlike the isochromats that separated during the experiment due to chemical shift differences, these vectors do not refocus at 2τ. If the time between the 90° pulse and 180° pulse in the spin echo is set to $\tau = 1/4J_{AX}$, the isochromats due to *J*-coupling will be oriented 180° opposite of each other in the **x**, **y** plane at the same time that the chemical shift isochromats refocus, and pulses applied along the axes where the chemical shift refocusing takes place *precisely at the time of the echo maximum* will affect only the *J*-separated isochromats (Figure 4.7). The spin echo thus provides a means of separating chemical shift effects from *J*-coupling effects, and virtually every heteronuclear NMR experiment employs a spin-echo pulse sequence somewhere to achieve exactly this result.

The reader is cautioned that while the model described here for the separation of chemical shift and *J*-coupling effects by the spin-echo experiment is a useful intuitive picture, it should be used with caution, because once the quantum mechanical descriptions of pulse NMR provided by product operators are introduced, we will find that the operator description of the spin echo is precisely the opposite of this picture (i.e. the chemical shift operator changes sign, and the *J*-coupling does not).

Mechanisms of nuclear spin relaxation in liquids and the spectral density function

A point that you may already be tired of hearing (but which is worth repeating nonetheless) is that NMR is different from most other forms of spectroscopy in that relaxation of nuclear-spin-excited states usually takes place on a much longer time

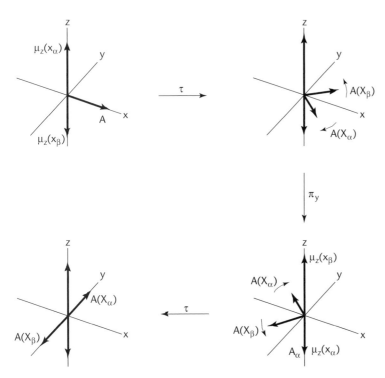

Figure 4.7 Pictorial model for J-coupling evolution during a spin echo. The components $\bar{\mu}_z(X)$ corresponding to the two spin states, α and β, of spin X are shown along the positive and negative **z** axis. The isochromats corresponding to the coupled spin A evolve in the transverse plane according to the torques exerted by the two spin X populations. After a time τ, the X populations are inverted by a π_y pulse (that also rotates the **x** components of both A vectors by 180°) and the directions of the torques on both A vectors are reversed, resulting in them being out of phase by 180° after a second time τ.

scale than is typical in optical spectroscopy. Besides the slow approach to equilibrium discussed above, slow relaxation in NMR is also related to the fact that under most circumstances the rates of the actual relaxation events (spin flips) are not very fast, at least when compared with other spectroscopic methods. We now consider what causes spin flips to occur and what effects are important in determining the rate at which they do.

The quantum mechanical description of a stationary state for a quantum system is provided by an eigenfunction of the Hamiltonian operator obtained by solving the time-independent Schrödinger equation. We have labeled those eigenfunctions for the stationary states of an $\mathbf{I} = 1/2$ nucleus as α and β. However, in order for the system to move from one stationary state to another, a "mixing" of those states must occur, that is, the correct description of the system will no longer be the time-independent α or β, but a linear combination of the stationary state descriptors of the form $\psi = c_1(t)\alpha + c_2(t)\beta$, where the $c_i(t)$ are coefficients that vary with time. In order for such a mixing of states to occur, some time-dependent perturbation must take place, and that perturbation *must occur at the frequency of the transition involved* in order for the transition to occur. Clearly, the application of an RF field at the frequency of the transition between α and β provides such a time-dependent perturbation. But what about relaxation? A perturbation at the correct (Larmor) frequency is required to mix the states in order also to get a transition back from the upper to the lower state. *These perturbations must originate from the local environment of the spins.* The random motion of nuclear dipoles in a magnetic field, such as occurs when molecules tumble in solution, results in the generation of electromagnetic "white noise", random perturbations in the local electromagnetic environment over a wide range of frequencies. To help picture this, imagine each nucleus to be a little radio transmitter/receiver. As the molecules containing the spins tumble in solution, the nuclei reorient in the applied magnetic field, generating local fluctuations in the electromagnetic environment. These fluctuations are detected by nearby nuclei, encouraging them to relax. Nuclei in an environment poor in local EMR fluctuations of the correct frequency will take longer to relax. A hypothetical nuclear spin isolated far from any source of EMR fluctuation would never relax from a higher energy state to a lower one because the environment did not present it with fluctuations in the electromagnetic environment of the correct frequency. (As you may have guessed by now, **spontaneous emission** is not an important pathway for nuclear spin relaxation. All nuclear spin transitions must be stimulated by magnetic field fluctuations of some sort.)

The more rapidly the molecule containing the spins tumbles, the higher the average frequency of the white noise generated. We can get an idea of the efficiency with which a particular transition will relax if we can characterize the "noise spectrum", i.e. the range of frequencies that can be found in the white noise generated by molecular (and hence nuclear) motions in solution. To do this, we need to describe the **autocorrelation function** for the fluctuations that give rise to the noise. An autocorrelation function $C(\tau)$ relates the probability of finding a system in a particular state at time $t + \tau$ as a function of the state it occupied at some earlier time t, and takes the form:

$$C(\tau) = \langle f(t) f(t + \tau) \rangle \qquad (4.7)$$

where the brackets indicate an average over the whole ensemble (i.e. one autocorrelation function is sufficient to describe the whole ensemble). If this sounds complicated, think of $C(\tau)$ as a measure of the "memory" of a system; if the system starts in a particular state, how long will it be before the system no longer "remembers" where it started? Consider a molecule oriented along the **z** axis in the laboratory frame of reference. The molecule undergoes random isotropic motion due to collisions, so it will not stay oriented along the **z** axis. However, if we wait only a short time before checking, chances are the molecule would not have strayed very far from its original orientation. The longer we wait, the more likely the molecule would have assumed an orientation very different from its starting position (the more likely "memory" of the original position is lost). In the simplest form, the decay (like relaxation) would be assumed to be exponential with respect to the absolute value of τ:

$$C(\tau) = \exp(-|\tau|/\tau_c) \qquad (4.8)$$

with a characteristic decay time of τ_c. τ_c, called the **correlation time**, is completely analogous to a relaxation time. If all the molecules started from the same orientation, τ_c would be the time it would take for the number of molecules still in the original orientation to decay from N_0 to N_0/e. In practice, one can think of the correlation time as the mean time required for a molecule to tumble 1 radian in any direction. In the context of nuclear spin relaxation, the correlation time is also a measure of the rate at which local magnetic fields change sign. Small molecules in nonviscous solvents tend to have correlation times of about 10^{-12} s, while for larger molecules (proteins, for example) correlation times range from 10^{-10} s upwards.

In order to extract a spectrum of the frequencies present in the white noise generated by random motion of nuclei in the ensemble, we use the Fourier transform of the time-domain autocorrelation function:

$$J(\omega) = \int_{-\infty}^{\infty} \exp(-|\tau|/\tau_c)\exp(i\omega\tau)d\tau \tag{4.9}$$

This yields the spectral density function $J(\omega)$:

$$J(\omega) = \frac{2\tau_c}{1 + \omega^2\tau_c^2} \tag{4.10}$$

The spectral density, when plotted as a function of ω, provides a **power spectrum** for the white noise generated by molecular motions (Figure 4.8). Note that the curves of $J(\omega)$ versus $\log_{10}\omega$ are flat from zero frequency to very close to the frequency that is the inverse of the correlation time τ_c, and then drops rapidly towards

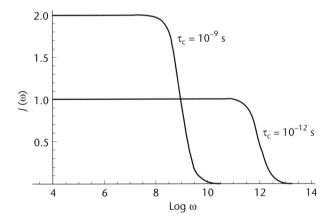

Figure 4.8 Plots of the power spectrum generated by the spectral density function (Equation 4.10). Frequency axis is given in $\log_{10}$ of the angular frequency ω. The spectral density $J(\omega)$ (vertical axis) is given in arbitrary units. Two curves are shown, one spectrum generated for a short correlation time, $\tau_c = 10^{-12}$ s, typical of a small molecule in a nonviscous solvent. Note that there is intensity from low frequencies out to the inverse of the correlation time at 10^{12} rad/s. The other spectrum is generated for $\tau_c = 10^{-9}$ s, and shows intensity only out to 10^9 rad/s. The areas under both curves are equal.

zero. This means that power required for enabling relaxation is equally available from the white noise for transitions with resonant frequencies ω over the range where $\omega^2\tau_c^2 \ll 1$. If all of the relevant transitions are equally represented in the power spectrum of the sample, this is described as the **extreme narrowing limit**. However, if there are frequencies of interest that are not represented in the power spectrum (frequencies much higher than τ_c^{-1}), the white noise no longer provides the necessary energy to stimulate those transitions, so relaxation of nuclear transitions with frequencies higher than τ_c^{-1} will be slow.

A useful analogy can be made between the spectral density and RF pulses. We have already seen that the shorter the time the RF pulse is applied (narrow pulse width), the wider the range of frequencies that are excited by that pulse. A long low-power pulse is used to excite a narrow range of frequencies, whereas a short high-power pulse is used for nonselective excitation. Now consider the origin of local fluctuations in the electromagnetic field, the motion of nuclear spins. Rapid tumbling, with short correlation times, produces short "pulses" and a correspondingly wide nonzero spectral density. Slow tumbling corresponds to long pulses, with a narrower frequency range of nonzero spectral density.

Dipolar relaxation and the nuclear Overhauser effect

From the previous discussion, we see that two nuclei that are close enough in space to detect each other's magnetic fields can **cross-relax**, i.e. stimulate relaxation in each other. We have also seen that two nuclei that are close enough to interact with each other by through-space interactions will be coupled via dipolar interactions (**D-coupled**). Recall that no splitting is observed in isotropic liquids due to D-coupling, since the D-coupling constant is modulated by the reorientation of the internuclear vector relative to the laboratory magnetic field, and in isotropic liquids, the time-average value of the D-coupling will be zero. However, the effects of D-coupling are still observed in solution in the form of cross-relaxation and a phenomenon known as the **nuclear Overhauser effect** (**NOE**). The NOE has become a topic of great importance in recent years, and only a general introduction to the concept will be provided here. The interested reader is directed to References 2 and 3.

If two nuclei are coupled, either by dipolar or scalar interactions, the proper description of the two-spin system is not given by the one-spin eigenfunctions (α and β) of the individual nuclear Hamiltonians, but rather by linear combinations of the eigenfunctions of the individual spins. These are eigenfunctions of the Hamiltonian appropriate for the two-spin system. We will discuss how one goes about deriving these eigenfunctions in the next chapter. For now, we simply represent the two-spin system (spin I and spin S) as a series of four states, with functions $\alpha\alpha$ (both spins in lower energy states), $\beta\alpha$ and $\alpha\beta$, (spin I in the higher energy state and S in the lower energy state and vice versa) and $\beta\beta$ (both spins in the higher energy state). An energy diagram of the four states and the transitions between them is shown in Figure 4.9. The rate constant for a particular transition is given by W_i. Multiplication of W_i by the population of the starting state N gives the rate for a particular transition. Three types of transitions are possible in this system. The four transitions marked with either W_{1I} or W_{1S} are **single-quantum** transitions, i.e. only

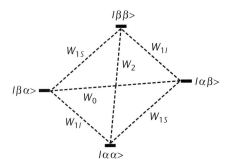

Figure 4.9 Energy level diagram for two dipolar-coupled spins *I* and *S* in an isotropic medium (i.e. no splitting due to dipolar coupling is observed).

one nucleus is changing spin state during the transition. The energy change involved in a single-quantum transition is given by either $\hbar\omega_I$ or $\hbar\omega_S$, depending upon which nucleus is undergoing the transition.

The ω_i are the resonance (Larmor) frequencies of spins *I* and *S*. The transition marked W_2 between the $\alpha\alpha$ and $\beta\beta$ states is a **double-quantum** transition, and the energy change required for this transition (either absorption or emission, depending on which direction the transition takes place) is the sum of the energies of the two single quantum transitions, $\hbar(\omega_I + \omega_S)$. Finally, the transition W_0 linking the $\beta\alpha$ and $\alpha\beta$ states is a **zero-quantum** transition, and the energy change involved is the difference between the energies of the two single-quantum transitions $\hbar(\omega_I + \omega_S)$. The sign of the energy change is determined by the direction of the transition. At thermal equilibrium, the relative populations of the four states are represented by a Boltzmann distribution. Since these are nuclear spin energy levels, the population differences will be relatively small. If we arbitrarily assign a population of *N* to the two states $\alpha\beta$ and $\beta\alpha$, that are very close in energy, then the higher energy state $\beta\beta$ will have a slightly lower population $N - \delta$, and the lowest energy state $\alpha\alpha$ will have a slightly larger population $N + \delta$, where δ is a small fraction of *N*.

Of the three types of transitions, only the single-quantum transitions give rise to directly observable coherence. This does not mean that the zero- and double-quantum transitions are unimportant, only that they cannot be detected directly. But since one can observe resonances corresponding to the W_{1I} and W_{1S} transitions, either of those transitions can be saturated by applying a weak RF field with the appropriate frequency. Imagine an experiment in which the two single quantum transitions in Figure 4.9 that result in a spin flip for nucleus *S* (W_{1S}) are irradiated with an RF field ω_S of sufficient power to saturate the transition, i.e. to equalize the populations of the two states that are connected by the W_{1S} transition. (For simplicity, assume that the two spins interact only via D-coupling, that is, there is no scalar coupling between *I* and *S*.) After irradiating for a time sufficient to reach steady state, the $\alpha\alpha$ and $\beta\alpha$ states will have (equal) populations of $N + \delta / 2$, while the $\alpha\beta$ and $\beta\beta$ states now each have populations of $N - \delta / 2$. If we were to now sample the signal intensity of the *I* transition, what would we expect to see? As the observed signal for the *I* transition is proportional to the population difference between the two states connected by W_{1I}, the signal intensity of the I resonance will be proportional to $(N + \delta/2) - (N - \delta/2) = \delta$. This is no different from what we calculate without saturation of the S transitions. However, if we turn on the D-coupling and consider the affect of W_0 and W_2, we do see an effect on the populations of the states connected by the saturated transitions. At equilibrium, the populations of the two states connected by the W_0 transition, $\alpha\beta$ and $\beta\alpha$, are equal (both to N). The tendency of the W_0 transition upon saturation of the S resonance will be to drive the now unequal populations of these two states back towards equality, i.e. the W_0 transition will drain excess population from the $\beta\alpha$ state (population of $N + \delta/2$) into the $\alpha\beta$ state (population of $N - \delta/2$). The net result of this process would be to reduce the population differences between the states connected by the W_{1I} transitions, for one of them by decreasing the population in the lower state ($\beta\alpha$) and for the other increasing the population in the upper state ($\alpha\beta$). Now the population differences across the I transitions would be less than δ and the signal intensity for resonance I would decrease relative to the equilibrium case.

What about the W_2 transition? The population difference between the two states connected by the double quantum transition ($\alpha\alpha$ and $\beta\beta$) is 2δ at equilibrium, but only δ after the S transitions are saturated. The tendency of the W_2 transition in this case would be to drain the excess population from the $\beta\beta$ state into the $\alpha\alpha$ state in order to re-establish the equilibrium 2δ population difference. This would have the net effect of increasing population differences across the W_{1I} transitions, by decreasing the population of the upper energy level of one of the W_{1I} transitions ($\beta\beta$), and increasing the population of the lower energy level ($\alpha\alpha$) of the other. The net result is a larger population difference across both of the W_{1I} transitions than at equilibrium, giving a signal intensity for resonance I greater than the equilibrium signal.

Obviously, the effects of the zero- and double-quantum transitions on the signal intensity of I are opposed to each other, and if they were perfectly balanced, no effect would be observed. However, the rate constants for these transitions depend on which frequency elements are present in the white noise provided by molecular motion. Without going into the derivations, the rate constants for the zero- and double-quantum transitions are clearly going to be functions of the spectral density at the frequency of the transition involved:

$$W_0 \propto r_{IS}^{-6} J(\omega_I - \omega_S) = \frac{2\tau_c}{r_{IS}^6 \left(1 + (\omega_I - \omega_S)^2 \tau_c^2\right)} \tag{4.11}$$

$$W_2 \propto 6 r_{IS}^{-6} J(\omega_I + \omega_S) = \frac{12\tau_c}{r_{IS}^6 \left(1 + (\omega_I + \omega_S)^2 \tau_c^2\right)} \tag{4.12}$$

$$W_1^I \propto r_{IS}^{-6} J(\omega_I) = \frac{3\tau_c}{r_{IS}^6 \left(1 + \omega_I^2 \tau_c^2\right)} \tag{4.13}$$

$$W_1^S \propto r_{IS}^{-6} J(\omega_S) = \frac{3\tau_c}{r_{IS}^6 \left(1 + \omega_S^2 \tau_c^2\right)} \tag{4.14}$$

where r_{IS}^{-6} is the inverse sixth power of the distance between nuclei I and S. At very short correlation times (small molecules in a nonviscous solvent), the spectral density at both the sum and difference frequencies will be of the same magnitude, and the double quantum term will predominate (because of the larger numerator term in Equation 4.12). This results in an *increase* in the signal intensity for I when the signal for S is saturated. As the correlation time increases (increasing molecular size and/or solvent viscosity), the zero-quantum term will increase in importance, since the denominator for Equation 4.11 is smaller than that of Equation 4.12 in all cases. When the correlation time has increased in length sufficiently that $\omega\tau_c = 1.118$ (assuming both nuclei have the same gyromagnetic ratio, so that $\omega_I + \omega_S \approx 2\omega$), the rate constants for both the zero- and double-quantum transitions will be the same, and there will be no net effect on the I signal when S is saturated. For very long correlation times, the zero-quantum transition will predominate, and saturation of the S spins will result in a *decrease* in signal intensity of spin I. The change in signal intensity at resonance I due to the saturation at resonance S is referred to as a nuclear Overhauser effect, or NOE. The word "nuclear" is to distinguish this phenomenon from the effect originally described by Overhauser, namely the result of electron-nuclear cross-relaxation.

The other important factor in the rate expressions is the inverse sixth power of the distance between nuclei I and S. The closer the two nuclei are to each other, the larger the effect. The NOE provides a valuable tool for determining when two nuclei are in close proximity, and is particularly valuable for examining macromolecules and molecular complexes in which nuclei in close proximity may not be part of the same bonding system and hence are not J-coupled to each other.

NOE measurements, indirect NOEs and saturation transfer

It can be shown (see Problem 4.4) that the maximum possible homonuclear NOE (NOE between nuclei with the same gyromagnetic ratio) in the extreme narrowing limit ($\omega^2\tau_c^2 \ll 1$) is +50%. In practice, NOEs observed between protons are smaller than this, rarely exceeding 20%. This is due in part to the fact that other relaxation mechanisms besides dipole–dipole interactions tend to short-circuit the NOE. Also, there are often multiple dipolar relaxation pathways operating for a given nuclear spin, and the overall relaxation of that spin is modulated by all of those pathways, not just the one of interest in the NOE experiment. This means that in a molecule containing multiple dipolar nuclei, there is a network of interconnected dipolar interactions, and when one member of the network is perturbed, this leads to a "ripple" effect on the other members of the network. Even spins that are not directly coupled to a perturbed spin can experience an **indirect NOE** (an NOE transmitted from the saturated spin to the observed spin via an intermediate spin). These will be small and negative (Figure 4.10) for molecules at the extreme narrowing limit. The opposite signs for direct and indirect NOEs can be rationalized by considering a diagram of the type shown in Figure 4.9 in which the population differences across the S transitions are increased relative to their equilibrium difference (that is, the S transition is experiencing a positive NOE) instead of being equalized by saturation. With time, the double quantum transition would tend to decrease the population differences across the I transitions, resulting in a smaller-

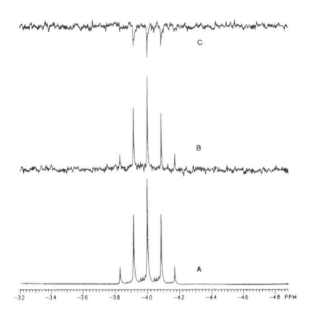

Figure 4.10 Examples of direct and indirect NOE. Spectrum A is the ^{11}B NMR spectrum of 0.1 M tetrabutylammonium borohydride (BH$_4^-$) in CDCl$_3$. Spectrum B is the NOE difference spectrum resulting from saturation of the 1-methylene ^{1}H resonances of the tetrabutylammonium ion. The direct interactions between the ^{11}B nucleus and the irradiated protons predominate, and a positive ^{11}B\{^{1}H\} NOE is observed (^{11}B signal enhancement). Spectrum C is the NOE difference spectrum observed upon saturation of the 2-methylene ^{1}H resonances of the tetrabutylammonium ion. In this case, the ^{11}B\{^{1}H\} NOE is indirect (negative), transmitted primarily through the directly bonded BH$_4^-$ protons to the boron nucleus. Note that this system is unusual in that NOEs are observed to a quadrupolar nucleus, ^{11}B. This is the result of the ^{11}B being in a high-symmetry (tetrahedral) site. The symmetry minimizes the electric field gradient at the ^{11}B nucleus, rendering the quadrupolar relaxation less efficient, so that the dipole–dipole processes that produce NOEs are competitive. [Reprinted with permission from *J. Am. Chem. Soc.* 112, 6714 (1990).]

than-equilibrium I signal. However, these effects are usually quite small, and in small molecules, indirect NOEs do not usually extend farther than one or two steps from the saturated nucleus. As such, NOE experiments on small molecules in non-viscous solvents are often performed by saturating the S transition until steady state is reached (saturation times on the order of five to 10 times T_1). This results in the maximum possible NOE. These are called **steady-state NOE** experiments, because the irradiation is continued until no further change is observed.

Steady-state NOE experiments are only practical at the extreme narrowing limit, where the double-quantum transitions (Equation 4.12) predominate. Away from the extreme narrowing limit (for large molecules and/or viscous solvents), when zero-quantum transitions (Equation 4.11) predominate, steady-state experiments are not practical. When the S transition is saturated under these conditions, eventually the population across the I transition will also become saturated. As the I spin becomes saturated, it in turn passes this saturation along to other nearby spins, in the manner of people in a bucket brigade passing buckets of water to put out a fire. This phenomenon is known as **saturation transfer**. By the time steady state is reached, saturation transfer may have reached quite some distance from the saturated nucleus, in which case the experiment would not provide much useful information about internuclear distances. A way around this problem is to avoid steady state. By measuring the NOE as a function of saturation time, it is possible to extract distance information from the experiment since it turns out that the initial *rate* of NOE buildup at nucleus I upon saturation of nucleus S is also proportional to r_{IS}^{-6}. This type of experiment is known as **truncated driven NOE** (Figure 4.11). Internuclear distances can be calculated from the rate of buildup of NOE between two nuclei that are a known distance apart, using this rate to determine the proportionality between r_{IS}^{-6} and NOE buildup. After saturation transfer has progressed to other spins in the area, however, back-transfer from those other spins to spin I complicates the analysis.

Another experiment that allows analysis of NOE data in the presence of saturation transfer is the **transient NOE** experiment (Figure 4.12). In this experiment, a selective 180° pulse (**selective inversion pulse**) is applied to nucleus S (*how* a selective inversion is accomplished will be discussed later), and the NOE detected at nucleus I is measured some time τ after the selective inversion of S. Again, the rate of initial NOE buildup as a function of time after the inversion will be proportional to r_{IS}^{-6}, but the NOE intensity curve will flatten and die back towards zero as τ increases. The one-dimensional transient NOE experiment provides the basis for the important two-dimensional NOESY experiment that will be discussed exten-

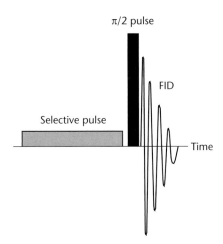

Figure 4.11 Pulse sequence for measuring selective steady-state or truncated driven NOEs. A low-power selective pulse, represented by the gray-shaded box, is applied at the resonance of interest. It is followed immediately by a nonselective read pulse and acquisition of the FID. The FID is then subtracted from a reference obtained without the selective pulse in order to obtain the difference spectrum.

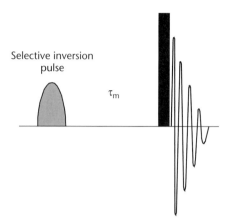

Figure 4.12 Pulse sequence for measuring selective transient NOEs. A low-power selective inversion (180° or π) pulse, represented by the gray-shaded parabola, is applied at the resonance of interest. The NOE develops during τ_m, followed by a nonselective read pulse and acquisition of the FID. The FID is then subtracted from a reference obtained without the selective inversion.

sively later. For both the truncated driven or transient NOE experiments, direct NOE and indirect NOE (saturation transfer) may be distinguished by the shape of the initial buildup curves (Figure 4.13). Direct effects between two nearby nuclei begin buildup immediately, while indirect effects exhibit a lag time, during which the intermediate spins begin to transfer saturation to nuclei further away from the site of the irradiated nucleus.

Heteronuclear NOE and the Solomon equation

The discussion of the NOE above did not make any assumptions concerning the types of nuclei involved, only that they could be separately irradiated (distinct resonance frequencies). Now we will be somewhat more specific about the differences between homo- and heteronuclear NOE (NOE involving nuclei of the same γ or different γ, respectively). We can set up the differential equations that relate the rate of change in state populations to how far the system is from equilibrium. If N_i is the population of state i and N_i^0 is the population of that state at equilibrium, then the rate that the system is returned to equilibrium dN_i / dt is a function of the rate constants connecting state i to other states. Let us look at state $\alpha\alpha$ for the IS spin system. In Figure 4.10, the $\alpha\alpha$ state is connected to all the other states in the system by the transitions W_{1I}, W_{1S} and W_2. The rate that the $\alpha\alpha$ state population is depleted by any of these transitions is determined by the difference between $N_{\alpha\alpha}$ and $N_{\alpha\alpha}^0$, while the rate that it repopulates is determined by the populations of the connected states. So the overall rate of change is given by:

$$
\begin{aligned}
\frac{dN_{\alpha\alpha}}{dt} = &-\left(W_{1I} + W_{1S} + W_2\right)\left(N_{\alpha\alpha} - N_{\alpha\alpha}{}^0\right) + W_2\left(N_{\beta\beta} - N_{\beta\beta}{}^0\right) \\
&+ W_{1I}\left(N_{\beta\alpha} - N_{\beta\alpha}{}^0\right) + W_{1S}\left(N_{\alpha\beta} - N_{\alpha\beta}{}^0\right)
\end{aligned}
$$

(4.15)

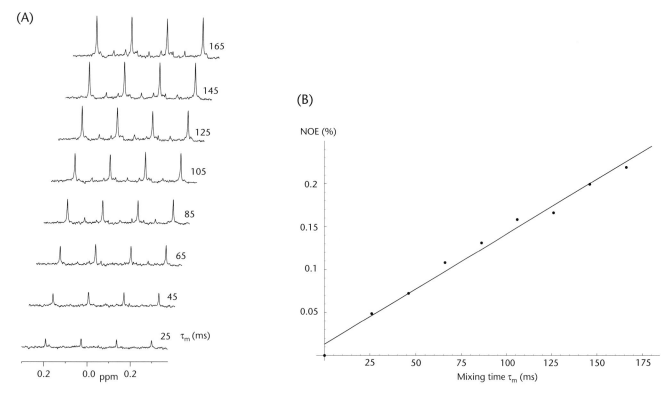

Figure 4.13 (A) Transient NOE measurement of interionic ^{1}H NOEs to the BH_4^- protons in the tight ion pair tetrabutylammonium borohydride (0.1 M in $CDCl_3$ at 11.74 T) upon selective inversion of the nearby 1-methylene protons. (B) Linear NOE buildup as a function of mixing time obtained from the signal intensities in A.

and similar equations can also be written for all of the other state populations (see Problem 4.8).

Now consider the steady-state NOE experiment described above in which spin S is saturated in order to observe an NOE at spin I. From the steady-state condition $dN_i/dt = 0$ and recalling that I, the signal intensity for nucleus I, is proportional to the population differences across both W_{1I} transitions, we get the following from the set of four equations of type 4.15:

$$\frac{I}{I_0} = 1 + \frac{S_0}{I_0}\frac{\left(W_2 - W_0\right)}{\left(W_0 + 2W_{1I} + W_2\right)} \qquad (4.16)$$

where the I_0 and S_0 are the equilibrium signal intensities for nuclei I and S, respectively. Since the equilibrium intensities of the two signals are proportional to the gyromagnetic ratios of the two nuclei, we can write:

$$\frac{I}{I_0} = 1 + \frac{\gamma_S}{\gamma_I}\frac{\left(W_2 - W_0\right)}{\left(W_0 + 2W_{1I} + W_2\right)} \qquad (4.17)$$

Equation 4.17 is known as the Solomon equation, and provides a tool for predicting steady-state NOEs as a function of correlation time (Reference 4). The numerator term $(W_2 - W_0)$ deals only with cross-relaxation between spins I and S, and is usually abbreviated simply as σ_{IS}. The denominator term $(W_0 + 2W_{1I} + W_2)$ deals with all contributions to relaxation of nucleus I and is usually given the symbol ρ_I. Since the single quantum transitions can be affected by mechanisms other than nuclear dipole–dipole interactions (see below), contributions other than dipolar to ρ_I are often noted separately as $\rho*$. The Solomon equation may now be rewritten as:

$$\frac{I}{I_0} = 1 + \frac{\gamma_S}{\gamma_I} \frac{\sigma_{IS}}{(\rho_I + \rho*)} \qquad (4.18)$$

For homonuclear NOE experiments the appropriate frequency components of the spectral density are very different for the zero-quantum and double-quantum transitions (near zero frequency for the zero-quantum term and double the resonance frequency for the double-quantum terms). However, they are much closer in frequency for typical heteronuclear spin systems. For example, for dipolar coupled ^{1}H–^{15}N nuclei, with a ^{1}H resonance frequency of 500 MHz (3.14×10^9 rad/s), ^{15}N resonates around 50 MHz (3.14×10^8 rad/s). The zero-quantum term depends upon the spectral density at 2.82×10^9 rad/s, while the double-quantum term depends on the spectral density at 3.54×10^9 rad/s. Unless the correlation time falls to about 3×10^{-9} s, the two spectral densities will be quite similar. A correlation time of 3×10^{-9} s is fairly long, observed for large molecules or in viscous solvents. A rough estimate of the correlation time for a spherical molecule of molecular weight M in a nonviscous solvent is given by $M \times 10^{-13}$. So for molecules of moderate size in a nonviscous isotropic solvent, the extreme narrowing condition applies to heteronuclear transitions. In this case (and assuming that the contributions to relaxation of nucleus I from mechanisms other than nuclear dipole–dipole are minimal), the Solomon equation for heteronuclear NOE experiments at steady state simplifies to:

$$\frac{I}{I_0} = 1 + \frac{\gamma_S}{2\gamma_I} \qquad (4.19)$$

Equation 4.19 gives the maximum possible NOE observable between nuclei with different γ at the extreme narrowing limit. Note that this expression for all practical purposes limits one to the irradiation of proton (nucleus S), and observing the effect on the heteronucleus I, since the ratio $\gamma_S / 2\gamma_I$ is otherwise too small to be useful. For ^{13}C when ^{1}H is saturated, this ratio is ~2, indicating that ^{13}C signal intensity can improve by a factor of three (200% NOE) if ^{1}H–^{13}C dipole–dipole interactions are the primary means of ^{13}C relaxation. This very significant increase is put to practical use in standard direct-observe ^{13}C NMR experiments. Typically, such experiments are acquired with broadband ^{1}H decoupling applied not only during the acquisition of the FID—so $J_{^1H^{13}C}$ splittings are not observed—but prior to the acquisition in order to build up the NOE and improve S/N. The application of the NOE to improving S/N in heteronuclear NMR experiments is far from general, however. For nuclei with negative gyromagnetic ratios (^{15}N and ^{29}Si, for example) the NOE obtained upon saturation of ^{1}H when observing these nuclei is negative and reduces signal intensity.

Other contributions to T_1 relaxation: chemical shift anisotropy, spin-rotation and paramagnetic effects

Although we have so far only discussed the contributions of nuclear dipole–dipole interactions to nuclear relaxation, it is important to remember that *any* time-dependent modulation of the local electromagnetic environment that contains frequency elements appropriate to the transition of interest will encourage relaxation. For $\mathbf{I} = 1/2$ nuclei other than ^{1}H, an important source of relaxation is *chemical shift anisotropy* (CSA), that was discussed earlier in relation to the directional nature of the electronic shielding that gives rise to chemical shift. The Zeeman splitting that gives rise to the energy difference between nuclear spin states depends on the orientation that the molecule occupies relative to the laboratory magnetic field. We noted earlier that in isotropic liquids, the molecular motion is usually rapid enough that the observed chemical shift is the scalar isotropic chemical shift σ_{iso}, the average of the three principal components of the shielding tensor, σ_{xx}, σ_{yy} and σ_{zz}. However, if there is a strong directional dependence of the shielding (σ_{xx}, σ_{yy} and/or σ_{zz} are very different from one another), the instantaneous chemical shift of the nucleus in question will fluctuate over a very wide range, and this fluctuation will encourage relaxation. CSA is not important for ^{1}H, since the electronic environment of the ^{1}H nucleus is spherically symmetric in the zero-order approximation, i.e. the electrons around the proton are in molecular orbitals that contain large contributions from the 1s atomic orbital. However, for nuclei such as ^{15}N and ^{13}C, that hybridize with orbital symmetries less than spherically symmetric, the shielding anisotropy is apt to be quite large, and chemical shift anisotropy a significant contributor to nuclear relaxation. Since chemical shift is field dependent, CSA becomes more important as magnetic field intensities become larger. (Note that this also is the reason why the chemical shift range for ^{1}H is so narrow relative to the chemical shift ranges seen for nuclei such as ^{13}C and ^{15}N.) Thus, the advantages that larger magnets bestow for ^{1}H NMR in terms of improved S/N and dispersion are often mixed blessings for NMR of other nuclei because of CSA contributions to relaxation.

Another contributor to the relaxation of $\mathbf{I} = 1/2$ nuclei is electron–nuclear dipole–dipole relaxation, which is observed in systems containing unpaired electrons (paramagnetic systems). Electron–nuclear spin interactions (*en*) work in much in the same way as nuclear dipole–dipole interactions in causing relaxation, but the electron–nuclear interactions are usually much more efficient. The resonances of nuclei near sites of unpaired electron spin density may exhibit extremely short relaxation times relative to other nuclei in the same molecule that are remote from the site of paramagnetism. The result can be spectral lines that are broadened, often to the point of being unobservable. The critical factor for determining the extent to which NMR lines are broadened by paramagnetic effects is the relaxation time τ_r of the unpaired electrons. If electronic relaxation is rapid ($\tau_r \sim 10^{-12}$ s), the nuclei detect only the average state of the electron and tend to show less paramagnetic relaxation effects. If electronic relaxation is comparatively slow (10^{-11} s$^{-1} < \tau_r < 10^{-9}$ s), nearby nuclei do not see just an average of the electronic states, but a flickering in the local electromagnetic field that provides very efficient nuclear relaxation. Note that in situations where electron–nuclear spin–spin interactions dominate the nuclear spin relaxation, the relevant correlation time is the electronic relaxation time τ_r rather than the molecular correlation time τ_c.

Less-commonly observed sources of relaxation for $\mathbf{I} = 1/2$ nuclei include modulated J-coupling and spin–rotation relaxation. Like the dipole–dipole coupling and chemical shift anisotropy, if the J-coupling between two nuclei is modulated on a time scale such that two coupling states are distinguishable, a "flicker" in the local electromagnetic environment occurs. Such flickering can be the result of chemical exchange (called "scalar relaxation of the first kind") or rapid T_1 relaxation of one of the coupling partners ("scalar relaxation of the second kind"). It turns out that this type of relaxation usually has a greater effect on T_2 than on T_1, because the time scale of the flicker must usually be quite slow in order for the states to be distinguishable. The J-coupling constants themselves are usually fairly small, and very rapid switching leads to the separate states being indistinguishable. As such, only the very-low-energy zero-quantum transitions (mutual spin flips or spin–spin interactions) are encouraged, resulting in no net change in the relative populations of states.

One can get the idea of what modulated J-coupling means by looking at the spectrum of a simple molecule such as n-hexane. This straight-chain hydrocarbon samples multiple conformations very rapidly. As the dihedral angles between individual protons change, so do the couplings between them. However, the motions are so fast that on the time scale of the NMR experiment these fluctuations are not detected, and the couplings between various protons appear to be constants. Now imagine a situation where the conformational changes occur at a frequency near to that of the coupling constant. This modulates the observed J-splitting, and leads to uncertainty broadening in the lines. However, the fluctuations are at such a low frequency that only zero-quantum ($\omega_I - \omega_S$) transitions are encouraged, not single-quantum (ω_I, ω_S) or double-quantum transitions ($\omega_I + \omega_S$), and no net loss of energy to the lattice takes place. This is an example of scalar relaxation of the first kind. Protons that are coupled to ^{14}N, a quadrupolar nucleus, provide an example of scalar relaxation of the second kind. Quadrupolar relaxation is fairly efficient (as we will see below), and protons bound to ^{14}N will generally see the ^{14}N nucleus changing spin states multiple times on the time scale of the coupling J_{HN}. The result is a broadened singlet for the proton resonance, since the proton detects the ^{14}N nucleus flickering between several states in a short period of time.

The final mechanism applicable to $\mathbf{I} = 1/2$ nuclei that we will consider is **spin–rotation** relaxation. When a functional group such as a methyl group rotates with respect to the rest of the molecule, the motion of charges produces a magnetic field, and nearby nuclei couple to that magnetic field, a phenomenon known as spin–rotation coupling. Since such rotational motion is quantized there are fixed energy levels corresponding to different rotational frequencies. When a collision occurs between molecules, this rotational motion can be disrupted, modulating the spin–rotation coupling. This random modulation of the spin–rotation coupling can lead to relaxation. The lifetime of the rotational states (related to the time between collisions) determines the efficiency of the relaxation. Interestingly, spin–rotation relaxation is the only relaxation mechanism that commonly becomes more efficient as temperature increases.

Quadrupolar relaxation

For quadrupolar nuclei ($\mathbf{I} > 1/2$), the most important contribution to relaxation is almost invariably due to modulation of quadrupolar coupling. Quadrupolar coupling arises from the interaction between the quadrupolar nucleus and an electric field gradient (dE/dr) at the nucleus. Collisions and molecular motions all modulate the electric field gradient, resulting in efficient relaxation of the quadrupolar nucleus. The quadrupolar relaxation is usually so efficient that it precludes the direct observation of spin–spin couplings to quadrupolar nuclei, because coupling partners detect only the average spin state of the nuclear spin quadrupole rather than discrete states. Quadrupolar relaxation also tends to short-circuit the NOE which depends on dipole–dipole interactions being an important relaxation pathway. The observation of direct spin–spin coupling and/or NOEs at a quadrupolar spin is typically limited to nuclei such as ^{6}Li and ^{7}Li, that have small quadrupolar moments, or to situations in which the quadrupolar spin is at a site of high symmetry that renders the electric field gradient at the nucleus zero on a time average (see Figure 4.10).

Selective and nonselective T_1 measurement and multi-exponential decay of coherence

The observed relaxation rate for a given nucleus is determined by the contributions of all of the mechanisms that contribute to that relaxation, and the rate constant k_{obs} for relaxation is given by:

$$k_{obs} = k_{dd} + k_{para} + k_q + k_{sr} + k_{CSA} + k_{scalar} \tag{4.20}$$

Going back to the definition of T_1 as the inverse of the rate constant for spin–lattice relaxation, we get:

$$\frac{1}{T_{1(obs)}} = \frac{1}{T_{1(dd)}} + \frac{1}{T_{1(para)}} + \frac{1}{T_{1(q)}} + \frac{1}{T_{1(sr)}} + \frac{1}{T_{1(CSA)}} + \frac{1}{T_{1(scalar)}} \tag{4.21}$$

Equations 20 and 21 can be cast more generally in terms of a multi-exponential decay:

$$A = A_0 \prod_1^i e^{-k_i t} \tag{4.22}$$

where the individual decay terms represent the i individual contributors to relaxation. Even if dd relaxation is the sole contributor to relaxation, the relaxation can be multi-exponential. If a spin is D-coupled with multiple other spins, each individual coupling contributes to the relaxation process as a separate component of Equation 4.22. The result is a multi-exponential decay that cannot be properly fit to a single exponential (although relaxation is quite often presented in just that way). Also, since the rate of cross-relaxation depends on state populations (that is, the relaxation rate depends on both populations and rate constants), the relaxation rate observed due to cross-relaxation will depend upon whether the populations of all of the D-coupled spins are perturbed simultaneously (**nonselective T_1** experiment) or only the resonance of interest is perturbed (**selective T_1** experiment).

One last point worth mentioning is that different relaxation processes are not completely independent of each other. In the end, all of the processes that lead to relaxation cause local electromagnetic field (EMF) fluctuations, and those fluctuations have both frequency and directionality (that is, the amplitude vector of the EMF). Like any wave phenomenon, this means that the fluctuations can interfere either constructively or destructively. Such effects are called **cross-correlation**, and are observed as either enhancements or slowing of relaxation rates. Often, such effects are opposite in sign for different lines of a coupling multiplet, that is, they have opposite effects for a given spin depending upon the spin state. Cross-correlation effects are often seen between CSA and *dd* relaxation mechanisms in heteronuclear spin systems, or between paramagnetic and *dd* relaxation in systems containing unpaired electrons. Cross-correlation will be discussed more thoroughly in Chapter 9.

Problems

*4.1   Figure 4.2 shows a series of transformed spectra from the inversion recovery experiment for sucrose in D_2O. The following table provides the measured line intensities in arbitrary units as a function of relaxation delay τ for two lines in the spectra shown, the anomeric proton and the solvent line (HDO). Based on these data, calculate T_1 and R_1 for each of the two protons.

Anomeric proton	HDO	τ (s)
−32.9	−132	0.1
−25.9	−126	0.2
−13.9	−117	0.4
3.72	−101	0.8
23.9	−70.7	1.6
38.1	−21.4	3.2
42.1	46.7	6.4
42.1	114	12.8
42.1	149	25.6

*4.2   Figure 4.5 shows a series of FIDs resulting from a spin-echo T_2 measurement for the ^{19}F signal from tetrabutylphosphonium tetrafluoroborate. The resulting maximum intensities as a function of relaxation time 2τ (total time of the echo sequence) are shown in the table below. Calculate T_2 for the ^{19}F resonance.

^{19}F intensity	2τ (s)
85.67	0.032
65.46	0.576
49.74	1.104
37.72	1.648
28.55	2.192
21.97	2.736
16.48	3.36
12.27	4.00
9.21	4.64
6.86	5.28

4.3   Show that the spectral density function is the Fourier transform of the auto-correlation function.

*4.4**   Show from the Solomon equation and relationships between the rate constants for zero-, single- and double-quantum transitions and the spectral density function that the maximum possible homonuclear NOE for a molecule in the extreme narrowing limit is +50%, and that for a large molecule with long correlation times it is −100%.

4.5   The maximum fractional theoretical NOE at nucleus S upon saturation of nucleus I is given by

$$f_s = 0.5 \frac{\gamma_I}{\gamma_S}$$

in the case when dd relaxation between I and S is the only efficient relaxation mechanism available to nucleus S. In the case where the relaxation is also modulated by other mechanisms, the equation modifies to:

$$f_s = 0.5 \frac{\gamma_I}{\gamma_S} \left(\frac{T_{1S(dd)}^{-1}}{T_{1S(dd)}^{-1} + T_{1S(other)}^{-1}} \right)$$

where $T_{1S(other)}^{-1}$ are the contributions to relaxation from other mechanisms. Assuming that only dd and quadrupolar mechanisms contribute to observed ^{11}B relaxation in the system shown in Figure 4.10, that a 0.31 fractional enhancement (f_s) is observed to ^{11}B upon saturation of all ^{1}H resonances in the sample, and that the observed ^{11}B relaxation rate, $T_{1(observed)}^{-1}$, is 0.355 s^{-1}, calculate the expected relative contributions of dd and quadrupolar mechanisms to the observed ^{11}B relaxation in units of inverse seconds (s^{-1}).

*4.6**   In the graph shown in Figure 4.8, indicate where the various transitions of importance for relaxation of ^{1}H signals at 11.74 T (500 MHz) occur. How does one recognize the extreme narrowing limit from this diagram?

*4.7**   Show from the definition of the spectral density function that the rate constants of the zero- and double-quantum transitions for two spins I and S will be equal when the term $\omega\tau_c = 1.118$ (assume that both nuclei have the same gyromagnetic ratio).

4.8   The rate equations which govern the relaxation of a two-spin system (I and S) are:

$$\frac{dN_{\alpha\alpha}}{dt} = -\left(k_1^I + k_1^S + k_2\right)\left(N_{\alpha\alpha} - N_{\alpha\alpha}{}^0\right) + k_2\left(N_{\beta\beta} - N_{\beta\beta}{}^0\right) + k_1^I\left(N_{\beta\alpha} - N_{\beta\alpha}{}^0\right) + k_1^S\left(N_{\alpha\beta} - N_{\alpha\beta}{}^0\right)$$

$$\frac{dN_{\beta\beta}}{dt} = -\left(k_1^I + k_1^S + k_2\right)\left(N_{\beta\beta} - N_{\beta\beta}{}^0\right) + k_2\left(N_{\alpha\alpha} - N_{\alpha\alpha}{}^0\right) + k_1^S\left(N_{\beta\alpha} - N_{\beta\alpha}{}^0\right) + k_1^I\left(N_{\alpha\beta} - N_{\alpha\beta}{}^0\right)$$

$$\frac{dN_{\alpha\beta}}{dt} = -\left(k_1^I + k_1^S + k_0\right)\left(N_{\alpha\beta} - N_{\alpha\beta}{}^0\right) + k_0\left(N_{\beta\alpha} - N_{\beta\alpha}{}^0\right) + k_1^I\left(N_{\beta\beta} - N_{\beta\beta}{}^0\right) + k_1^S\left(N_{\alpha\alpha} - N_{\alpha\alpha}{}^0\right)$$

$$\frac{dN_{\beta\alpha}}{dt} = -\left(k_1^I + k_1^S + k_0\right)\left(N_{\beta\alpha} - N_{\beta\alpha}{}^0\right) + k_0\left(N_{\alpha\beta} - N_{\alpha\beta}{}^0\right) + k_1^S\left(N_{\beta\beta} - N_{\beta\beta}{}^0\right) + k_1^I\left(N_{\alpha\alpha} - N_{\alpha\alpha}{}^0\right)$$

Remembering that at steady state, the change in all populations $dN_i/dt = 0$ and that saturation equalizes populations across the saturated transition, show that the change in signal intensity for nucleus I upon saturation of the S spin is given by:

$$\frac{I}{I_0} = 1 + \frac{S_0}{I_0} \frac{\left(k_2 - k_0\right)}{\left(k_0 + 2k_1^I + k_2\right)}$$

where I_0 and S_0 are the equilibrium signal intensities for nuclei I and S, respectively. (Don't forget that the signal intensities are proportional to population differences across *all* transitions corresponding to that signal).

References

1. A. G. Redfield, *Adv. Magn. Reson.* 1, 1 (1965).
2. J. H. Noggle and R. E. Schirmer, *The Nuclear Overhauser Effect: Chemical Applications.* Academic Press, New York (1971).
3. D. Neuhaus and M. Williamson, *The NOE in Structural and Conformational Analysis.* 2nd edition, Wiley, New York (2000).
4. I. Solomon, *Phys. Rev.* 99, 559 (1955).

5

CLASSICAL AND QUANTUM DESCRIPTIONS OF NMR EXPERIMENTS IN LIQUIDS

The classical approach: the Bloch equations of motion for macroscopic magnetization

We have now considered everything needed to describe the behavior of macroscopic magnetization in bulk matter, and it is time to assemble these disparate pieces into a concise set of laws governing such behavior. Physicists call laws governing time-dependent behavior "equations of motion", and the laws governing the time evolution of magnetization in condensed matter are the Bloch equations of motion (Reference 1). As discussed in Chapter 2, the torque $\vec{N}_i$ (the rate of change in angular momentum) exerted by a magnetic field $\vec{B}$ upon an individual magnetic dipole is given by:

$$\vec{N}_i = \frac{d\vec{p}_i}{dt} = -\vec{\mu}_i \times \vec{B} \tag{5.1}$$

where $\vec{p}_i$ is the angular momentum of the ith nuclear dipole $\vec{\mu}_i$. The torque exerted on bulk magnetization $\vec{M}$ (i.e. the vector sum of the individual dipoles) is given by:

$$\vec{N} = \frac{d}{dt}\vec{P} = \left(\frac{1}{\gamma}\right)\frac{d}{dt}\vec{M} = -\sum_i \vec{\mu}_i \times \vec{B} = -\vec{M} \times \vec{B} \tag{5.2}$$

The derivative of $\vec{M}$ was obtained using the relationship $\vec{M} = \gamma\vec{P}$. Rearrangement gives:

$$\frac{d}{dt}\vec{M} = -\gamma\vec{M}\times\vec{B} \tag{5.3}$$

Equation 5.3 provides an expression for the time dependence of the bulk magnetization $\vec{M}$ under the influence of a magnetic field $\vec{B}$. The $\vec{B}$ field can be either static or time dependent, that is $\vec{B} = \vec{B}(t)$. If $\vec{M}$ is displaced from equilibrium by a time-dependent $\vec{B}(t)$, relaxation provides a restoring force that tends to drive the system back to equilibrium. Since relaxation will affect the magnetization only until the system returns to equilibrium, the rate of relaxation is determined by how far the system is from equilibrium, i.e. the difference $\vec{M}(t) - \vec{M}_0$. As long as this difference is nonzero, relaxation will result in a time-dependent change in magnetization according to the relationship:

$$\frac{d}{dt}\vec{M} = \hat{\mathbf{R}}(\vec{M}(t) - \vec{M}_0) \tag{5.4}$$

where $\hat{\mathbf{R}}$ is a matrix operator that contains as diagonal elements the relaxation rate constants for components of magnetization along each of the three Cartesian axes of the rotating frame:

$$\hat{\mathbf{R}} \equiv \begin{pmatrix} T_2^{-1} & 0 & 0 \\ 0 & T_2^{-1} & 0 \\ 0 & 0 & T_1^{-1} \end{pmatrix} \tag{5.5}$$

This representation can be interpreted in the following way. Relaxation along the **x** and **y** axes is governed by T_2 processes, while that along **z** is governed by T_1. The diagonal elements of $\hat{\mathbf{R}}$ are thus the appropriate relaxation rate constants for each axis in the rotating frame. The magnetization terms $\vec{M}(t)$ and $\vec{M}_0$ can be represented by vectors containing the **x**, **y** and **z** components of the magnetization:

$$\vec{M}(t) \equiv \begin{pmatrix} M_x(t) \\ M_y(t) \\ M_z(t) \end{pmatrix} \tag{5.6}$$

$$\vec{M}_0 \equiv \begin{pmatrix} 0 \\ 0 \\ M_0 \end{pmatrix} \tag{5.7}$$

so that substitution of Matrices 5.6 and 5.7 into Equation 5.4 for $\vec{M}(t) - \vec{M}_0$ followed by vector multiplication will yield the same expressions for relaxation described previously, except that the **x** and **y** components of the transverse magnetization are now explicitly separated.

Combining Equations 5.3 and 5.4 provides expressions for the overall time dependence of the magnetization under the influence of a perturbing torque (usually from an applied RF field) and a restoring torque provided by relaxation. These expressions are known as the Bloch equations. Using matrix representations for the equations, and remembering that the two torques oppose each other, and so are opposite in sign, the Bloch equations can be stated succinctly as:

$$\frac{d}{dt}\bar{M}(t) = \gamma\bar{M}(t)\times\bar{B}(t) - \hat{\mathbf{R}}(\bar{M}(t) - \bar{M}_0) \tag{5.8}$$

The only part of Equation 5.8 that still needs clarification is the form that $\bar{B}(t)$ takes. In the laboratory frame, $\bar{B}(t)$ is represented by:

$$\bar{B}(t) = B_1[\cos(\omega_{RF}t + \phi)\bar{i} + \sin(\omega_{RF}t + \phi)\bar{j}] + B_0\bar{k} \tag{5.9}$$

Equation 5.9 is the sum of the static field B_0 and the oscillating B_1 field of the applied RF (or at least the circularly polarized component of B_1 that is assumed to be fixed in the rotating frame). That component precesses in **x**, **y** of the laboratory frame with a frequency ω_{RF} and an arbitrary phase ϕ. The $\bar{i}, \bar{j}$ and $\bar{k}$ are unit vectors in the **x**, **y** and **z** directions, respectively. However, in the rotating frame (assumed to rotate at ω_{RF}), the components of $\bar{B}_1$ are fixed in the **x**, **y** plane, and $\bar{B}(t)$ takes a simpler form:

$$\bar{B}(t) = B_1[(\cos\phi)\bar{i}_r + (\sin\phi)\bar{j}_r] - (\Omega/\gamma)\bar{k}_r \tag{5.10}$$

In this representation, $\bar{B}_1$ is independent of time and offset from the **x** axis by the arbitrary phase factor ϕ. Alternatively, one can think of the time dependence of the field as implicit in the magnitude of $\bar{B}_1$: for an RF pulse, at $t < 0$, $B_1 = 0$. While the pulse is applied ($t = 0$ to $t = t_p$), B_1 is of the magnitude determined by the pulse power, and $t > t_p$, B_1 is once again zero. The $\bar{i}, \bar{j}$ and $\bar{k}$ now have a subscript r to indicate that they are unit vectors along the three axes of the rotating frame. The magnitude of the static magnetic field is replaced by Ω/γ, which is the **chemical shift operator** divided by the gyromagnetic ratio. The chemical shift operator is the difference between the RF frequency and the Larmor frequency of the magnetization of interest, $\Omega = -\gamma H_0 - \omega_{RF}$. The reason for this substitution is simple. The laboratory field $\bar{B}_0$ causes magnetization to precess in the laboratory frame at the Larmor frequency. However, in the rotating frame, the observed precessional frequency is the *difference* between the Larmor frequency and the frequency at which the frame precesses. As a result, the chemical shift operator can be represented as a vector along **z** in the rotating frame around which the magnetization precesses at an angular frequency Ω. The chemical shift operator will also be used in the discussions of the density matrix and product operator formalisms in Chapter 6.

For a more complete discussion of the transformation of $\bar{B}(t)$ from the laboratory to the rotating frame, see Reference 2. However, for the present it is sufficient to take the result in Equation 5.10, and use it to expand the expression of the Bloch equations (Equation 5.8). $\bar{B}(t)$ can be represented as a vector with components along three axes in the rotating frame:

$$\bar{B}(t) \equiv \begin{pmatrix} B_1 \cos\phi \\ B_1 \sin\phi \\ -\Omega/\gamma \end{pmatrix} \tag{5.11}$$

Taking this expression, substituting into Equation 5.8, and performing the appropriate vector multiplications, we obtain the expanded forms of the Bloch equations in the rotating frame:

$$\frac{d}{dt} M_x = \gamma \left[M_y(-\Omega/\gamma) - M_z B_1 \sin\phi \right] - M_x/T_2 \tag{5.12}$$

$$\frac{d}{dt} M_y = \gamma \left[M_z B_1 \cos\phi - M_x(-\Omega/\gamma) \right] - M_y/T_2 \tag{5.13}$$

$$\frac{d}{dt} M_z = \gamma \left[M_x B_1 \sin\phi - M_y B_1 \cos\phi \right] - (M_z - M_0)/T_1 \tag{5.14}$$

Classical description of a pulsed NMR experiment

The Bloch equations contain all the information needed to describe the time evolution of the macroscopic magnetization after application of an RF pulse (a one-pulse NMR experiment), and with them we can predict what the FID should look like as well as the transformed spectrum. As an example, we will predict the appearance of a spectrum containing a single resonance with a chemical shift of Ω rad/s and a transverse relaxation time T_2. After the application of a 90° pulse ($\phi = \pi/2$) with $\bar{B}_1$ along $+\mathbf{y}$ to the equilibrium magnetization $\bar{M}_0$ (along $+\mathbf{z}$), the magnetization will lie along $+\mathbf{x}$ as evolution begins ($t = 0$, $M_x = M_0$, $M_y = M_z = 0$). Integration of Equations 5.12–5.14 yields expressions for the time-dependent intensities of M_x, M_y and M_z. Since the RF pulse ($\bar{B}_1$) is off during the evolution of the time-dependent signal, terms containing $\bar{B}_1$ are not included in the integration. Furthermore, as we are only interested in observable magnetization (coherence in the $\mathbf{x}$, $\mathbf{y}$ plane) we need not consider the integration of Equation 5.14 that describes the return to equilibrium. The simultaneous solution to the two differential equations 5.12 and 5.13:

$$\frac{d}{dt} M_x = -M_y \Omega - M_x/T_2$$
$$\frac{d}{dt} M_y = M_x \Omega - M_y/T_2 \tag{5.15}$$

yields the following expressions for the instantaneous magnetization along $\mathbf{x}$ and $\mathbf{y}$ at time t:

$$M_x(t) = M_0 \left[\frac{\exp(-i\Omega t) + \exp(i\Omega t)}{2} \right] \exp(-t/T_2)$$

$$M_y(t) = iM_0 \left[\frac{\exp(-i\Omega t) - \exp(i\Omega t)}{2} \right] \exp(-t/T_2) \tag{5.16}$$

For illustration purposes, it is useful to replace the complex exponential terms using the trigonometric substitutions:

$$\cos z = \frac{\exp(iz) + \exp(-iz)}{2}$$

$$\sin z = \frac{\exp(iz) - \exp(-iz)}{2i} \tag{5.17}$$

This yields the simple expressions for the transverse components of the magnetization:

$$M_x(t) = M_0 (\cos \Omega t) \exp(-t/T_2)$$

$$M_y(t) = M_0 (\sin \Omega t) \exp(-t/T_2) \tag{5.18}$$

Referring to the discussion of quadrature detection in Chapter 3, a plot of either of these functions versus time will provide the projection of the FID that one is accustomed to seeing on the spectrometer display.

The expressions for $M_x(t)$ and $M_y(t)$ given in Equations (5.18) can be combined in two ways to give a complex number representation of the FID:

$$S^+(t) = M_x(t) + iM_y(t) = M_0 \exp(i\Omega t - t/T_2) \tag{5.19}$$

$$S^-(t) = M_x(t) - iM_y(t) = M_0 \exp(-i\Omega t - t/T_2) \tag{5.20}$$

$S^+(t)$ gives rise to clockwise rotation of the signal in the complex plane, while $S^-(t)$ gives counterclockwise rotation. Intuitively, one realizes that after Fourier transformation of $S^+(t)$, a signal centered on a frequency $+\Omega$ relative to the center of the spectrum would be observed, while transformation of $S^-(t)$ would give a signal centered on $-\Omega$. By choosing one or the other of the two expressions, we arrive at the desired result of quadrature detection, that is, selection of only one of the two frequencies for display in the frequency domain. If the two RF receiver channels are not perfectly balanced, the linear combination of $M_x(t)$ and $M_y(t)$ would not result in exact cancellation of the $-\Omega$ frequency term, and a quadrature image would be observed at $-\Omega$ (assuming that $+\Omega$ was the true frequency).

The Fourier transform $Q(\omega)$ of a function $q(t)$ is given by:

$$Q(\omega) = \int_{-\infty}^{\infty} q(t) \exp(-i\omega t)dt \tag{5.21}$$

The appropriate expression for the FT of $S^+(t)$ of a causal function (i.e. one that is zero when $t < 0$) is the Laplace transform:

$$\int_0^{\infty} M_0 \exp(i\Omega t - t/T_2) \exp(-i\omega t)dt \tag{5.22}$$

which rearranges to give:

$$\int_0^\infty M_0 \exp\big(i(\Delta\omega)t\big)\exp(-t/T_2)\,dt \qquad (5.23)$$

where $\Delta\omega = \omega - \Omega$. Integration of Expression 5.23 yields:

$$S(\Delta\omega) = M_0 T_2 \frac{\big(1 + i\Delta\omega T_2\big)}{1 + \Delta\omega^2 T_2^2} \qquad (5.24)$$

After rearrangement, Equation 5.24 can be separated into real and imaginary components:

$$\mathrm{Re}[S(\Delta\omega)] = M_0 \frac{T_2^{-1}}{T_2^{-2} + \Delta\omega^2} \qquad (5.25)$$

$$\mathrm{Im}[S(\Delta\omega)] = M_0 \frac{\Delta\omega}{T_2^{-2} + \Delta\omega^2} \qquad (5.26)$$

Note that Equations 5.25 and 5.26 are the expressions for the frequency-dependent absorptive and dispersive line shapes discussed in Chapter 3 (see Equations 3.7 and 3.8). Now we have come full circle, using the Bloch equations of motion for macroscopic magnetization to describe the NMR spectrum of a single resonance.

A quantum mechanical description of NMR of a single spin in an isotropic liquid

The analysis of the Bloch equations above shows that quantum mechanics is not necessary in order to describe the NMR spectrum of a simple ensemble in which all of the spins are chemically equivalent with no *J*-coupling. So why invoke a quantum description of NMR at all? We will see later that many phenomena, particularly multiple quantum coherence, defy description by classical analogy, and a quantum description is necessary.

We will begin with the quantum description of a single $\mathbf{I} = 1/2$ spin in an isotropic medium. In order to arrive at this description, we first need to recall the basic premise of quantum mechanics: that the *i*th stationary state of the system is described by the wave function ψ_i, which is an eigenfunction of the Hamiltonian operator $\hat{H}$ in the time-independent Schrödinger equation:

$$E_i\psi_i = \hat{E}\psi_i \qquad (5.27)$$

The constant E_i is the eigenvalue associated with ψ_i, and represents the energy of the *i*th stationary state of the system. In order to come up with a quantum description of the NMR experiment, we must first determine the correct form of the Hamiltonian operator $\hat{E}$, and then generate a set of eigenvectors ψ_i that correctly describe the states of the system. Finally, we calculate the energies E_i associated with the individual states.

A general expression for a nuclear spin Hamiltonian for a single spin in an isotropic fluid (usually a liquid) must take into account two factors that together determine the energy of a particular state, chemical shift and J-coupling. (We specify that the system is isotropic because this makes chemical shift a scalar quantity and removes explicit dipole coupling from the calculation.) Recall that the Larmor frequency of the ith spin is given (in angular frequency) by the expression $\omega_i = \gamma B_0(1 - \sigma_i)$, where σ_i is a scalar quantity describing the local electronic shielding of spin i. The energy corresponding to that transition is $\hbar\omega_i = \hbar\gamma B_0(1 - \sigma_i)$. The Larmor frequency is proportional to the energy difference between the lower energy α state ($\mathbf{m} = +1/2$) and higher energy β spin state ($\mathbf{m} = -1/2$). We can specify that the zero-point of the energy is halfway between the upper and lower spin states (where it would be if the spin states were degenerate, i.e. in the absence of an external magnetic field). Now the energy of the upper state is given by $+1/2(\hbar\gamma B_0(1 - \sigma_i))$ and that of the lower energy state is $-1/2(\hbar\gamma B_0(1 - \sigma_i))$. Note that these results may also be written as $-\mathbf{m}_i(\hbar\gamma B_0(1 - \sigma_i))$.

The reader should be aware that this description is for a *single spin*, not an ensemble of identical spins. In the absence of an ensemble treatment, we cannot account for relaxation behavior quantum mechanically. For that we require a statistical mechanical approach that will be discussed in Chapter 11. For the moment, we will assume that relaxation of the ensemble can be treated classically, and use the quantum mechanical approach only to predict the positions of resonance lines in the spectrum, not line shape.

A quantum mechanical description of NMR of coupled spins in an isotropic liquid

J-coupling between two magnetically non-equivalent spins results in the splitting of each resonance into two lines separated by a frequency J, where J is the coupling constant. If spin A is coupled to spin X by a coupling J_{AX}, the energy of spin A is changed relative to the uncoupled state by an amount $J_{AX}\mathbf{m}_A\cdot\mathbf{m}_X$. This corresponds $\pm(1/4)J_{AX}$, with the sign depending on the relative signs of the spin quantum numbers of the coupled spins and the sign of their coupling constant. Thus, the energy of unpaired spin states ($\mathbf{m}_A$ and $\mathbf{m}_X$ are both $+1/2$ or are both $-1/2$) will be raised by $(1/4)J_{AX}$, while the energy of spin paired states ($\mathbf{m}_A$ and $\mathbf{m}_X$ are $+1/2$ and $-1/2$, respectively, or vice versa) will be lowered by $(1/4)J_{AX}$ (assuming J_{AX} is positive) (see Figure 5.1). Since only single quantum events are detected, or **allowed** ($\mathbf{m}_A$ changes from $+1/2$ to $-1/2$ or vice versa, while $\mathbf{m}_X$ remains unchanged), the observed transitions for spin A in a simple NMR experiment will be between states

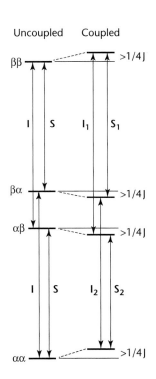

Figure 5.1 Uncoupled (left) and coupled (right) **IS** spin systems. Splitting for two **I** transitions are equal on the left (uncoupled), but are shifted in the coupled case by $\pm J/4$, so that two separate lines **I₁** and **I₂** are observed, separated by **J**. A similar situation is observed for spin **S**.

with the same value of $\mathbf{m}_X$. In one case this results in an increase by $(1/2)J_{AX}$ in the transition frequency over that due to Zeeman splitting alone, whereas in the other case, the transition frequency is reduced by $(1/2)J_{AX}$ from that due only to the Zeeman splitting. The net result is just what you expect when you look at the spectrum of the AX spin system: a splitting of J_{AX} between the two resonance lines of spin A. A similar result is observed for spin X.

The total energy of a coupled spin system in a particular state can be described by:

$$E_T = -\sum_i \mathbf{m}_i \left(\gamma B_0 (1 - \sigma_i) \right) + h \sum_{i<j} J_{ij} \mathbf{m}_i \cdot \mathbf{m}_j \qquad (5.28)$$

with summation over all of the spins in the system. Division of Equation 5.28 by Planck's constant gives each term in units of frequency (hertz). This is convenient as coupling constants are typically described in hertz.

$$\frac{E_T}{h} = -\sum_i \mathbf{m}_i \left(\frac{1}{2\pi} \gamma B_0 (1 - \sigma_i) \right) + \sum_{i<j} J_{ij} \mathbf{m}_i \cdot \mathbf{m}_j \qquad (5.29)$$

In order to bring this classical description of the energy of the system into a form usable for quantum mechanical calculations, it must be converted into a Hamiltonian operator that yields the correct energy for a given spin state when acting upon the wave description of that state. We start by remembering that the angular momentum of a nuclear spin is quantized. The magnitude of the spin angular momentum vector (fixed for a given spin) is given by $\hbar[\mathbf{I}(\mathbf{I}+1)]^{\frac{1}{2}}$ where $\mathbf{I}$ is the spin angular momentum quantum number for the spin. The quantum number $\mathbf{m}$ indicates the spin state occupied, and can be taken as indicating the direction of the angular momentum vector the magnitude of which is specified by $\hbar[\mathbf{I}(\mathbf{I}+1)]^{\frac{1}{2}}$ (see Figure 5.2). The $\mathbf{z}$ component of the angular momentum has a magnitude of $\hbar\mathbf{m}$. The cosine of the angle between the two vectors is given by $\mathbf{m}[\mathbf{I}(\mathbf{I}+1)]^{\frac{1}{2}}$ (see Equation 2.8 for the classical equivalent). The value of $\mathbf{m}$ can range from $-\mathbf{I}$ to $+\mathbf{I}$ in increments of 1.

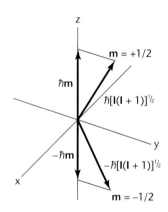

Figure 5.2 Quantization of angular momentum of a nuclear spin represented vectorially. The magnitude of the angular momentum is given by $\hbar[\mathbf{I}(\mathbf{I}+1)]^{\frac{1}{2}}$, while the direction is specified by $\mathbf{m}$. Note that the magnitude of both the total spin angular momentum and the $\mathbf{z}$ component can be specified simultaneously.

Since the magnitudes of both the total spin angular momentum and the $\mathbf{z}$ component of that angular momentum are fixed, one might suspect that they are eigenvalues returned by the action of the appropriate operators on eigenfunctions of those operators that describe the stationary spin states of the spin. And so we will try to find operators that, when acting upon the wave descriptions of the system, return eigenvalues that are the spin angular momenta of the system. *Both the total spin angular momentum and the $\mathbf{z}$ component of the angular momentum can be determined simultaneously for a given stationary state of the system.* As such, we should be able to find a set of wave descriptions of the system that are simultaneously eigenfunctions of the total spin angular momentum operator and the operator for the $\mathbf{z}$ component of the spin angular momentum. When acted on by the former, the appropriate eigenfunction should generate the eigenvalue $\hbar[\mathbf{I}(\mathbf{I}+1)]^{\frac{1}{2}}$, and when acted on by the operator for the $\mathbf{z}$ component of the spin angular momentum it should generate the value of $\hbar\mathbf{m}$. In general, if one can find a set of functions that are simultaneously eigenfunctions of two operators, those operators are said to **commute**, which means it does not matter in what order the two operators are applied to a function. The operators for the total spin angular momentum are represented by $\hat{\mathbf{I}}^2$, as it turns out to be more convenient to describe the square of the total angular momentum. The operator for the **z** component of the angular momentum ($\hat{\mathbf{I}}_z$) and $\hat{\mathbf{I}}^2$ should (and do) commute, since we expect to find a set of wave descriptions of the system that are simultaneously eigenvectors of both $\hat{\mathbf{I}}^2$ and $\hat{\mathbf{I}}_z$.

Now it is possible to make some substitutions into Equation 5.29 that will yield a Hamiltonian operator appropriate for calculating stationary state energies in this coupled spin system. First we turn to the terms dealing with chemical shift:

$$-\sum_i \mathbf{m}_i \left(\frac{1}{2\pi} \gamma B_0 (1-\sigma_i) \right) \tag{5.30}$$

Recall that the magnetic dipole of the *i*th spin, $\vec{\mu}_i$, is directly proportional to the spin angular momentum, which is now described by an operator $\hat{\mathbf{I}}_i$, with the gyromagnetic ratio γ as the proportionality constant (Equation 2.5). In turn, the **z** component of $\vec{\mu}_i$, μ_z, will have the same proportionality with $\hat{\mathbf{I}}_z$. Furthermore, the potential energy of the state generated by the interaction of the nuclear magnetic dipole with an applied field $\vec{B}_0$ along the **z** axis is given by $-\vec{\mu} \cdot \vec{B}_0 = -\mu_z B_0$ (see Equation 2.6). Substituting $\gamma \hat{\mathbf{I}}_z$ for μ_z, we obtain the operator for the energy of a nuclear spin in terms of the quantized **z** component of the spin angular momentum. Including the shielding terms from Expression 5.30, one obtains:

$$-\sum_i \left(\frac{1}{2\pi} \gamma B_0 (1-\sigma_i) \right) \hat{\mathbf{I}}_{iz} \tag{5.31}$$

Operation on this expression by the appropriate spin wave functions will yield energy terms the same as in the classical descriptions of the stationary state energies in Expression 5.30. The term for the couplings between spins is straightforward:

$$\sum_{i<j} J_{ij} \hat{\mathbf{I}}_i \cdot \hat{\mathbf{I}}_j \tag{5.32}$$

And now the complete form of the Hamiltonian operator can be given:

$$\frac{\hat{\mathbf{E}}}{h} = -\sum_i \left(\frac{1}{2\pi} \gamma B_0 (1-\sigma_i) \right) \hat{\mathbf{I}}_{iz} + \sum_{i<j} J_{ij} \hat{\mathbf{I}}_i \cdot \hat{\mathbf{I}}_j \tag{5.33}$$

The next step is to determine the appropriate eigenfunctions for use in Equation 5.27. Before we do so, we need to review some basic concepts and nomenclature. A proper wave description ψ of a quantum state should be **normalized**, i.e. the integral of that wave function multiplied by its complex conjugate over all space should be equal to 1. [The **complex conjugate** of a function is defined by changing the sign of the square root of -1 (i) where it is found. Thus the complex conjugate of $f(t) = \exp(i\omega t)$ is given by $f^*(t) = \exp(-i\omega t)$.] Thus, if α represents an eigenfunction of the spin angular momentum operator $\hat{\mathbf{I}}_z$ with the associated eigenvalue $\hbar \mathbf{m} = \hbar(1/2)$, α must satisfy the following relationship:

$$\int_{-\infty}^{\infty} \alpha * \cdot \alpha dq = 1 \tag{5.34}$$

meaning that integration of $\alpha^* \cdot \alpha$ over all space has unit value. This requirement is often short-handed using Dirac notation as:

$$\langle \alpha | \alpha \rangle = 1 \tag{5.35}$$

Equations 5.34 and 5.35 mean exactly the same thing. The left-hand term in the Dirac notation $\langle \alpha |$ is referred to as the **bra**, while the right hand term $| \alpha \rangle$ is the **ket** (from the hyphenation *bra-cket*, with *cket* shortened to *ket*). An eigenfunction of an operator is also sometimes called an **eigenket** of that operator.

It is worth noting that eigenfunctions represent vectors in an n-dimensional space called a **Hilbert space**, with the dimensionality of that space defined by the number of stationary states available to the system. For a single spin, the dimensionality of the space is 2 ($\mathbf{m}$ = +1/2, $\mathbf{m}$ = –1/2). One can think of the eigenfunctions as representing the unit vectors of that space, in complete analogy to unit vectors that represent each of the three dimensions in Cartesian coordinates. As any point in Cartesian space can be represented by a vector from the origin to that point with components $\mathbf{x}$, $\mathbf{y}$ and $\mathbf{z}$, so any possible state of the quantized system (including states that are not stationary, but time dependent) can be represented by a linear combination of the complete set of eigenfunctions that describe that system. For this reason, eigenfunctions are often referred to as **eigenvectors**. A good general introduction to these concepts and nomenclature can be found in Reference 3.

The action of $\hat{\mathbf{I}}_z$ on α is described by the expression:

$$\hbar(1/2)|\alpha\rangle = \hbar \mathbf{m}_\alpha |\alpha\rangle = \hat{\mathbf{I}}_z |\alpha\rangle \tag{5.36}$$

We can apply the same requirements to a function β that describes the upper energy spin state of the spin, and will return a value of $\hbar \mathbf{m} = \hbar(-1/2)$ when operated on by $\hat{\mathbf{I}}_z$:

$$\hbar(-1/2)|\beta\rangle = \hbar \mathbf{m}_\beta |\beta\rangle = \hat{\mathbf{I}}_z |\beta\rangle \tag{5.37}$$

$$\langle \beta | \beta \rangle = 1 \tag{5.38}$$

The action of an operator on an eigenket yields a constant (the eigenvalue) multiplied by the eigenket (Equation 5.37). Post-multiplication by the complex conjugate of the eigenket followed by integration yields the eigenvalue without the eigenket:

$$\langle \alpha | \hat{\mathbf{I}}_z | \alpha \rangle = \hbar \mathbf{m}_\alpha = \hbar / 2$$
$$\langle \beta | \hat{\mathbf{I}}_z | \beta \rangle = \hbar \mathbf{m}_\beta = -\hbar / 2 \tag{5.39}$$

Note that the order of operation in Equations 5.37–5.39 is from right to left.

A second requirement made upon the eigenvectors describing the spin states of a spin 1/2 spin, α and β, is that they be orthogonal to each other, a condition defined by the expression:

$$\int_{-\infty}^{\infty} \alpha * \cdot \beta dq = \int_{-\infty}^{\infty} \beta * \cdot \alpha dq = 0 \tag{5.40}$$

or in Dirac notation:

$$\langle \alpha | \beta \rangle = \langle \beta | \alpha \rangle = 0 \tag{5.41}$$

We now have all of the information we need about the functions describing the quantized spin states of a spin 1/2 nucleus. Notice that we have not precisely defined the forms of the spin functions α and β, nor do we need to. We only need to define their properties, and go under the blissful assumption that functions fitting this description actually exist. We can obtain all of the important information about the eigenstates of the spin system by assuming that α and β are normalized functions that are orthogonal to each other (called **orthonormality**), and are simultaneously eigenvectors of both $\hat{\mathbf{I}}^2$ and $\hat{\mathbf{I}}_z$. By analogy, we need not know any more about the unit vectors $\mathbf{x}$, $\mathbf{y}$ and $\mathbf{z}$ than that they are orthonormal in order to use those vectors to define any point in Cartesian space.

Equations 5.36 and 5.37 describe the effects of the operation of $\hat{\mathbf{I}}_z$ on α and β, and the action of $\hat{\mathbf{I}}^2$ on both functions is described by:

$$\hbar^2 [\mathbf{I}(\mathbf{I}+1)] | \beta \rangle = \hbar^2 [\mathbf{I}(\mathbf{I}+1)] | \alpha \rangle = \hat{\mathbf{I}}^2 | \beta \rangle = \hat{\mathbf{I}}^2 | \alpha \rangle \tag{5.42}$$

We now consider the properties of the angular momentum operators. Recalling from classical physics that the angular momentum of a particle around a point is the cross-product of the linear momentum of that particle $\vec{\mathbf{u}}$ and the vector $\vec{\mathbf{r}}$ connecting the particle to that point. Representing the vectors in three-dimensional space as 3×1 matrices and cross-multiplying those vectors yields:

$$\vec{\mathbf{r}} = \begin{bmatrix} x \\ y \\ z \end{bmatrix}, \qquad \vec{\mathbf{u}} = \begin{bmatrix} u_x \\ u_y \\ u_z \end{bmatrix}, \qquad \vec{\mathbf{I}} = \vec{\mathbf{r}} \times \vec{\mathbf{u}} = \begin{bmatrix} u_z y - u_y z \\ u_x z - u_z x \\ u_y x - u_x y \end{bmatrix} \tag{5.43}$$

In order to generate the quantum mechanical operator equivalents of the classical angular momentum, we go to the basic postulate of quantum mechanics, namely, the variables representing basic properties of a particle, or **observables** (e.g. position, energy and momentum) are represented by operators that act upon the wave descriptions of the system. The operator for the observable energy is the Hamiltonian. The position operators in Cartesian coordinates are given as $\hat{\mathbf{x}}$, $\hat{\mathbf{y}}$ and $\hat{\mathbf{z}}$. The operators representing the linear momentum along the $\mathbf{x}$, $\mathbf{y}$ and $\mathbf{z}$ axes respectively are given by:

$$-i\left(\frac{d}{dx}\right)$$

$$-i\left(\frac{d}{dy}\right)$$

$$-i\left(\frac{d}{dz}\right) \tag{5.44}$$

Substitution of the quantum mechanical operators from Equation 5.44 into the vectorial expression of the angular momentum in Equation 5.43 yields the following vector representation for the total angular momentum of the system:

$$\hat{\mathbf{I}} = \begin{bmatrix} \hat{\mathbf{I}}_x \\ \hat{\mathbf{I}}_y \\ \hat{\mathbf{I}}_z \end{bmatrix} = -i\hbar \begin{bmatrix} \hat{\mathbf{y}}\left(\dfrac{d}{dz}\right) - \hat{\mathbf{z}}\left(\dfrac{d}{dy}\right) \\ \hat{\mathbf{z}}\left(\dfrac{d}{dx}\right) - \hat{\mathbf{x}}\left(\dfrac{d}{dz}\right) \\ \hat{\mathbf{x}}\left(\dfrac{d}{dy}\right) - \hat{\mathbf{y}}\left(\dfrac{d}{dz}\right) \end{bmatrix} \tag{5.45}$$

In order to determine other fundamental properties of the angular momentum operators, we need the **commutators** of the individual angular momentum operators. The commutator of two operators $\hat{\mathbf{A}}$ and $\hat{\mathbf{B}}$ is defined as:

$$\left[\hat{\mathbf{A}}, \hat{\mathbf{B}}\right] = \hat{\mathbf{A}}\hat{\mathbf{B}} - \hat{\mathbf{B}}\hat{\mathbf{A}} \tag{5.46}$$

If the action of a commutator of two operators on a function returns a value of zero, the two operators are said to **commute**, and one can find sets of functions that are eigenfunctions of both operators simultaneously. The following commutator expressions are important:

$$\left[\hat{\mathbf{I}}_x, \hat{\mathbf{I}}_y\right] = i\hbar\hat{\mathbf{I}}_z$$

$$\left[\hat{\mathbf{I}}_z, \hat{\mathbf{I}}_x\right] = i\hbar\hat{\mathbf{I}}_y$$

$$\left[\hat{\mathbf{I}}_y, \hat{\mathbf{I}}_z\right] = i\hbar\hat{\mathbf{I}}_x \tag{5.47}$$

The fact that the commutators for the individual components of the angular momentum are nonzero means that we cannot find a set of functions that are simultaneously eigenfunctions of two of the Cartesian operators of the angular momentum. Translated to what was discussed earlier, this means that one can never simultaneously get values for the angular momentum in more than one direction at a time. However, the scalar operator $\hat{\mathbf{I}}^2$, defined from the Pythagorean relationship $\hat{\mathbf{I}}^2 = \hat{\mathbf{I}} \cdot \hat{\mathbf{I}} = \hat{\mathbf{I}}_x^2 + \hat{\mathbf{I}}_y^2 + \hat{\mathbf{I}}_z^2$, does commute with the three component operators, so one can simultaneously obtain eigenvalues for $\hat{\mathbf{I}}^2$ and one component $\hat{\mathbf{I}}_i$ when the system is in a stationary state. In analogy to the classical picture, where the equilibrium magnetization M_z^0 is taken to lie along the $\mathbf{z}$ axis, the one known component is usually chosen to be $\hat{\mathbf{I}}_z$, with α and β chosen to be eigenfunctions of $\hat{\mathbf{I}}_z$. The qualification "in a stationary state" is required because time-dependent states obtained by superposition of the stationary states of the system are used to describe the generation of coherence, which results in known but fluctuating values for transverse components of the angular momentum. This topic will be discussed in Chapter 6.

One last set of useful operators comprises the **raising** and **lowering** operators. These connect the upper and lower spin states represented by $|\alpha\rangle$ ($\mathbf{m}=+1/2$) and $|\beta\rangle$ ($\mathbf{m}=-1/2$):

$$\hat{\mathbf{I}}_- = \hat{\mathbf{I}}_x - i\hat{\mathbf{I}}_y, \quad \hat{\mathbf{I}}_+ = \hat{\mathbf{I}}_x + i\hat{\mathbf{I}}_y \qquad (5.48)$$

$$|\alpha\rangle = \hat{\mathbf{I}}_+ |\beta\rangle \qquad (5.49)$$

$$|\beta\rangle = \hat{\mathbf{I}}_- |\alpha\rangle \qquad (5.50)$$

$$0 = \hat{\mathbf{I}}_+ |\alpha\rangle \qquad (5.51)$$

$$0 = \hat{\mathbf{I}}_- |\beta\rangle \qquad (5.52)$$

The raising and lowering operators represent what we usually think of as observable coherence in NMR. They connect two states (and so represent a spectroscopic transition), and are analogous to M_x and M_y in the classical description from earlier in this chapter. The connection with the classical picture is now complete: $\hat{\mathbf{I}}_z$ correlates with the relative populations of states M_z, while the linear combinations of $\hat{\mathbf{I}}_x$ and $\hat{\mathbf{I}}_y$ represented by the raising and lowering operators $\hat{\mathbf{I}}_+$ and $\hat{\mathbf{I}}_-$ correlate with the amplitudes of the transverse (detectable) coherences M_x and M_y.

We can now tackle the solution to the time-independent form of the Schroedinger equation 5.27 as it applies to systems of $\mathbf{I} = 1/2$ spins. In the simplest case, that of a single spin $\mathbf{I} = 1/2$, one obtains this version of the Hamiltonian operator:

$$\hat{\mathbf{E}} = -\hbar \gamma B_0 (1 - \sigma_i)\hat{\mathbf{I}}_{iz} \qquad (5.53)$$

where i is an index labeling the spin. When placed into Equation 5.27 along with the appropriate wave function for a single spin, this form for $\hat{\mathbf{E}}$ will produce the energy eigenvalues for $\pm\hbar\gamma\mathcal{B}_0(1-\sigma_i)/2$ for $\mathbf{m} = \mp 1/2$. Equation 5.53 has been rearranged in order to yield the eigenvalue in energy units rather than in frequency units.

More complicated situations arise when more than one spin is present and there is coupling between spins. For the coupled two-spin case, the Hamiltonian operator obtained by expanding Equation 5.33 and rearranging to get units of energy is:

$$\hat{\mathbf{E}} = -\hbar\gamma_i\left(1-\sigma_i\right)\hat{\mathbf{I}}_{iz}B_z - \hbar\gamma_j\left(1-\sigma_j\right)\hat{\mathbf{I}}_{jz}B_z + hJ_{ij}\left(\hat{\mathbf{I}}_{ix}\hat{\mathbf{I}}_{jx} + \hat{\mathbf{I}}_{iy}\hat{\mathbf{I}}_{jy} + \hat{\mathbf{I}}_{iz}\hat{\mathbf{I}}_{jz}\right) \quad (5.54)$$

Note that the Hamiltonian in Equation 5.54 is not in terms of operators for which eigenfunctions have been defined other than $\hat{\mathbf{I}}_z$. In order to get the Hamiltonian in terms of operators for which eigenfunctions are defined, the product operators $(\hat{\mathbf{I}}_{ix}\hat{\mathbf{I}}_{jx} + \hat{\mathbf{I}}_{iy}\hat{\mathbf{I}}_{jy} + \hat{\mathbf{I}}_{iz}\hat{\mathbf{I}}_{jz})$ can be replaced by $\hat{\mathbf{I}}_{iz}\hat{\mathbf{I}}_{jz} + \frac{1}{2}(\hat{\mathbf{I}}_{i+}\hat{\mathbf{I}}_{j-} + \hat{\mathbf{I}}_{i-}\hat{\mathbf{I}}_{j+})$ where the products of raising and lowering operators are called the "**flip-flop**" operators, because they connect coupled spins by simultaneously raising the state of one spin and lowering the state of the other. Now the Hamiltonian is in terms of operators for which we have defined eigenfunctions, the $\hat{\mathbf{I}}_z$. We can use products of the eigenvectors $|\alpha_i\rangle$, $|\beta_i\rangle$, $|\alpha_j\rangle$ and $|\beta_j\rangle$, that describe the stationary states of the isolated spins, as a **basis set** for determining the eigenvectors and eigenvalues of the Hamiltonian that describe the coupled spins. The elements of the **Hamiltonian matrix**, $\langle j|\langle i|\hat{\mathbf{E}}|i\rangle|j\rangle$, are obtained by operation of $\hat{\mathbf{E}}$ on the product ket $|i\rangle|j\rangle$, where $|i\rangle$ is either $|\alpha_i\rangle$ or $|\beta_i\rangle$ and $|j\rangle$ is either $|\alpha_j\rangle$ or $|\beta_j\rangle$. This is followed by operation with the appropriate bra. For example, the upper left-hand diagonal element of the Hamiltonian matrix (see Expression 5.59) expands to:

$$\langle\alpha_i|\langle\alpha_j|\hat{\mathbf{E}}|\alpha_i\rangle|\alpha_j\rangle = \langle\alpha_i|\langle\alpha_j|\left(-\hbar\gamma_i(1-\sigma_i)\hat{\mathbf{I}}_{iz}B_z\right)|\alpha_i\rangle|\alpha_j\rangle +$$

$$\langle\alpha_i|\langle\alpha_j|\left(-\hbar\gamma_j(1-\sigma_j)\hat{\mathbf{I}}_{jz}B_z\right)|\alpha_i\rangle|\alpha_j\rangle + \langle\alpha_i|\langle\alpha_j|\left(hJ_{ij}\hat{\mathbf{I}}_{iz}\hat{\mathbf{I}}_{jz}\right)|\alpha_i\rangle|\alpha_j\rangle +$$

$$\frac{1}{2}\langle\alpha_i|\langle\alpha_j|\left(hJ_{i+}(\hat{\mathbf{I}}_{iz}\hat{\mathbf{I}}_{j-} + \hat{\mathbf{I}}_{i-}\hat{\mathbf{I}}_{j+})\right)|\alpha_i\rangle|\alpha_j\rangle \quad (5.55)$$

However, each operator acts only on the portion of the product function that is an eigenfunction of that operator, so Equation 5.55 simplifies to:

$$\langle\alpha_i|\langle\alpha_j|\hat{\mathbf{E}}|\alpha_i\rangle|\alpha_j\rangle = \frac{1}{2}\left(-\hbar\gamma_i(1-\sigma_i)B_z\right) + \frac{1}{2}\left(-\hbar\gamma_j(1-\sigma_j)B_z\right) + \frac{1}{4}\left(hJ_{ij}\right) \quad (5.56)$$

Note that the flip-flop operators always return a value of 0 for diagonal elements of the matrix. On the other hand, in off-diagonal elements (that represent connections between states), *only* the flip-flop operators give nonzero terms, owing to the requirements of orthonormality that have been imposed on the original eigenvectors. For example, element 3,2 is derived as follows:

$$\langle\alpha_i|\langle\beta_j|\hat{\mathbf{E}}|\beta_i\rangle|\alpha_j\rangle = \langle\alpha_i|\langle\beta_j|\left(-\hbar\gamma_i(1-\sigma_i)\hat{\mathbf{I}}_{iz}B_z\right)|\beta_i\rangle|\alpha_j\rangle +$$

$$\langle\alpha_i|\langle\beta_j|\left(-\hbar\gamma_j(1-\sigma_j)\hat{\mathbf{I}}_{jz}B_z\right)|\beta_i\rangle|\alpha_j\rangle + \langle\alpha_i|\langle\beta_j|\left(\hbar2\pi J_{ij}\hat{\mathbf{I}}_{iz}\hat{\mathbf{I}}_{jz}\right)|\beta_i\rangle|\alpha_j\rangle +$$

$$\frac{1}{2}\langle\alpha_i|\langle\beta_j|\left(hJ_{ij}(\hat{\mathbf{I}}_{i+}\hat{\mathbf{I}}_{j-} + \hat{\mathbf{I}}_{i-}\hat{\mathbf{I}}_{j+})\right)|\beta_i\rangle|\alpha_j\rangle \quad (5.57)$$

which yields:

$$\langle \alpha_i | \langle \beta_j | \hat{\mathbf{E}} | \beta_i \rangle | \alpha_j \rangle = hJ_{ij} / 2 \qquad (5.58)$$

The complete Hamiltonian matrix is:

	$| \alpha_i \rangle | \alpha_j \rangle$	$| \alpha_i \rangle | \beta_j \rangle$	$| \beta_i \rangle | \alpha_j \rangle$	$| \beta_i \rangle | \beta_j \rangle$
$\langle \alpha_i | \langle \alpha_j |$	$\dfrac{1}{2}\left(-\varepsilon_i - \varepsilon_j + \dfrac{hJ_{ij}}{2}\right)$	0	0	0
$\langle \alpha_i | \langle \beta_j |$	0	$\dfrac{1}{2}\left(-\varepsilon_i + \varepsilon_j - \dfrac{hJ_{ij}}{2}\right)$	$\dfrac{hJ_{ij}}{2}$	0
$\langle \beta_i | \langle \alpha_j |$	0	$\dfrac{hJ_{ij}}{2}$	$\dfrac{1}{2}\left(\varepsilon_i - \varepsilon_j - \dfrac{hJ_{ij}}{2}\right)$	0
$\langle \beta_i | \langle \beta_j |$	0	0	0	$\dfrac{1}{2}\left(\varepsilon_i + \varepsilon_j + \dfrac{hJ_{ij}}{2}\right)$

$$(5.59)$$

where $\varepsilon_i = \hbar\gamma_i(1 - \sigma_i)B_z$ and $\varepsilon_j = \hbar\gamma_j(1 - \sigma_j)B_z$, and the basis set of bras and kets is shown along the upper and left-hand edges.

In the case of heteronuclear coupling, where $\gamma_i \neq \gamma_j$, the energies represented by off-diagonal terms will always be smaller than the difference between the diagonal terms, which can be taken as representing the energies of the stationary states of the system. The spectrum will be first-order (see Chapter 2); that is, the coupling is much smaller than the difference in the chemical shift of the two spins. Allowed transitions are ones where the total change in spin angular momentum is $\Delta\mathbf{m} = \pm 1$ (spin i undergoes a transition while spin j remains in the same state, or vice versa). If diagonal elements are numbered from upper left to lower right as 1, 2, 3 and 4, the allowed transitions are thus between 1 and 2, 3 and 4, 1 and 3, and 2 and 4 for a total of four distinguishable transitions. The frequencies of the transition lines are proportional to the differences between the energies of each state as always.

For homonuclear coupled spins, where $\gamma_i = \gamma_j$, the situation is more complicated. In this case, the differences between transition energies are not necessarily much larger than the coupling energies (non-first order spectra). In this case we will need to **diagonalize** the Hamiltonian matrix in order to derive a correct description of the spin system; the simple product functions used to approximate the wave functions of the weakly coupled system will not be good enough to describe the stationary states or calculate the energies of those states of the more strongly coupled system. Consider this problem in the following way. If two spins perturb each other only weakly, they behave for the most part as if the other spin was not there, and the weak coupling approximation (that the stationary states of the spins can be described using the wave functions that describe their uncoupled behavior) is good enough. However, when the coupling energies are on the same order of magnitude as the differences between Zeeman transition energies, the behavior of one spin becomes important in determining the behavior of the other; that is, the appropriate wave descriptions of the stationary states of the system are no longer the wave functions of the isolated spins. The wave functions used to build the Hamiltonian matrix, that are eigenfunctions of the $\hat{\mathbf{I}}_{zi}$ angular momentum operators, are clearly *not* eigenfunctions of the Hamiltonian of the two-spin system, since the orthonormality requirements for eigenfunctions do not apply. However, we can use linear combinations of the original basis set to generate a new set of functions that *are* eigenfunctions of the Hamiltonian operator $\hat{\mathbf{E}}$.

In order to understand how this is done, we return to the analogy between Cartesian space and Hilbert spaces presented earlier in this chapter. The original set of functions proposed to be the eigenvectors of the coupled two-spin system, $|\alpha_i\rangle|\alpha_j\rangle$, $|\alpha_i\rangle|\beta_j\rangle$, $|\beta_i\rangle|\alpha_j\rangle$ and $|\beta_i\rangle|\beta_j\rangle$, can be considered as unit vectors in a Hilbert space with as many dimensions as there are eigenfunctions for the spin system (in the present case, four). If this set is complete (as the unit vectors $\vec{x}$, $\vec{y}$ and $\vec{z}$ **span** Cartesian space), linear combination of these four vectors can be used to generate a new set of **eigenvectors** that are normalized and orthogonal to each other. This is exactly analogous to the description of any vector in three-dimensional Cartesian space by a linear combination of the orthogonal $\vec{x}$, $\vec{y}$ and $\vec{z}$ unit vectors. In matrix representation, a vector with the x, y and z coordinates a, b and $\vec{c}$ in Cartesian space can be represented as a 1×3 matrix as follows:

$$\begin{pmatrix} a \\ b \\ c \end{pmatrix} \Rightarrow a\vec{x} + b\vec{y} + c\vec{z}$$

(5.60)

The choice of coordinate system is completely arbitrary, and we could easily imagine rotating the frame of reference so that the same vector described in Expression 5.60 lay along only one of the three principal axes (let us say the new $\vec{x}'$ axis). This transformation (called a **change of basis**) is accomplished by the action of a **rotation operator** $\hat{\mathbf{R}}$ on the original description of the system, giving a description in the new frame of reference:

$$\begin{pmatrix} a \\ b \\ c \end{pmatrix} \hat{\mathbf{R}} = \begin{pmatrix} a' \\ 0 \\ 0 \end{pmatrix} \Rightarrow a'\vec{x}'$$

(5.61)

Problem 5.14 presents a simple example of a basis change. At present, it is not necessary to understand exactly how this operation is accomplished; it is only necessary to understand that it *can* be accomplished. This sort of operation goes on all the time in the physical sciences: a problem can be simplified by finding the correct frame of reference to describe the behavior. We have already had some experience with this in going from the laboratory to the rotating frame of reference in order to simplify descriptions of precessing magnetization. We now do the same type of operation in order to find the eigenvalues that describe the stationary states of the coupled system. By rotating the frame of reference in Hilbert space so that the wave functions that describe stationary states of the coupled spin system represent the principal axes of the space in the new coordinate system, they become the new set of eigenvectors of the system, and the eigenvalues generated using the Hamiltonian are the energies of those stationary states. The Hamiltonian matrix in this coupled basis set would be diagonal [all elements $E_{ij}(i \neq j)=0$], since all the stationary states of the system are "pure", i.e. each is represented by a single eigenvector. The new Hamiltonian matrix would have the following appearance:

	$\lvert\alpha'_i\rangle\lvert\alpha'_j\rangle$	$\lvert\alpha'_i\rangle\lvert\beta'_j\rangle$	$\lvert\beta'_i\rangle\lvert\alpha'_j\rangle$	$\lvert\beta'_i\rangle\lvert\beta'_j\rangle$
$\langle\alpha'_i\rvert\langle\alpha'_j\rvert$	A	0	0	0
$\langle\alpha'_i\rvert\langle\beta'_j\rvert$	0	B	0	0
$\langle\beta'_i\rvert\langle\alpha'_j\rvert$	0	0	C	0
$\langle\beta'_i\rvert\langle\beta'_j\rvert$	0	0	0	D

$$(5.62)$$

where A, B, C and D are the eigenvalues (energies) of the stationary states represented by the eigenvectors along the top and left-hand sides of the matrix. The primes (′) in the bras and kets indicate that this is a new basis set, made up from linear combinations of the old basis set. The task now is to determine the values of A, B, C and D, which are the energies of the stationary states of the coupled system, and the form of the eigenvectors. The latter are simply linear combinations of the starting basis set spin functions that will provide the basis set for the diagonalized Hamiltonian. If we define this new set as $\tilde{\psi}$, each of the four eigenkets has the form $\lvert\psi_k\rangle = c_{1k}\lvert\alpha_i\rangle\lvert\alpha_j\rangle + c_{2k}\lvert\alpha_i\rangle\lvert\beta_j\rangle + c_{3k}\lvert\beta_i\rangle\lvert\alpha_j\rangle + c_{4k}\lvert\beta_i\rangle\lvert\beta_j\rangle$, and the goal is to determine the coefficients c_k.

In the present case, we only need to diagonalize the submatrix obtained by leaving off the first and last rows and columns of Matrix 5.59, where all the off-diagonal elements are already zero. In other words, $\lvert\alpha_i\rangle\lvert\alpha_j\rangle$ and $\lvert\beta_i\rangle\lvert\beta_j\rangle$ are two of the four $\lvert\psi_k\rangle$ eigenkets, and the other two are produced by mixing $\lvert\alpha_i\rangle\lvert\beta_j\rangle$ and $\lvert\beta_i\rangle\lvert\alpha_j\rangle$. The 2×2 matrix thus obtained now needs to be converted from the form:

$$\bar{M} = \begin{bmatrix} \dfrac{1}{2}\left(-\varepsilon_i + \varepsilon_j - \dfrac{hJ_{ij}}{2}\right) & \dfrac{hJ_{ij}}{2} \\[2ex] \dfrac{hJ_{ij}}{2} & \dfrac{1}{2}\left(\varepsilon_i - \varepsilon_j - \dfrac{hJ_{ij}}{2}\right) \end{bmatrix} \tag{5.63}$$

to a form:

$$\bar{M}' = \begin{bmatrix} B & 0 \\ 0 & C \end{bmatrix} \tag{5.64}$$

There are general methods for accomplishing diagonalization using linear algebra (see Appendix B for details). This involves finding the rotation operator $\hat{\mathbf{R}}$ that will make the off-diagonal elements of the submatrix zero, i.e.:

$$\hat{\mathbf{R}}^{\perp}\vec{M}\hat{\mathbf{R}} = \vec{M}' \tag{5.65}$$

where $\hat{\mathbf{R}}^{\perp}$ is the adjoint of $\hat{\mathbf{R}}$, obtained by transposing (interchanging rows and columns) and taking the complex conjugates of the elements of $\hat{\mathbf{R}}$. This process is called "finding the eigenvalues" of $\overline{M}$. We need not go into the mathematical details of how this is done (for the interested reader, any linear algebra text describes the process). Applying the same rotation operator $\hat{\mathbf{R}}$ to the original basis set $\vec{\zeta}\,(\hat{\mathbf{R}}\vec{\zeta})$ yields a new set of orthogonal functions $\vec{\psi}$ that are eigenkets of $\hat{\mathbf{E}}$. These eigenkets are such that $M'\vec{\psi}=\hat{\mathbf{E}}\vec{\psi}$, M' representing the diagonal elements of the Hamiltonian matrix. The individual eigenvalues will form the entries of the matrix represented in Equation 5.64. The new eigenvectors will be linear combinations of the starting basis set of the form $\psi_k=c_{ki}|\alpha\rangle|\beta\rangle+c_{kj}|\beta\rangle|\alpha\rangle$.

For matrix 5.62 (the submatrix of the Hamiltonian), diagonalization yields the following result:

$$\overline{M}' = \begin{bmatrix} -\dfrac{hJ_{ij}}{4} - \dfrac{1}{2}\left((\varepsilon_i - \varepsilon_j)^2 + (hJ_{ij})^2\right)^{1/2} & 0 \\[2em] 0 & -\dfrac{hJ_{ij}}{4} + \dfrac{1}{2}\left((\varepsilon_i - \varepsilon_j)^2 + (hJ_{ij})^2\right)^{1/2} \end{bmatrix} \tag{5.66}$$

The diagonal entries are the eigenvalues of the matrix, and represent the energies of the two stationary states of the system linked by the flip-flop operators. The rotation matrix that yielded this result is:

$$\hat{\mathbf{R}} = \begin{bmatrix} -\dfrac{1}{hJ_{ij}}\left(\varepsilon_i - \varepsilon_j + \dfrac{1}{2}\left(\left(\varepsilon_i - \varepsilon_j\right)^2 + \left(hJ_{ij}\right)^2\right)^{1/2}\right) & 1 \\[2em] -\dfrac{1}{hJ_{ij}}\left(\varepsilon_i - \varepsilon_j - \dfrac{1}{2}\left(\left(\varepsilon_i - \varepsilon_j\right)^2 + \left(hJ_{ij}\right)^2\right)^{1/2}\right) & 1 \end{bmatrix} \tag{5.67}$$

and two linear combinations of the original spin operators that are eigenvectors of the Hamiltonian operator are given by:

$$\hat{\mathbf{R}}\begin{bmatrix} |\alpha_i\rangle|\beta_j\rangle \\ |\beta_i\rangle|\alpha_j\rangle \end{bmatrix} = \begin{bmatrix} \left(\left(-\dfrac{1}{hJ_{ij}}\left(\varepsilon_i - \varepsilon_j + \dfrac{1}{2}\left(\left(\varepsilon_i - \varepsilon_j\right)^2 + \left(hJ_{ij}\right)^2\right)^{1/2}\right)\right)|\alpha_i\rangle|\beta_j\rangle + |\beta_i\rangle|\alpha_j\rangle \\[2em] \left(\left(-\dfrac{1}{hJ_{ij}}\left(\varepsilon_i - \varepsilon_j - \dfrac{1}{2}\left(\left(\varepsilon_i - \varepsilon_j\right)^2 + \left(hJ_{ij}\right)^2\right)^{1/2}\right)\right)|\alpha_i\rangle|\beta_j\rangle + |\beta_i\rangle|\alpha_j\rangle \end{bmatrix} \tag{5.68}$$

Although the functions in Equation 5.68 are eigenvectors of the submatrix Equation 5.66, they are not normalized. We therefore need to find scaling factors that will make the sum of the squares of the coefficients equal to 1. The coefficients of the basis functions in 5.68 provide the proportionality c_{2i} / c_{3i}, and orthonormality requires that $c_{2i}^2 + c_{3i}^2 = 1$ and that $c_{22} = -c_{33}$ and $c_{23} = c_{32}$. The coefficients must obey simple mathematical relationships based on the trigonometric identity $\sin^2\theta + \cos^2\theta = 1$, so $-c_{22} = c_{33} = \sin\theta$ and $c_{23} = c_{32} = \cos\theta$. If the value of θ is given as $\tan\theta = J_{ij}/\Delta\nu_{ij}$ where J_{ij} is the coupling between spin i and j and $\Delta\nu_{ij} = (\varepsilon_i - \varepsilon_j)/h$ is the difference in their transition frequencies (in hertz), then the value of θ varies as a function of the state energies and coupling constants. Therefore, the contributions of the uncoupled basis spin functions to the mixed states depend upon the size of the coupling constants relative to the state energies. At the limit of very weak coupling ($J << (\varepsilon_2 - \varepsilon_3)/h$), θ is very small and the eigenvectors are essentially the same as the uncoupled basis. This is nearly always true for heteronuclear spin systems, where i and j have different γ, and the chemical shift difference between the coupled spins is orders of magnitude larger than the coupling between them. When $\varepsilon_2 - \varepsilon_3 = 0$ (chemical shifts of the two nuclei are the same), $\sin\theta = \cos\theta$, and both uncoupled basis functions contribute equally to the eigenvector representing the states. Both of these situations (and intermediate cases as well) are shown in Figure 5.3.

This analysis of the two-spin system yields not only the energies of the stationary states of the system and the eigenvectors that represent those states, but can also give us the relative intensities of the lines that make up the NMR spectrum. Why? Recall that in the NMR experiment we observe coherences that connect two states of the system. These coherences are represented quantum mechanically by the raising and lowering operators discussed earlier, $\hat{\mathbf{I}}_+$ and $\hat{\mathbf{I}}_-$. The **selection rules** for changes in spin state require that the total change in spin angular momentum in the system be confined to $\Delta\mathbf{m} = \pm 1$ during a transition. The selection rules will be discussed in more detail in the next section, but for now, we just need to note that double quantum ($\alpha\alpha \to \beta\beta$, $\beta\beta \to \alpha\alpha$, $\Delta\mathbf{m}_{total} = \pm 2$) and zero quantum ($\alpha\beta \to \beta\alpha$, $\beta\alpha \to \alpha\beta$, $\Delta\mathbf{m}_{total} = 0$) are forbidden in the first approximation. The allowed transitions are therefore $\alpha\alpha \to \alpha\beta$, $\alpha\alpha \to \beta\alpha$, $\beta\alpha \to \beta\beta$, $\alpha\beta \to \beta\beta$ and their inverses. As expected, this gives four lines in the NMR spectrum. The intensities of these lines come from the transition probabilities, which are calculated as follows.

If we generate a lowering operator by the addition of the operators of the individual spins $\hat{\Gamma}_- = \hat{\mathbf{I}}_{-i} + \hat{\mathbf{I}}_{-j}$, this operator can be applied to the eigenvectors representing one stationary state in order to connect to another state. For example, the lowering operator connects the state $c_{23}|\alpha\beta\rangle + c_{32}|\beta\alpha\rangle$ to the highest energy state $|\beta\beta\rangle$. This transition is represented by:

$$\hat{\Gamma}_-(\cos\theta\,|\,\alpha\beta\rangle + \sin\theta\,|\,\beta\alpha\rangle) = (\hat{\mathbf{I}}_{-i} + \hat{\mathbf{I}}_{-j})(\cos\theta\,|\,\alpha\beta\rangle + \sin\theta\,|\,\beta\alpha\rangle) \qquad (5.69)$$

From the definition of the lowering operator (Equation 5.51), and the results of its operation on the individual spin functions, this expression yields:

$$\hat{\Gamma}_-(\cos\theta\,|\,\alpha\beta\rangle + \sin\theta\,|\,\beta\alpha\rangle) = (\cos\theta + \sin\theta)\,|\,\beta\beta\rangle \qquad (5.70)$$

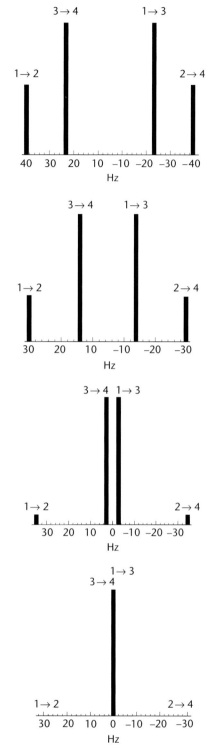

Figure 5.3 Non-first-order transitions in a two-spin system as a function of $\Delta\delta$ (chemical shift difference) and coupling constant J. The coupling (J) is fixed at 16 Hz. The chemical shift differences are 50, 30, 10 and 0 Hz (top to bottom). Lines correspond to allowed $\Delta m = \pm 1$ transitions: 1→2 is $|\alpha\alpha\rangle\rightarrow(c_{22}|\alpha\beta\rangle + c_{23}|\beta\alpha\rangle)$, 3→4 is $(c_{32}|\alpha\beta\rangle + c_{33}|\beta\alpha\rangle) \rightarrow|\beta\beta\rangle$, 1→3 is $|\alpha\alpha\rangle\rightarrow (c_{32}|\alpha\beta\rangle + c_{33}|\beta\alpha\rangle)$, and 2→4 is $(c_{22}|\alpha\beta\rangle + c_{23}|\beta\alpha\rangle) \rightarrow|\beta\beta\rangle$. Line heights correspond to the probability for a particular transition, and are calculated as shown in Equations 5.74 and 5.75.

Treatment of this with the bra for the connected $\langle\beta\beta|$ gives:

$$\langle\beta\beta| \,\hat{\Gamma}_-(\cos\theta\,|\alpha\beta\rangle + \sin\theta\,|\beta\alpha\rangle) = (\cos\theta + \sin\theta) \tag{5.71}$$

The result gives the probability for the transition between the states represented by the eigenvectors $|\beta\beta\rangle$ and $c_{32}|\alpha\beta\rangle + c_{33}|\beta\alpha\rangle$. The result is squared to remove the sign dependence as one squares a wave function in order to find spatial distributions of orbitals. This gives:

$$(\cos\theta + \sin\theta)^2 = \cos^2\theta + 2\cos\theta\sin\theta + \sin^2\theta \tag{5.72}$$

The probabilities for the other transitions are derived in the same fashion:

$$
\begin{aligned}
|\alpha\alpha\rangle &\to c_{22}|\alpha\beta\rangle + c_{23}|\beta\alpha\rangle, & \sin^2\theta - 2\cos\theta\sin\theta + \cos^2\theta \\
|\alpha\alpha\rangle &\to c_{22}|\alpha\beta\rangle + c_{23}|\beta\alpha\rangle, & \sin^2\theta + 2\cos\theta\sin\theta + \cos^2\theta \\
|\beta\beta\rangle &\to c_{22}|\alpha\beta\rangle + c_{23}|\beta\alpha\rangle, & \sin^2\theta - 2\cos\theta\sin\theta + \cos^2\theta
\end{aligned}
\tag{5.73}
$$

The time-dependent nuclear spin Hamiltonian operator and solutions to the time-dependent Schrödinger equation

Although the derivation shown above yields the energies of the stationary states of the system as well as the relative intensities of lines, it does not yield an expression for the line shape. In order to completely predict the appearance of the spectrum as was done in the classical case using the Bloch equations, we must consider time-dependent contributions to the Hamiltonian operator. If an RF field with an amplitude $B_1(\cos\omega t)$ is applied along the **x** axis of the laboratory frame, the Hamiltonian aquires a time dependence defined by:

$$\hat{\mathbf{E}}'(t) = -\gamma\hat{\mathbf{I}}\cdot\vec{B}_1\cos\omega t = -\gamma\hat{\mathbf{I}}_x\vec{B}_1\cos\omega t \tag{5.74}$$

To deal with the time-dependent Hamiltonian, we require the time-dependent version of the Schrödinger equation, which is a complete chapter in itself (see Appendix A). However, the results of the time-dependent treatment can be summarized as follows. The probability P_{kj} per unit time that a transition between the kth and jth states will occur is given by:

$$P_{kj}(\omega, t) = \gamma^2 B_1^2 \left\langle \psi_k \mid \hat{\mathbf{I}}_x \mid \psi_j \right\rangle^2 \left[\frac{\sin^2((\omega_{kj} - \omega)t/2)}{4(\omega_{kj} - \omega)^2} \right] \tag{5.75}$$

The bracketed term $\sin^2((\omega_{kj} - \omega)t/2)/4(\omega_{kj} - \omega)^2$ gives a line shape. The ψ_i are the time-independent eigenvectors representing the stationary states of the system that are connected by the transition. At the limit of $\omega = \omega_{kj}$, the function $P_{kj}(\omega, t)$ reaches a maximum, with the height of that maximum proportional to the square of the time over which the probability was measured. On the other hand, the line width is inversely proportional to the time over which the probability is measured, as expected from the uncertainty principle (note that we still have not brought relaxation into the expression). The square of the gyromagnetic ratio and the magnitude of B_1 also contribute to the transition probability, as does the off-diagonal term of the Hamiltonian matrix representing the connectivity between the kth and jth states (associated with the eigenkets of the time-independent Hamiltonian operators ψ_k and ψ_j). We can rewrite this term using the definition of the raising and lowering operators as:

$$\left\langle \psi_k \mid \hat{\mathbf{I}}_x \mid \psi_j \right\rangle = \left\langle \psi_k \mid \hat{\mathbf{I}}_+ + \hat{\mathbf{I}}_- \mid \psi_j \right\rangle \qquad (5.76)$$

Since the raising and lowering operators change the spin state **m** by +1 and –1, respectively (see Appendix A), the term in Equation 5.76 is nonzero only if the two connected states differ by $\Delta\mathbf{m} = \pm 1$.

The previous discussion is necessarily incomplete: we still have not arrived at a complete quantum mechanical description of the ensemble. In other words, we are not yet dealing with spin *populations* or *coherences* quantum mechanically, and the impact of relaxation on populations and coherences has yet to be considered. However, treatment of the isolated single-spin and coupled two-spin cases has allowed us to arrive at a basic description of an NMR spectrum using the principles of quantum mechanics. For a more rigorous treatment of this material, interested students should consult Reference 4.

Problems

5.1 Show that Equation 5.53 can be expanded to give Equation 5.54.

5.2 Using vector multiplication, show that Equation 5.8 can be used to generate Equations 5.12–5.14.

5.3 Show the conversion of Equations 5.15 to 5.18, as well as the conversion of 5.23 to 5.24.

*5.4 What would the relaxation operator **R** look like for a spin with a $T_1 = 2.1$ s and a $T_2 = 1.5$ s?

5.5 A Lorentzian line shape is described by the expression $M_0 T_2 [1/1 + \Delta\omega^2 T_2^2]$, which is obtained by rearranging Equation 5.25. The integrated area of the line described by this expression is independent of T_2. However, the height of the line depends on the value of $M_0 T_2$. Based on this expression, show that $\Delta\nu$, the line width in hertz of a peak at half height, is equal to T_2^{-1}/π.

*5.6 A line in a ^{1}H spectrum shows strongly temperature-dependent narrowing (more so than other signals in the same spectrum, so the effect is not due to changes in solvent viscosity). Explain this effect if (a) the signal is a carbon-attached proton in a diamagnetic sample; (b) the molecule contains a paramagnetic center (i.e. a radical or paramagnetic metal ion) or (c) the proton is attached to a ^{14}N atom.

*5.7 What is the chemical shift operator Ω for a ^{1}H spin resonating at 1 ppm with the carrier frequency set at 4.8 ppm in a 500 MHz NMR?

*5.8 Two lines in a ^{1}H NMR spectrum are believed to be from the same spin in a compound undergoing an exchange reaction of the form: $A \underset{k_r}{\overset{k_f}{\rightleftharpoons}} B$.

(a) By integration, the line A has twice the area of line B. What are their relative heights?

(b) Line A is 3 Hz wide at half-height. How wide is line B?

(c) If a 10°C increase in temperature doubles the line width of A, what is the activation enthalpy for the interconversion of A and B?

5.9 Given that $|\alpha\rangle$ and $|\beta\rangle$ are orthonormal spin functions for an $I = 1/2$ nucleus, give values for the following:

(a) $\langle\beta|\alpha\rangle$

(b) $\langle\beta|\hat{I}_+|\alpha\rangle$

(c) $\langle\alpha|\hat{I}_-|\alpha\rangle$

(d) $\langle\beta|\beta\rangle$

(e) $\langle\alpha|\hat{I}^2|\alpha\rangle$

5.10 The Pauli spin matrices are matrix representations of the interactions of the various spin operators on the orthonormal spin functions for an $I = 1/2$ nucleus, $|\alpha\rangle$ and $|\beta\rangle$. For an operator $\hat{O}$, they take the form:

$$\begin{bmatrix} \langle\alpha|\hat{O}|\alpha\rangle & \langle\alpha|\hat{O}|\beta\rangle \\ \langle\beta|\hat{O}|\alpha\rangle & \langle\beta|\hat{O}|\beta\rangle \end{bmatrix}$$

Calculate the appropriate forms of the spin matrices for each of the following operators:

$\hat{I}^2, \hat{I}_+, \hat{I}_-, \hat{I}_x, \hat{I}_y, \hat{I}_z$.

*5.11 Determine in each case whether the functions given are eigenfunctions of the operators $\dfrac{d}{dt}$ and/or $\dfrac{d^2}{dt^2}$.

(a) $e^{\omega t}$.

(b) t^2.

(c) $\sin 2t$.

*5.12 Calculate the commutators of the following pairs of operators.

(a) $[\hat{I}_z, \hat{I}^2]$.

(b) $[x, d/dx]$.

(c) $[\sqrt{\ }, (\ )^2]$.

5.13 Show that the expressions in Equation 5.47 are correct.

*5.14 The concept of deriving a basis set of orthonormal functions by linear combination of elements in a nonorthogonal basis set is exactly analogous to the idea of rotating a coordinate system in space so as to simplify calculations involving vectors and points within that space. In fact, a Hilbert space is exactly that: a space of N dimensions that must be described by a set of N orthogonal normalized vectors. In Cartesian space, we have three orthonormal vectors, the unit vectors $\vec{x}$, $\vec{y}$ and $\vec{z}$, that can be used to describe any vector in that space. In the rotating frame of reference, we convert from a static Cartesian space to a rotating space that simplifies mathematics involving rotating vectors. As a simpler example, consider a change in basis for a two-dimensional space with axes x and y. If we rotate the x and y axes by an angle α to generate a new set of axes x' and y', the relationship between the location of a point at the end of a vector (x,y) in the old coordinate system and in the new coordinate system is given by:

$x' = x \cos \alpha + y \sin \alpha$

$y' = -x \sin \alpha + y \cos \alpha$

(a)   Given two vectors $\vec{V}$ between the origin and a point at $x = 2$, $y = 4$ and $\vec{V}'$ between the origin and a point $x = -2$, $y = 3$, give the coordinates for the same points if the coordinate system is rotated through α so that the x' axis is now collinear with $\vec{V}$.

(b)   Write both vectors $\vec{V}$ and $\vec{V}'$ in terms of sums of unit vectors $\vec{x}$, $\vec{y}$ and $\vec{x}'$, $\vec{y}'$ (the eigenvectors) in both coordinate systems.

(c)   Show that the eigenvectors within each coordinate system are orthonormal to each other.

(d)   Imagine that as one went from the x,y coordinate system to the x',y' coordinate system, the x component doubled in magnitude, i.e. $x' = 2x$. Determine whether the operations of rotation and multiplication commute with each other (i.e. does one get the same result if the multiplication occurs before the rotation or after the rotation).

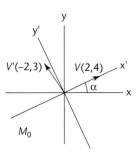

References

1.   F. Bloch, *Phys. Rev.* 70, 460 (1946).
2.   R. R. Ernst, G. Bodenhausen, and A. Wokaun, *Principles of Nuclear Magnetic Resonance in One and Two Dimensions*, Clarendon Press, New York (1991), p. 118.
3.   J. D. Macomber, *The Dynamics of Spectroscopic Transitions*, Chapter 2. John Wiley & Sons, New York (1976).
4.   C. Cohen-Tannoudji, B. Diu, and F. Laloe, *Quantum Mechanics*, Volume 1 (English translation). Hermann, Wiley-Interscience, Paris (1977).

6

DENSITY OPERATOR AND PRODUCT OPERATOR DESCRIPTIONS OF NMR EXPERIMENTS IN LIQUIDS

An ensemble of identical spins at equilibrium: an introduction to the density matrix formalism

In the NMR experiment, one typically deals with ensembles of spins rather than single particles, and ensembles must be handled statistically. The uncertainty principle teaches us that we cannot predict the precise time-dependent behavior of a single quantized particle. However, given a reasonable description of any time-dependent perturbations acting upon it, one can predict the statistical (likely) behavior of a single spin or several coupled spins over a long time. The transition from single spins or spin systems to statistical treatments of ensembles is made simpler by the **ergodic hypothesis**, that observation of a single system for a very long time is the same as observing many identical systems for a short time. In essence, one takes the description for a single spin or spin system and extends it to the ensemble by considering the average effect of time-dependent perturbations upon it.

We first consider the case of an ensemble of identical spins ($\mathbf{I}$ = 1/2). In the absence of spin–spin coupling, the spins at equilibrium will be distributed between two states, described by the spin functions $|\alpha\rangle$ and $|\beta\rangle$. The time-average probability of finding a given spin in either state is determined by the relative energies of the two states, and is given by Equation 2.9. The total time-average probability of finding a spin in either state must total to one. The overall state of the ensemble (or of a single spin observed for a long time) can be described using a wave function of the form:

$$\psi = c_1 \,|\, \alpha \rangle + c_2 \,|\, \beta \rangle \tag{6.1}$$

The c_i must meet the requirements of orthonormality, i.e.:

$$\psi * \psi = \sum_k c_k * c_k = 1 \tag{6.2}$$

It is convenient to represent the spin eigenfunctions as vectors:

$$|\alpha\rangle = \begin{pmatrix} 1 \\ 0 \end{pmatrix} \quad |\beta\rangle = \begin{pmatrix} 0 \\ 1 \end{pmatrix} \tag{6.3}$$

We now consider how to represent an ensemble of spins using the **density matrix** formalism. First, we recall (or learn) that the **expectation value** $\langle O \rangle$ for an observable represented by a quantum mechanical operator $\hat{\mathbf{O}}$ in a state of the system described by the wave function ψ is given by:

$$\langle \boldsymbol{O} \rangle = \langle \psi | \hat{\mathbf{O}} | \psi \rangle = \int_{-\infty}^{\infty} \psi * \hat{\mathbf{O}} \psi \, dq \tag{6.4}$$

The standard form for the expression is shown on the right of Equation 6.4, while the Dirac notation is shown in the middle. For an ensemble of identical spins $I = 1/2$, with spin states represented by $|\alpha\rangle$ and $|\beta\rangle$, we can use Expression 6.4 to generate matrix representations of the various spin angular momentum operators of the form:

$$\begin{pmatrix} \langle \alpha | \hat{\mathbf{O}} | \alpha \rangle & \langle \alpha | \hat{\mathbf{O}} | \beta \rangle \\ \langle \beta | \hat{\mathbf{O}} | \alpha \rangle & \langle \beta | \hat{\mathbf{O}} | \beta \rangle \end{pmatrix} \tag{6.5}$$

$$\hat{\mathbf{I}}^2 = \frac{3}{4}\begin{pmatrix} 1 & 0 \\ 0 & 1 \end{pmatrix} \quad \hat{\mathbf{I}}_+ = \begin{pmatrix} 0 & 1 \\ 0 & 0 \end{pmatrix} \quad \hat{\mathbf{I}}_- = \begin{pmatrix} 0 & 0 \\ 1 & 0 \end{pmatrix} \quad \hat{\mathbf{I}}_\alpha = \begin{pmatrix} 1 & 0 \\ 0 & 0 \end{pmatrix}$$

$$\hat{\mathbf{I}}_z = \frac{1}{2}\begin{pmatrix} 1 & 0 \\ 0 & -1 \end{pmatrix} \quad \hat{\mathbf{I}}_y = \frac{1}{2i}\begin{pmatrix} 0 & 1 \\ -1 & 0 \end{pmatrix} \quad \hat{\mathbf{I}}_x = \frac{1}{2}\begin{pmatrix} 0 & 1 \\ 1 & 0 \end{pmatrix} \quad \hat{\mathbf{I}}_\beta = \begin{pmatrix} 0 & 0 \\ 0 & 1 \end{pmatrix} \tag{6.6}$$

The matrix representations of the spin operators in Equations 6.6 are known as the **Pauli spin matrices**. Both of the pure states shown in Equations 6.3 are eigenfunctions of $\hat{\mathbf{I}}_z$ that commute with the total angular momentum operator $\hat{\mathbf{I}}^2$ and the time-independent Hamiltonian operator. The $\hat{\mathbf{I}}_z$ and $\hat{\mathbf{I}}^2$ operators are **population operators**, i.e. they return eigenvalues relating to the populations of spin states when operating on the spin functions in Equations 6.3. However, the pure-state representations in Equation 6.3 are not eigenfunctions of $\hat{\mathbf{I}}_x$ or $\hat{\mathbf{I}}_y$. The eigenfunctions of $\hat{\mathbf{I}}_x$ are given by:

$$\frac{1}{\sqrt{2}}\begin{pmatrix} 1 \\ 1 \end{pmatrix} \text{ and } \frac{1}{\sqrt{2}}\begin{pmatrix} 1 \\ -1 \end{pmatrix} \tag{6.7}$$

while the eigenfunctions of $\hat{\mathbf{I}}_y$ are:

$$\frac{1}{\sqrt{2}}\begin{pmatrix} 1 \\ i \end{pmatrix} \text{ and } \frac{1}{\sqrt{2}}\begin{pmatrix} 1 \\ -i \end{pmatrix} \tag{6.8}$$

Problem 6.9 is designed to convince you that this is true. Note that these are **mixed** states (i.e. they contain contributions from both pure states represented by Equations 6.3), and fit with the idea that the transverse operators $\hat{\mathbf{I}}_x$ and $\hat{\mathbf{I}}_y$ represent **coherences**, or connections between states.

The commutator relationships can be used to determine a number of important properties of spin operators. These are as follows:

$$\begin{aligned}
\left[\hat{\mathbf{I}}_x, \hat{\mathbf{I}}_y\right] &= i\hat{\mathbf{I}}_z & \hat{\mathbf{I}}_z|\alpha\rangle &= \tfrac{1}{2}|\alpha\rangle \\
\left[\hat{\mathbf{I}}_y, \hat{\mathbf{I}}_z\right] &= i\hat{\mathbf{I}}_x & \hat{\mathbf{I}}_z|\beta\rangle &= \tfrac{1}{2}|\beta\rangle \\
\left[\hat{\mathbf{I}}_z, \hat{\mathbf{I}}_x\right] &= i\hat{\mathbf{I}}_y & \hat{\mathbf{I}}_+|\alpha\rangle &= 0 & \langle\beta|\hat{\mathbf{I}}_z|\beta\rangle &= -\tfrac{1}{2} \\
\left[\hat{\mathbf{I}}_+, \hat{\mathbf{I}}_y\right] &= -\hat{\mathbf{I}}_+ & \hat{\mathbf{I}}_-|\beta\rangle &= 0 & \langle\alpha|\hat{\mathbf{I}}_z|\alpha\rangle &= \tfrac{1}{2} \\
\left[\hat{\mathbf{I}}_-, \hat{\mathbf{I}}_z\right] &= \hat{\mathbf{I}}_- & \hat{\mathbf{I}}_-|\alpha\rangle &= |\beta\rangle \\
\left[\hat{\mathbf{I}}_+, \hat{\mathbf{I}}_-\right] &= 2\hat{\mathbf{I}}_z & \hat{\mathbf{I}}_+|\beta\rangle &= |\alpha\rangle
\end{aligned} \tag{6.9}$$

Having defined the matrix forms of the **Cartesian operators** that represent the Cartesian components of spin angular momentum ($\hat{\mathbf{I}}_x$, $\hat{\mathbf{I}}_y$ and $\hat{\mathbf{I}}_z$), the total spin angular momentum ($\hat{\mathbf{I}}^2$), and operators representing connectivity between the stationary states of the system ($\hat{\mathbf{I}}_+$ and $\hat{\mathbf{I}}_-$), we can now define the **density matrix**. The density matrix allows one to calculate ensemble averages of observables based on the properties of the spin operators. There is no easy way to write down the idea, as a wave function, that 51% of all spins are α, while 49% are β (a situation that can only result from very unusual circumstances!). But the density matrix formalism allows us to do that. More importantly, the density matrix formalism and the shorthand known as **product operator formalism** allow one to follow the generation and evolution of coherences during the course of a multipulse NMR experiment.

The elements of the density matrix $\bar{\sigma}$ are given by $\sigma_{ij} = \langle c_i c_j^* \rangle$ where the c_i are the coefficients of the basis functions as defined in Equation 6.1 for the two-spin system, and the brackets indicate an ensemble average. For an ensemble of identical $\mathbf{I} = 1/2$ spins in the absence of coupling, the density matrix is a 2×2 matrix of the form:

$$\begin{pmatrix} \sigma_{11} & \sigma_{12} \\ \sigma_{21} & \sigma_{22} \end{pmatrix} \tag{6.10}$$

The diagonal terms of the density matrix ($i = j$) are just the probabilities of finding the spins in either the α or β states over a time average. The off-diagonal elements,

σ_{12} and σ_{21}, represent coherences between the two states. At thermal equilibrium, the off-diagonal elements average to 0, since the instantaneous values of $c_i c_j^*$ $i \neq j$ "flicker", that is they change with time, and average to zero. Note also that the density matrix is Hermitian, i.e. $(c_i c_j) = (c_j c_i)^*$.

The alert student will notice that there must be some time-dependent term in the elements of the density matrix in order for this "flickering" to occur, and to account for changes in spin populations when the system is perturbed (e.g. by an RF pulse). In Appendix A, it is shown (Equation A.7) that the time-dependence of $\psi(s,t)$, the wave function describing an arbitrary state of the system, is given by:

$$\psi_k(s,t) = \varsigma_k(s)\varphi_k(0)\exp\left[\frac{-iE_k t}{\hbar}\right] \tag{A.7}$$

where $\varphi_k(0)$ is the time-dependent portion of the wave function of the kth state at time zero and E_k is the energy of that state. If coherence is generated as the result of some time-dependent perturbation, the off-diagonal elements of the density matrix become nonzero, implying that the time-dependent portion of Equation A1.7 corresponds to the c_i in Equation 6.1. So Equation 6.1 can be written as:

$$\psi = c_1(t)|\alpha\rangle + c_2(t)|\beta\rangle \tag{6.11}$$

where each of the $c_k(t) = \varphi_k(0)\exp\left[\frac{-iE_k t}{\hbar}\right] = \varphi_k(0)\exp(-i\omega_k t)$ and ω_i is the

Larmor frequency of the spins in the ensemble. In the homonuclear two-spin case, the frequency terms are similar to those seen in the example of the two-spin system discussed in Chapter 5.

In reality, very few spins in the ensemble are in "pure" states at any given time: there are constantly local perturbations in the electromagnetic environment that give rise to mixed or **superposition** states. The frequency distribution of such fluctuations is given by the spectral density function, as described in Chapter 4. The presence of superposition states permits both the generation of coherence and spin relaxation. A more complete discussion of relaxation as applied to the density matrix is given in Appendix B.

Expansion of the density matrix for an uncoupled spin in terms of Cartesian angular momentum operators

The time-weighted average $\langle\langle O \rangle\rangle$ of an observable represented by the operator $\hat{\mathbf{O}}$ over the ensemble is called the **expectation value** of $\hat{\mathbf{O}}$ and is given by:

$$\langle\langle O \rangle\rangle = \left\langle \sum_{ij} c_j^* \mathbf{O}_{ji} c_i \right\rangle = \sum_{ij} \mathbf{O}_{ji} \left\langle c_j^* c_i \right\rangle$$

$$= \sum_{ij} \sigma_{ij} \mathbf{O}_{ji} = \sum_i (\sigma\mathbf{O})_{ii} = Tr\hat{\sigma}\hat{\mathbf{O}} \tag{6.12}$$

The $\mathbf{O}_{ij}$ are elements of the matrix representation of the operator $\hat{\mathbf{O}}$. The double arrows in the last two statements indicate multiplication of the $\hat{\sigma}$ and $\hat{\mathbf{O}}$ matrices to yield a new matrix, the sum of the diagonal elements of which (i.e. the **trace**) yields the desired expectation value. (The trace of the density matrix itself must be unity. Why?)

The density matrix for an ensemble of identical $\mathbf{I} = 1/2$ spins can be written as:

$$\bar{\sigma} = \begin{pmatrix} a+b & c+id \\ c-id & a-b \end{pmatrix} \tag{6.13}$$

As the trace of the matrix must be 1, $a = \frac{1}{2}$, and the matrix is Hermitian, the matrix can be expanded as follows:

$$\bar{\sigma} = a\begin{pmatrix} 1 & 0 \\ 0 & 1 \end{pmatrix} + b\begin{pmatrix} 1 & 0 \\ 0 & -1 \end{pmatrix} + c\begin{pmatrix} 0 & 1 \\ 1 & 0 \end{pmatrix} + id\begin{pmatrix} 0 & 1 \\ -1 & 0 \end{pmatrix} \tag{6.14}$$

where the first matrix in the expansion is the identity matrix and the last three matrices in the expansion are the $\hat{\mathbf{I}}_z$, $\hat{\mathbf{I}}_x$ and $\hat{\mathbf{I}}_y$ operators, respectively (assuming that the appropriate constants from Equation 6.6 are included in a, b and c). The coefficients for each term are the expectation values for the corresponding operator, obtained from the trace of the product of the density matrix and the appropriate operator matrix. For example, multiplication of the $\hat{\mathbf{I}}_x$ matrix by Expansion 6.14 gives all zeros except for the third term. Thus c equals the trace of the product matrix $\bar{\sigma} \times \hat{\mathbf{I}}_x$, or the ensemble average of the expectation value for $\hat{\mathbf{I}}_x$, $\langle\langle\hat{\mathbf{I}}_x\rangle\rangle$. The same operations performed with the $\hat{\mathbf{I}}_y$ and $\hat{\mathbf{I}}_z$ operators yields:

$$\bar{\sigma} = \frac{1}{2}\begin{pmatrix} 1 & 0 \\ 0 & 1 \end{pmatrix} + \langle\langle I_z\rangle\rangle\begin{pmatrix} 0 & 0 \\ 1 & -1 \end{pmatrix} + \langle\langle I_x\rangle\rangle\begin{pmatrix} 0 & 1 \\ 1 & 0 \end{pmatrix} + \langle\langle I_y\rangle\rangle\begin{pmatrix} 0 & i \\ -i & 0 \end{pmatrix} \tag{6.15}$$

Thus, the state of the ensemble, as represented by the density matrix, can be described in terms of the Cartesian operators that represent populations and observable coherences. Equation 6.15 is an expansion of the density matrix in the **Cartesian operator basis**. Furthermore, the Cartesian operators can be related to the **single-element operator basis** using the relationships described in Equations 6.9 and the following identities:

$$\hat{\mathbf{I}}_z = \frac{1}{2}\left(\hat{\mathbf{I}}_\alpha - \hat{\mathbf{I}}_\beta\right)$$
$$\hat{\mathbf{I}}_x = \frac{1}{2}\left(\hat{\mathbf{I}}_+ + \hat{\mathbf{I}}_-\right)$$
$$\hat{\mathbf{I}}_y = \frac{1}{2i}\left(\hat{\mathbf{I}}_+ - \hat{\mathbf{I}}_-\right)$$
$$\frac{1}{2}\hat{\mathbf{I}} = \frac{1}{2}\left(\hat{\mathbf{I}}_\alpha + \hat{\mathbf{I}}_\beta\right) \tag{6.16}$$

One can think of the $\hat{\mathbf{I}}_z$ operator as representing the population distribution of spins, in the same way as the $\bar{M}_z$ vector does in the classical picture. $\hat{\mathbf{I}}_x$ and $\hat{\mathbf{I}}_y$, on the other hand, being linear combinations of the raising and lowering operators, can be thought of as representing coherences between states (and hence observable magnetization in some cases). The identity operator 1 represents the sum of all the populations of the stationary states and is therefore invariant to changes in the system.

The important lesson to be learned from this discussion is that NMR observables can be represented by expansions of the density matrix. In fact, *any NMR experiment can be described by applying the correct series of operators to the density matrix* and calculating the results of each operation in terms of the expansion shown in Equation 6.15. The appropriate operators for RF pulses, chemical shift and coupling, the most typical perturbations in NMR experiments, are discussed below.

Weakly coupled ensembles and the weak-coupling approximation

The statistical treatment becomes more complicated when coupling is introduced. For a system of n coupled spins ($\mathbf{I} = 1/2$), there is a total of 2^n states, each of which has a corresponding eigenfunction (eigenvector). The simplest case, $n = 2$, yields four states. For a system of 20 mutually coupled spins, the total number of states is 1048576, and diagonalization to obtain eigenfunctions and energies becomes unwieldy, at least for back-of-the-envelope calculations. In order to keep the problem tractable for multiple pulse NMR sequences, spectroscopists use the two-spin weak-coupling approximation when they can. As discussed in Chapter 5, in the weak-coupling approximation the products of eigenvectors from the uncoupled case can be used to represent the coupled spin functions, and each pair of coupled spins can be dealt with independently. This also means that spin transformations (changes in coherence and/or population) can be considered sequentially, even if they are occurring simultaneously. For example, if spins are evolving under the influence of both coupling and chemical shift, the coupling evolution can be considered first, and then shift evolution determined (or vice versa).

In the weak-coupling approximation, basis sets used for constructing representations of complicated spin systems can be products of the eigenfunctions of a single element operator, usually $\hat{\mathbf{I}}_{zi}$ for each type of spin i. If the eigenfunctions of $\hat{\mathbf{I}}_{zi}$ are represented by $|\alpha_i\rangle$ and $|\beta_i\rangle$, the product states for two spins i and j are represented by $|\alpha_i\alpha_j\rangle$, $|\alpha_i\beta_j\rangle$, $|\beta_i\alpha_j\rangle$ and $|\beta_i\beta_j\rangle$. In an ensemble of two coupled spins described by these product functions, there is some probability that one might find some fraction of the ensemble in one particular state, and the total probability of finding the two-spin system in one of the four states again must add to unity over a time average. The overall state of the ensemble (or of a single two-spin system observed for a long time) can be described using a wave function of the form:

$$\psi = c_i\left|\alpha_i\alpha_j\right\rangle + c_2\left|\alpha_i\beta_j\right\rangle + c_3\left|\beta_i\alpha_j\right\rangle + c_4\left|\beta_i\beta_j\right\rangle \tag{6.17}$$

The representation of the state of the system shown in Equation 6.17 is not necessarily the most convenient one to use, since the operators that will be used to transform the system all have matrix representations, and matrix algebra provides a convenient format for calculating the transformations of the system under the influence of those operators. The operators representing the pure product states can be represented by column vectors of the form:

$$|\alpha_i \alpha_j\rangle = \begin{pmatrix} 1 \\ 0 \\ 0 \\ 0 \end{pmatrix} \quad |\alpha_i \beta_j\rangle = \begin{pmatrix} 0 \\ 1 \\ 0 \\ 0 \end{pmatrix} \quad |\beta_i \alpha_j\rangle = \begin{pmatrix} 0 \\ 0 \\ 1 \\ 0 \end{pmatrix} \quad |\beta_i \beta_j\rangle = \begin{pmatrix} 0 \\ 0 \\ 0 \\ 1 \end{pmatrix} \tag{6.18}$$

The ensemble state can be represented as a vector using the matrices in 6.18 as a basis set:

$$\psi = \begin{pmatrix} c_1 \\ c_2 \\ c_3 \\ c_4 \end{pmatrix} \tag{6.19}$$

Normalization requires that:

$$\psi * \psi = \sum_k c_k * c_k = 1 \tag{6.20}$$

and the probability of finding the system in any one product state k is given by $c_k * c_k$. As was the case for Equation 6.11, c_k in Equation 6.20 can be replaced with the time-dependent portion of the wave function $\varphi_k(0)$ and the exponential term to obtain:

$$\psi = \begin{pmatrix} c_1 \\ c_2 \\ c_3 \\ c_4 \end{pmatrix} = \begin{pmatrix} c_1(t) \\ c_2(t) \\ c_3(t) \\ c_4(t) \end{pmatrix} \tag{6.21}$$

The Hamiltonian operator for a system of coupled nuclear spins can be written as a sum of contributions from the Zeeman term $\hat{E}_{Zeeman}$ (that includes chemical shift), the scalar coupling term $\hat{E}_{scalar}$ and time-dependent terms $\hat{E}_{relaxation}$ and $\hat{E}_{excitation}$. The first two terms were described in Chapter 5, while the excitation due to an RF field $\hat{E}_{excitation}$ was dealt with at the end of Chapter 5 (Equation 5.76). A more complete description of time-dependent perturbation is given in Appendix A.

For protons in an 11.74 T magnetic field, the top and bottom states differ by approximately twice the resonance frequency, $\omega_1 \approx -\omega_4 \approx 2\pi(500 \times 10^6)$ rad/s, while the middle two states differ only by the chemical shift difference between the two spins, $\omega_2 \approx -\omega_3 \approx 0$. In the rotating frame, these "absolute" frequencies, ω_i, are replaced by the difference frequencies $\Omega_i = (\omega_i - \omega_R)$ where ω_R is the carrier frequency (assumed to be the same as the frequency of the rotating frame). Note that the middle two terms are unchanged, since the carrier frequencies cancel out in the difference expressions. In the heteronuclear case, all of the frequency spacings are relatively large, so that $\omega_1 \approx 1/2(\omega_A + \omega_B)$, $\omega_4 \approx -1/2(\omega_A + \omega_B)$, $\omega_2 \approx 1/2(\omega_A - \omega_B)$ and $\omega_3 \approx 1/2(\omega_A - \omega_B)$. In practice, one uses two rotating frames that are considered separately, one for each type of nucleus.

Single-element operators for a two-spin system

How does one go about representing a multiple-spin system using spin operator expansions of the density matrix? The density matrix presentation becomes more complicated with increasing numbers of spins. (Note that for the 20-spin system discussed above, the density matrix will have $>10^{12}$ elements!) For analyzing systems with more than one coupled spin, it is usually convenient to use the product operator formalism which is derived from the single-element basis set in order to calculate the effects of pulses and evolutions on one type of spin in the ensemble. Product operators extend the concept of the single-element operators and Cartesian operators to ensembles of weakly coupled spins. We will find the product operator formalism extremely useful for describing multidimensional NMR experiments.

The single element operators and their product operators are defined in terms of how they affect the various spin functions. The spin functions that form the basis set for the product operators are shown in Equation 6.18. The single-element operators have the following matrix representations for a two-spin ensemble (spin 1 and spin 2):

$$\hat{\mathbf{I}}_{1x} = \frac{1}{2}\begin{pmatrix} 0 & 0 & 1 & 0 \\ 0 & 0 & 0 & 1 \\ 1 & 0 & 0 & 0 \\ 0 & 1 & 0 & 0 \end{pmatrix}, \hat{\mathbf{I}}_{1y} = \frac{1}{2i}\begin{pmatrix} 0 & 0 & 1 & 0 \\ 0 & 0 & 0 & 1 \\ -1 & 0 & 0 & 0 \\ 0 & -1 & 0 & 0 \end{pmatrix},$$

$$\hat{\mathbf{I}}_{1z} = \frac{1}{2}\begin{pmatrix} 1 & 0 & 0 & 0 \\ 0 & 1 & 0 & 0 \\ 0 & 0 & -1 & 0 \\ 0 & 0 & 0 & -1 \end{pmatrix}, \hat{\mathbf{I}}_{1+} = \frac{1}{2}\begin{pmatrix} 0 & 0 & 1 & 0 \\ 0 & 0 & 0 & 1 \\ 0 & 0 & 0 & 0 \\ 0 & 0 & 0 & 0 \end{pmatrix}, \hat{\mathbf{I}}_{1-} = \frac{1}{2}\begin{pmatrix} 0 & 0 & 0 & 0 \\ 0 & 0 & 0 & 0 \\ 1 & 0 & 0 & 0 \\ 0 & 1 & 0 & 0 \end{pmatrix} \quad (6.22)$$

$$\hat{\mathbf{I}}_{2x} = \frac{1}{2}\begin{pmatrix} 0 & 1 & 0 & 0 \\ 1 & 0 & 0 & 0 \\ 0 & 0 & 0 & 1 \\ 0 & 0 & 1 & 0 \end{pmatrix}, \hat{\mathbf{I}}_{2y} = \frac{1}{2i}\begin{pmatrix} 0 & 1 & 0 & 0 \\ -1 & 0 & 0 & 0 \\ 0 & 0 & 0 & 1 \\ 0 & 0 & -1 & 0 \end{pmatrix},$$

$$\hat{\mathbf{I}}_{2z} = \frac{1}{2}\begin{pmatrix} 1 & 0 & 0 & 0 \\ 0 & -1 & 0 & 0 \\ 0 & 0 & 1 & 0 \\ 0 & 0 & 0 & -1 \end{pmatrix}, \hat{\mathbf{I}}_{2+} = \frac{1}{2}\begin{pmatrix} 0 & 1 & 0 & 0 \\ 0 & 0 & 0 & 0 \\ 0 & 0 & 0 & 1 \\ 0 & 0 & 0 & 0 \end{pmatrix}, \hat{\mathbf{I}}_{2-} = \frac{1}{2}\begin{pmatrix} 0 & 0 & 0 & 0 \\ 1 & 0 & 0 & 0 \\ 0 & 0 & 0 & 0 \\ 0 & 0 & 1 & 0 \end{pmatrix} \quad (6.23)$$

Note that the nonzero off-diagonal terms in the matrix representations of the transverse operators are those that connect the appropriate states, i.e. the nonzero terms in the $\hat{\mathbf{I}}_{1i}$ connect states in which spin 1 changes state while spin 2 is unchanged (6.22), and vice versa for $\hat{\mathbf{I}}_{2i}$ (6.23). The population operators for spins 1 and 2 in the two-spin system are:

$$\hat{\mathbf{I}}_{1\beta} = \begin{pmatrix} 0 & 0 & 0 & 0 \\ 0 & 0 & 0 & 0 \\ 0 & 0 & 1 & 0 \\ 0 & 0 & 0 & 1 \end{pmatrix}, \hat{\mathbf{I}}_{1\alpha} = \begin{pmatrix} 1 & 0 & 0 & 0 \\ 0 & 1 & 0 & 0 \\ 0 & 0 & 0 & 0 \\ 0 & 0 & 0 & 0 \end{pmatrix}$$

$$\hat{\mathbf{I}}_{2\alpha} = \begin{pmatrix} 1 & 0 & 0 & 0 \\ 0 & 0 & 0 & 0 \\ 0 & 0 & 1 & 0 \\ 0 & 0 & 0 & 0 \end{pmatrix}, \hat{\mathbf{I}}_{2\beta} = \begin{pmatrix} 0 & 0 & 0 & 0 \\ 0 & 1 & 0 & 0 \\ 0 & 0 & 0 & 0 \\ 0 & 0 & 0 & 1 \end{pmatrix} \qquad (6.24)$$

Equations 6.22–6.24 are the coupled spin equivalents of the Pauli spin matrices. From these matrix representations, all of the commutator relationships between the operators can be derived. Matrix multiplication of the Cartesian and single-element operators shown in Equations 6.22–6.24 yields the **product operators** for the two-spin system. Although we will not formally present any of the product operators in the following sections using their matrix representations, it is important to remember that such representations exist and exhibit the same properties as described below for their nonmatrix representations. Several problems at the end of this chapter will deal with the explicit representation of product operators as matrices. A simple introduction to the density matrix formalism and its explicit use in the calculation of the results of NMR experiments can be found in Reference 1. A review of the product operator formalism as used in two-dimensional NMR was published in Reference 2.

Interconversion between the single-element and the Cartesian operator bases

As we will find, the single element and product operator bases are useful for predicting spectral appearance, while the Cartesian product basis is appropriate for describing the behavior of coupled spins interacting with a nonselective RF pulse and evolution of spins under the influence of chemical shift and coupling. The Cartesian operators for a two-spin system can be expanded in terms of the single element and product operators as follows:

$$\tfrac{1}{2}\hat{\mathbf{I}} = 2\left(\tfrac{1}{2}\hat{\mathbf{I}}_1\right)\left(\tfrac{1}{2}\hat{\mathbf{I}}_2\right) = \tfrac{1}{2}\left(\hat{\mathbf{I}}_{1\alpha} + \hat{\mathbf{I}}_{1\beta}\right)\left(\hat{\mathbf{I}}_{2\alpha} + \hat{\mathbf{I}}_{2\beta}\right)$$

$$\hat{\mathbf{I}}_{1z} = 2\left(\hat{\mathbf{I}}_{1z}\right)\left(\tfrac{1}{2}\hat{\mathbf{I}}_2\right) = \tfrac{1}{2}\left(\hat{\mathbf{I}}_{1\alpha} - \hat{\mathbf{I}}_{1\beta}\right)\left(\hat{\mathbf{I}}_{2\alpha} + \hat{\mathbf{I}}_{2\beta}\right)$$

$$\hat{\mathbf{I}}_{2z} = 2\left(\tfrac{1}{2}\hat{\mathbf{I}}_1\right)\left(\hat{\mathbf{I}}_{2z}\right) = \tfrac{1}{2}\left(\hat{\mathbf{I}}_{1\alpha} + \hat{\mathbf{I}}_{1\beta}\right)\left(\hat{\mathbf{I}}_{2\alpha} - \hat{\mathbf{I}}_{2\beta}\right)$$

$$\hat{\mathbf{I}}_{1x} = 2\left(\hat{\mathbf{I}}_{1x}\right)\left(\tfrac{1}{2}\hat{\mathbf{I}}_2\right) = \tfrac{1}{2}\left(\hat{\mathbf{I}}_{1+} + \hat{\mathbf{I}}_{1-}\right)\left(\hat{\mathbf{I}}_{2\alpha} + \hat{\mathbf{I}}_{2\beta}\right)$$

$$2\mathbf{I}_{1z}\mathbf{I}_{2z} = \tfrac{1}{2}\left(\hat{\mathbf{I}}_{1\alpha} - \hat{\mathbf{I}}_{1\beta}\right)\left(\hat{\mathbf{I}}_{2\alpha} - \hat{\mathbf{I}}_{2\beta}\right)$$

$$2\mathbf{I}_{1y}\mathbf{I}_{2x} = \tfrac{1}{2i}\left(\hat{\mathbf{I}}_{1+} - \hat{\mathbf{I}}_{1-}\right)\left(\hat{\mathbf{I}}_{2+} + \hat{\mathbf{I}}_{2-}\right) \qquad (6.25)$$

As with the single-element operators, we can ascribe physical significance to the expanded Cartesian operators. Consider the expansion $\hat{\mathbf{I}}_{1y} = \frac{1}{2i}(\hat{\mathbf{I}}_{1+}\hat{\mathbf{I}}_{2\alpha} + \hat{\mathbf{I}}_{1+}\hat{\mathbf{I}}_{2\beta} - \hat{\mathbf{I}}_{1-}\hat{\mathbf{I}}_{2\alpha} - \hat{\mathbf{I}}_{1-}\hat{\mathbf{I}}_{2\beta})$. This operator describes coherence along the y axis resulting from the two single-quantum transitions of spin 1, $\hat{\mathbf{I}}_{1-}$ and $\hat{\mathbf{I}}_{1+}$. The population terms for spin 2 indicate that spin 1 will evolve under the influence of the populations of spin 2. After grouping together terms that describe spin 1 coherences for a given spin 2 population, we see that for the $\hat{\mathbf{I}}_{1y}$ operator, both terms have the same sign: $\hat{\mathbf{I}}_{1y} = \frac{1}{2i}((\hat{\mathbf{I}}_{1+}\hat{\mathbf{I}}_{2\alpha} - \hat{\mathbf{I}}_{1-}\hat{\mathbf{I}}_{2\alpha}) + (\hat{\mathbf{I}}_{1+}\hat{\mathbf{I}}_{2\beta} - \hat{\mathbf{I}}_{1-}\hat{\mathbf{I}}_{2\beta}))$. The coherence that this operator describes will be in phase for spin 1, i.e. their time-dependent evolution will have the same origin and will be detected by the receiver circuit as having the same phase. They will transform to a spectrum with an in-phase doublet for spin 1 split by the coupling J_{12} (see Figure 6.1).

In the case of the operator $2\hat{\mathbf{I}}_{1z}\hat{\mathbf{I}}_{2y} = \frac{1}{2i}((\hat{\mathbf{I}}_{1\alpha}\hat{\mathbf{I}}_{2+} - \hat{\mathbf{I}}_{1\alpha}\hat{\mathbf{I}}_{2-}) - (\hat{\mathbf{I}}_{1\beta}\hat{\mathbf{I}}_{2+} - \hat{\mathbf{I}}_{1\beta}\hat{\mathbf{I}}_{2-}))$, the terms associated with $\hat{\mathbf{I}}_{1\alpha}$ are opposite in sign from those associated with $\hat{\mathbf{I}}_{1\beta}$, and so will lead to a transformed spectrum for spin 2 in that the line associated with $\hat{\mathbf{I}}_{1\alpha}$ is antiphase with respect to the line associated with $\hat{\mathbf{I}}_{1\beta}$ (Figure 6.2).

Why are $\hat{\mathbf{I}}_{1z}$ operators needed to describe the evolution of single-quantum coherences for spin 2? This is because the spin-2 coherences are evolving under the

$$\hat{\mathbf{I}}_{1y} = \frac{1}{2i}((\hat{\mathbf{I}}_{1+}\hat{\mathbf{I}}_{2\alpha} - \hat{\mathbf{I}}_{1-}\hat{\mathbf{I}}_{2\alpha}) + (\hat{\mathbf{I}}_{1+}\hat{\mathbf{I}}_{2\beta} - \hat{\mathbf{I}}_{1-}\hat{\mathbf{I}}_{2\beta}))$$

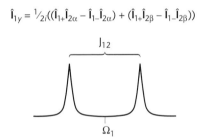

Figure 6.1 In-phase doublet corresponding to the resonance of spin 1 resulting (after FT) from a coherence in which the terms modulated by the two spin states of spin 2 have the same sign (see text).

$$2\hat{\mathbf{I}}_{1z}\hat{\mathbf{I}}_{2y} = \frac{1}{2i}((\hat{\mathbf{I}}_{1\alpha}\hat{\mathbf{I}}_{2+} - \hat{\mathbf{I}}_{1\alpha}\hat{\mathbf{I}}_{2-}) - (\hat{\mathbf{I}}_{1\beta}\hat{\mathbf{I}}_{2+} - \hat{\mathbf{I}}_{1\beta}\hat{\mathbf{I}}_{2-}))$$

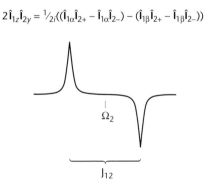

Figure 6.2 Antiphase doublet corresponding to the resonance of spin 2 resulting (after FT) from a coherence in which the terms modulated by the two spin states of spin 1 have the opposite sign (see text).

influence of the $\hat{\mathbf{I}}_{1z}$ operator in order to produce either clockwise or counterclockwise rotations of the coherences. This can be visualized as we did earlier with macroscopic magnetizations, in that the half of the transverse magnetization of spin 2 that sees spin 1 in the α state evolves in one direction, while the half that sees spin 1 in the β state evolves in the other direction.

Evolution of Cartesian operators under the influence of pulses, chemical shift and *J*-coupling

The beauty of the product operator formalism is that it can be used to describe the time evolution of coupled spin systems during the course of a multipulse NMR experiment. It allows one to predict which types of coherences will be generated at what stage of the experiment, and then to choose which ones to select. (This selection is done using phase cycling or pulsed field gradients and will be discussed later.) Let us see how one describes the effects of a nonselective RF pulse on an ensemble of two-spin systems.

A pulse is represented by $\beta\hat{\mathbf{I}}_\varphi$, where β is the flip angle around the φ axis in the **x**, **y** plane. A pulse is represented by a rotation operator $\hat{R}_\varphi$, which has the form for rotation through an angle β around the x axis:

$$\begin{pmatrix} \cos(\beta/2) & i\sin(\beta/2) \\ i\sin(\beta/2) & \cos(\beta/2) \end{pmatrix} = \hat{R}_{\beta x} \tag{6.26}$$

The rotation operator is applied to the spin operator as $\hat{R}_\varphi^{-1}\hat{\mathbf{I}}\hat{R}_\varphi$. Problem 6.4 is designed to convince you of this. Examples of the effects of pulses on various types of operators are:

$$\hat{\mathbf{I}}_x \xrightarrow{\beta\hat{\mathbf{I}}_y} \hat{\mathbf{I}}_x \cos\beta - \hat{\mathbf{I}}_z \sin\beta \tag{6.27}$$

$$\hat{\mathbf{I}}_y \xrightarrow{\beta\hat{\mathbf{I}}_y} \hat{\mathbf{I}}_y \tag{6.28}$$

$$\hat{\mathbf{I}}_z \xrightarrow{\beta\hat{\mathbf{I}}_y} \hat{\mathbf{I}}_z \cos\beta + \hat{\mathbf{I}}_x \sin\beta \tag{6.29}$$

Chemical shift is represented as **z-directed** rotation operator $\Omega t\hat{\mathbf{I}}_z$ that influences the time evolution of coherence:

$$\hat{\mathbf{I}}_x \xrightarrow{\Omega t\hat{\mathbf{I}}_z} \hat{\mathbf{I}}_x \cos\Omega t + \hat{\mathbf{I}}_y \sin\Omega t \tag{6.30}$$

$$\hat{\mathbf{I}}_y \xrightarrow{\Omega t\hat{\mathbf{I}}_z} \hat{\mathbf{I}}_y \cos\Omega t - \hat{\mathbf{I}}_x \sin\Omega t \tag{6.31}$$

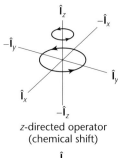

z-directed operator (chemical shift)

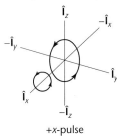

+x-pulse

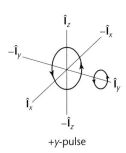

+y-pulse

Figure 6.3 Evolution of Cartesian operators under the influence of chemical shift (top), +x (middle) and +y pulses (bottom).

As expected, population operators are invariant to chemical shift (or other z-directed operators):

$$\hat{\mathbf{I}}_z \xrightarrow{\Omega \hat{\mathbf{I}}_z} \hat{\mathbf{I}}_z \tag{6.32}$$

Figure 6.3 summarizes the effects of pulses and chemical shift evolution on the Cartesian operators. A more extensive discussion of product operators is given in Reference 3.

How are product operators affected by nonselective pulses and chemical shift evolution? For nonselective pulses, the spins are transformed independently, so the outcomes are easy to predict:

$$2\hat{\mathbf{I}}_{1z}\hat{\mathbf{I}}_{2z} \xrightarrow{\pi/2(x)} 2\hat{\mathbf{I}}_{1y}\hat{\mathbf{I}}_{2y} \tag{6.33}$$

$$2\hat{\mathbf{I}}_{1x}\hat{\mathbf{I}}_{2z} \xrightarrow{\pi/2(x)} -2\hat{\mathbf{I}}_{1x}\hat{\mathbf{I}}_{2y} \tag{6.34}$$

$$2\hat{\mathbf{I}}_{1y}\hat{\mathbf{I}}_{2z} \xrightarrow{\pi/2(x)} -2\hat{\mathbf{I}}_{1z}\hat{\mathbf{I}}_{2y} \tag{6.35}$$

The signs of the product terms depend on the products of the signs of the individual terms. In Expressions 6.32 and 6.33, the pulse transforms the initial operators into operators in which both terms are transverse (ones that, if converted to single-element operators, will contain only zero- and double-quantum terms) so no signal would be observed after the pulse. In Expression 6.35, single-quantum coherence on spin 1 is converted into single-quantum coherence on spin 2. This sort of transformation provides the basis of spin–spin correlation in multidimensional NMR.

Evolution of operators with weak J-coupling

In multidimensional NMR experiments, we are usually trying to identify connectivity between spins, and so an important type of product operator evolution is that due to J-coupling (at least in solution NMR experiments). This is also where the most complicated events take place, and several operator spaces must be defined in order to predict what happens. J-coupling between spins 1 and 2 is represented by the rotation operator $\exp(i2\pi J_{12}t\hat{\mathbf{I}}_{1z}\hat{\mathbf{I}}_{2z})$, which expands to $\cos[(1/2)\pi J_{12}t\,\hat{\mathbf{I}}) + i\,\sin(1/2)\pi J_{12}t(4\hat{\mathbf{I}}_{1z}\hat{\mathbf{I}}_{2z})]$. Application of this rotation operator (often abbreviated as $\pi J_{12}\hat{\mathbf{I}}_{1z}\hat{\mathbf{I}}_{2z}$) to the product operators (again, following counterclockwise rotation) results in the evolutions shown in Figure 6.4.

As expected, $\hat{\mathbf{I}}_\mathbf{z}$ is invariant to J-coupling evolution (Expression 6.36), because there is no **z** component to drive the evolution.

$$\hat{\mathbf{I}}_{1z} \xrightarrow{\pi J_{12}\hat{\mathbf{I}}_{1z}\hat{\mathbf{I}}_{2z}} \hat{\mathbf{I}}_{1y} \tag{6.36}$$

Product operators in which both components are transverse (**x** or **y**) are also invariant to the **active coupling**, i.e. the coupling between the two spins represented by the coupling operator:

$$2\hat{\mathbf{I}}_{1x}\hat{\mathbf{I}}_{2y} \xrightarrow{\ \pi J_{12}\hat{\mathbf{I}}_{1z}\hat{\mathbf{I}}_{2z}\ } 2\hat{\mathbf{I}}_{1x}\hat{\mathbf{I}}_{2y} \qquad (6.37)$$

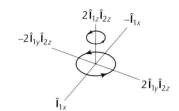

However, single transverse operators evolve with all couplings:

$$\hat{\mathbf{I}}_{1x} \xrightarrow{\ \pi J_{12}\hat{\mathbf{I}}_{1z}\hat{\mathbf{I}}_{2z}\ } \hat{\mathbf{I}}_{1x}\ \cos\pi J_{12}t + 2\hat{\mathbf{I}}_{1y}\hat{\mathbf{I}}_{2z}\ \sin\pi J_{12}t \qquad (6.38)$$

$$\hat{\mathbf{I}}_{1y} \xrightarrow{\ \pi J_{12}\hat{\mathbf{I}}_{1z}\hat{\mathbf{I}}_{2z}\ } \hat{\mathbf{I}}_{1y}\ \cos\pi J_{12}t - 2\hat{\mathbf{I}}_{1x}\hat{\mathbf{I}}_{2z}\ \sin\pi J_{12}t \qquad (6.39)$$

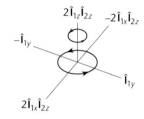

Product operator terms that evolve from single transverse operators (e.g. $2\hat{\mathbf{I}}_{1x}\hat{\mathbf{I}}_{2y}$) are not directly observable, but can evolve into observable coherence with time. Thus, new operators are created from old ones under the influence of *J*-coupling, and old ones are *annihilated*. If the evolution time t in Expressions 6.38 and 6.39 is appropriately set, one can select for one operator exclusively. Thus, after time $t = \tfrac{1}{2}J_{12}$, a pure $\hat{\mathbf{I}}_{1x}$ term evolves completely into $2\hat{\mathbf{I}}_{1y}\hat{\mathbf{I}}_{2z}$, while a pure $2\hat{\mathbf{I}}_{1x}\hat{\mathbf{I}}_{2z}$ term evolves completely into $\hat{\mathbf{I}}_{1y}$. As we will see in Chapter 7, this is a very common feature of multipulse NMR experiments. One sets an **evolution time** between pulses appropriately so as to exclusively create or annihilate a particular coupling operator.

Figure 6.4 Evolution of transverse coherence on spin 1 under the influence of *J*-coupling with spin 2, as represented by the operator $\pi J_{12}\hat{\mathbf{I}}_{1z}\hat{\mathbf{I}}_{2z}$.

It is important to note that the various evolution operators (chemical shift and coupling) can be applied to the product operators sequentially, even though all of the various evolutions are occurring simultaneously. Furthermore, it does not matter in what order the various operators are applied (i.e. the operators *commute*). This is only true in the weak-coupling case, and this assumption must be valid for the formalism to be properly applied.

We can now summarize how product operators evolve under the influence of various evolution and rotation operators.

(a) Product operators evolve independently with chemical shift (i.e. each term evolves separately) and $\hat{\mathbf{I}}_z$ does not evolve with chemical shift. Some examples of this are:

$$2\hat{\mathbf{I}}_{1x}\hat{\mathbf{I}}_{2z} \xrightarrow{\ \Omega_1\hat{\mathbf{I}}_{1z}+\Omega_2\hat{\mathbf{I}}_{2z}\ } 2\hat{\mathbf{I}}_{1x}\hat{\mathbf{I}}_{2z}\ \cos\Omega_1 t + 2\hat{\mathbf{I}}_{1y}\hat{\mathbf{I}}_{2z}\ \sin\Omega_1 t$$

$$2\hat{\mathbf{I}}_{1x}\hat{\mathbf{I}}_{2y} \xrightarrow{\ \Omega_1\hat{\mathbf{I}}_{1z}+\Omega_2\hat{\mathbf{I}}_{2z}\ } 2\left(\hat{\mathbf{I}}_{1x}\ \cos\Omega_1 t + \hat{\mathbf{I}}_{1y}\ \sin\Omega_1 t\right)\left(\hat{\mathbf{I}}_{2y}\ \cos\Omega_2 t - \hat{\mathbf{I}}_{2x}\ \sin\Omega_2 t\right) \quad (6.40)$$

(b) The evolution of Cartesian operators under the influence of scalar coupling results in the creation and annihilation of operators containing terms for both spins involved in the coupling. For example, the $2\hat{\mathbf{I}}_{1x}\hat{\mathbf{I}}_{2z}$ operator evolves due to the coupling between spins 1 and 2 (the active coupling) to generate a term containing only the $\hat{\mathbf{I}}_{1y}$ operator (Expression 6.41). Conversely, the $\hat{\mathbf{I}}_{1y}$ operator will evolve to generate $2\hat{\mathbf{I}}_{1x}\hat{\mathbf{I}}_{2z}$ (Expression 6.42). These operations are shown in Figure 6.4.

$$2\hat{\mathbf{I}}_{1x}\hat{\mathbf{I}}_{2z} \xrightarrow{\pi J_{12}\hat{\mathbf{I}}_{1z}\hat{\mathbf{I}}_{2z}} 2\hat{\mathbf{I}}_{1x}\hat{\mathbf{I}}_{2z}\ \cos\pi J_{12}t + \hat{\mathbf{I}}_{1y}\ \sin\pi J_{12}t \qquad (6.41)$$

$$\hat{\mathbf{I}}_{1y} \xrightarrow{\pi J_{12}\hat{\mathbf{I}}_{1z}\hat{\mathbf{I}}_{2z}} \hat{\mathbf{I}}_{1y}\ \cos\pi J_{12}t - 2\hat{\mathbf{I}}_{1x}\hat{\mathbf{I}}_{2z}\ \sin\pi J_{12}t \qquad (6.42)$$

(c) In the case of product operators evolving under the influence of a coupling in which only one of the two spins is involved (i.e. coupling between spins 1 and 3 in a product operator with terms for spins 1 and 2), only that operator evolves:

$$2\hat{\mathbf{I}}_{1x}\hat{\mathbf{I}}_{2z} \xrightarrow{\pi J_{13}\hat{\mathbf{I}}_{1z}\hat{\mathbf{I}}_{3z}} 2\hat{\mathbf{I}}_{1x}\hat{\mathbf{I}}_{2z}\ \cos\pi J_{13}t + 4\hat{\mathbf{I}}_{1y}\hat{\mathbf{I}}_{2z}\hat{\mathbf{I}}_{3z}\ \sin\pi J_{13}t \qquad (6.43)$$

(d) The product operators of two transverse (**x** or **y**) operators or two longitudinal operators (**z**, –**z**) do not evolve with the active J-coupling.

$$2\hat{\mathbf{I}}_{1x}\hat{\mathbf{I}}_{2x} \xrightarrow{\pi J_{12}\hat{\mathbf{I}}_{1z}\hat{\mathbf{I}}_{2z}} 2\hat{\mathbf{I}}_{1x}\hat{\mathbf{I}}_{2x}$$
$$2\hat{\mathbf{I}}_{1x}\hat{\mathbf{I}}_{2y} \xrightarrow{\pi J_{12}\hat{\mathbf{I}}_{1z}\hat{\mathbf{I}}_{2z}} 2\hat{\mathbf{I}}_{1x}\hat{\mathbf{I}}_{2y}$$
$$2\hat{\mathbf{I}}_{1z}\hat{\mathbf{I}}_{2z} \xrightarrow{\pi J_{12}\hat{\mathbf{I}}_{1z}\hat{\mathbf{I}}_{2z}} 2\hat{\mathbf{I}}_{1z}\hat{\mathbf{I}}_{2z} \qquad (6.44)$$

(e) If the product of two transverse operators evolves under the influence of a third spin ($\hat{\mathbf{I}}_{3z}$), the operators evolve independently:

$$2\hat{\mathbf{I}}_{1x}\hat{\mathbf{I}}_{2y} \xrightarrow{\pi J_{13}\hat{\mathbf{I}}_{1z}\hat{\mathbf{I}}_{3z}} 2\hat{\mathbf{I}}_{1x}\hat{\mathbf{I}}_{2y}\ \cos\pi J_{13}t + 4\hat{\mathbf{I}}_{1y}\hat{\mathbf{I}}_{2y}\hat{\mathbf{I}}_{3z}\ \sin\pi J_{13}t \xrightarrow{\pi J_{23}\hat{\mathbf{I}}_{2z}\hat{\mathbf{I}}_{3z}}$$
$$2\hat{\mathbf{I}}_{1x}\hat{\mathbf{I}}_{2y}\ \cos\pi J_{13}t\ \cos\pi J_{23}t + 4\hat{\mathbf{I}}_{1y}\hat{\mathbf{I}}_{2y}\hat{\mathbf{I}}_{3z}\ \sin\pi J_{13}t\ \cos\pi J_{23}t$$
$$-4\hat{\mathbf{I}}_{1x}\hat{\mathbf{I}}_{2x}\hat{\mathbf{I}}_{3z}\ \cos\pi J_{13}t\ \sin\pi J_{23}t - 2\hat{\mathbf{I}}_{1y}\hat{\mathbf{I}}_{2x}\ \sin\pi J_{13}t\ \sin\pi J_{23}t \qquad (6.45)$$

Note that in Expressions 6.45, despite the fact that two transverse terms are present in the parent operator, both evolve under the influence of the operator not represented by a transverse term (**passive coupling**). This evolution implies that spins 1 and 2 are both coupled to spin 3. If not, only the spin that is coupled to spin 3 evolves (see Expression 6.44).

In deciding what operators will result in observable coherence, the following rules apply:

(1) Only operators with *one* transverse operator represent observable coherence (i.e. $\hat{\mathbf{I}}_{ix}$ or $\hat{\mathbf{I}}_{iy}$). Operators with one transverse term and any number of longitudinal terms ($\hat{\mathbf{I}}_{iz}$) can evolve into observable coherence. All others are not directly detectable by standard NMR methods nor will they evolve into observable

(single-quantum) coherence. This is equivalent to saying that only single-quantum coherence can be detected directly in an NMR experiment. Some examples of single-quantum operators that are observable or will evolve into observable coherence include:

$$\hat{\mathbf{I}}_{1\beta}\hat{\mathbf{I}}_{2+}$$
$$\hat{\mathbf{I}}_{1\alpha}\hat{\mathbf{I}}_{2\beta}\hat{\mathbf{I}}_{3-}$$
$$\hat{\mathbf{I}}_{1+} \qquad (6.46)$$

(2) All product operators that contain more than one transverse component are not observable and will not evolve into observable coherence. This category includes multiple-quantum coherences that are represented by operators containing more than one raising operator or more than one lowering operator (i.e. products of transverse operators). However, such coherences can be detected *indirectly* by observing the modulation that they exert on single-quantum coherences generated from them. This will be discussed in more detail in the next chapter. $\hat{\mathbf{I}}_{1-}\hat{\mathbf{I}}_{2-}$ and $\hat{\mathbf{I}}_{1+}\hat{\mathbf{I}}_{2+}$ are double-quantum operators, while $\hat{\mathbf{I}}_{1-}\hat{\mathbf{I}}_{2-}\hat{\mathbf{I}}_{3-}$ is a triple-quantum coherence operator. Zero-quantum coherence operators contain one raising and one lowering operator, and like the multiple-quantum operators, can only be detected by their modulation of single-quantum terms. Examples of zero-quantum coherence operators include:

$$\hat{\mathbf{I}}_{1-}\hat{\mathbf{I}}_{2+}$$
$$\hat{\mathbf{I}}_{1+}\hat{\mathbf{I}}_{2-} \qquad (6.47)$$

Analysis of a simple NMR spectrum using product operators

A particularly convenient aspect of the product operator formalism is that, at a glance, one can learn a great deal about the expected results of a multipulse NMR experiment by inspection of the operators that evolve. We first analyze a simple NMR experiment in terms of the product operators. Consider an ensemble of spins 1, with a chemical shift Ω_1. Spin 1 is coupled to two other spins, 2 and 3, with coupling constants J_{12} and J_{13}, respectively. Using the product operator formalism, we can watch the evolution of the $\hat{\mathbf{I}}_{1x}$ term to account for the evolution of chemical shift and both couplings:

$$\hat{\mathbf{I}}_{1x} \to \hat{\mathbf{I}}_{1x}\cos\Omega_1 t\,\cos\pi J_{12}t\,\cos\pi J_{13}t + \hat{\mathbf{I}}_{1y}\sin\Omega_1 t\,\cos\pi J_{12}t\,\cos\pi J_{13}t + \dots \quad (6.48)$$

There are eight terms altogether in Expression 6.48 (from cos cos cos to sin sin sin). Although some of these terms do not represent observable coherence, it is helpful in this case to calculate them all (see Problem 6.5). The sine and cosine coefficients determine the time-dependent modulation of the operators. This means that cosine-modulated terms start with maximum amplitude at $t = 0$ and reach zero amplitude when $t = (1/2J)$. Conversely, sine-modulated terms start with zero amplitude at $t = 0$ and reach maximum amplitude when $t = (1/2J)$. It is important to note that the magnitude of each term (and the signal intensity that it might represent) can never be greater than the magnitude of the original coherence represented by the parent operator, in this case, $\hat{\mathbf{I}}_{1x}$. Even if all of the original magnetization ends up in terms representing observable coherence, they will only add up to the amplitude represented by the parent operator. Many of the gymnastics of pulse programming in NMR are attempts to insure that as much as possible of the original signal intensity ends up where you want it, in the product terms of interest. Deciding which terms those are and understanding how to collect them will be discussed in the Chapter 7. First, we need to understand how the coefficients of the operators that give rise to observable coherence determine the appearance of the spectrum.

Although we determine the evolution of the Cartesian operators due to chemical shift and coupling independently in Expressions 6.47, the Fourier transformation (FT) of these terms acts on the products of the coefficients, and for this we recall the trigonometric relationships between products of sine and cosine functions:

$$\cos\theta\cos\phi = \tfrac{1}{2}\cos(\theta - \phi) + \tfrac{1}{2}\cos(\theta + \phi)$$
$$\sin\theta\cos\phi = \tfrac{1}{2}\sin(\theta - \phi) + \tfrac{1}{2}\sin(\theta + \phi)$$
$$\sin\theta\sin\phi = \tfrac{1}{2}\cos(\theta - \phi) - \tfrac{1}{2}\cos(\theta + \phi) \qquad (6.49)$$

The FT of a product of trigonometric functions will report the sum and difference frequencies as shown in the right sides of Expressions 6.49. Although the absolute phase of the peaks in the transformed spectrum is arbitrary, the relative phases of those peaks can be predicted using the following rules. The relative phases of $\hat{\mathbf{I}}_x$ coherence with only cosine factors and $\hat{\mathbf{I}}_x$ coherence with n sine-modulated terms is given by $\pi n/2$ (in radians). For example, if **x** coherence with only cosine factors is phased pure positive absorptive, as in Figure 6.5A, then $\hat{\mathbf{I}}_x$ terms with one sine and one cosine factor will be $\pi/2$ radians out of phase with the cosine term (positive dispersive). Note that two frequency terms are generated upon Fourier transformation, one at half of J_{12} (in hertz) greater than the shift frequency Ω_1 and one at half of J_{12} less than the shift frequency (as expected from Expressions 6.49). Thus FT of this term will yield two peaks, a doublet separated by J_{12} and centered on Ω_1. Since both terms are sine-modulated and have the same sign, they are dispersive and in phase with each other (see Figure 6.5B).

The presence of two sine terms along with a cosine term will yield the results shown in Figure 6.5C. Once again, two cosine terms are generated, one at half of J_{12} (in hertz) greater than Ω_1 and one at half of J_{12} less than Ω_1. Since both terms are cosine-modulated, they will be absorptive, but having opposite signs, they are antiphase with respect to one other.

In the last example (Figure 6.5D), a second cosine term is present along with the two sine terms shown above, giving four terms (the original coherence is modulated by two couplings). After transformation, those peaks separated by the cosine-modulated coupling constant J_{13} are in phase with respect to each other, while those

(A)

$$\mathbf{I}_{1x} \cos \Omega_1 t \overset{FT}{\Rightarrow}$$

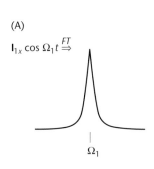

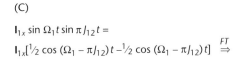

(B)

$$\mathbf{I}_{1x} \sin \Omega_1 t \cos \pi J_{12}) t =$$

$$\mathbf{I}_{1x}[\tfrac{1}{2} \sin (\Omega_1 - \pi J_{12}) t + \tfrac{1}{2} \sin (\Omega_1 - \pi J_{12}) t] \overset{FT}{\Rightarrow}$$

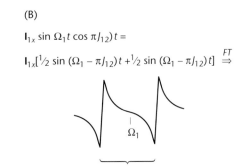

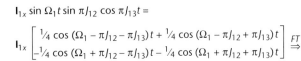

(C)

$$\mathbf{I}_{1x} \sin \Omega_1 t \sin \pi J_{12} t =$$

$$\mathbf{I}_{1x}[\tfrac{1}{2} \cos (\Omega_1 - \pi J_{12}) t - \tfrac{1}{2} \cos (\Omega_1 - \pi J_{12}) t] \overset{FT}{\Rightarrow}$$

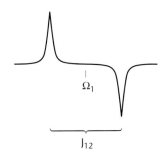

(D)

$$\mathbf{I}_{1x} \sin \Omega_1 t \sin \pi J_{12} \cos \pi J_{13} t =$$

$$\mathbf{I}_{1x} \begin{bmatrix} \tfrac{1}{4} \cos (\Omega_1 - \pi J_{12} - \pi J_{13}) t + \tfrac{1}{4} \cos (\Omega_1 - \pi J_{12} + \pi J_{13}) t \\ -\tfrac{1}{4} \cos (\Omega_1 + \pi J_{12} - \pi J_{13}) t - \tfrac{1}{4} \cos (\Omega_1 + \pi J_{12} + \pi J_{13}) t \end{bmatrix} \overset{FT}{\Rightarrow}$$

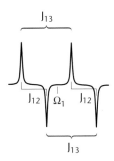

Figure 6.5 (A) FT of the term resulting from cosine-modulated $\hat{\mathbf{I}}_{1x}$. (B) FT of the term resulting from $\hat{\mathbf{I}}_{1x} \sin\Omega_1 t \cos\pi J_{12} t$ (sin–cos modulation), resulting in dispersive character relative to pure absorptive character for a single-cosine term as in (A). (C) FT of a two-sine term $\hat{\mathbf{I}}_{1x} \sin\Omega_1 t \sin\pi J_{12} t$, which results in a sum and difference of cosine terms, giving pure absorptive antiphase signals relative to a pure-cosine term. (D) FT of a two-sine one-cosine term $\hat{\mathbf{I}}_{1x} \sin\Omega_1 t \sin\pi J_{12} t \cos\pi J_{13} t$, which results in sums and differences of cosine terms, giving pure absorptive in-phase signals split by coupling J_{13} and antiphase absorptive signals split by J_{12}.

separated by the sine-modulated coupling J_{12} are antiphase. All terms are of the cosine form and therefore all peaks are absorptive. The antiphase pairing is directly apparent from the signs of the terms.

Problems

*6.1   Using their matrix representations, show that the following commutator relationships are true:

$$\left[\hat{\mathbf{I}}_+, \hat{\mathbf{I}}_z \right] = -\hat{\mathbf{I}}_+$$

$$\left[\hat{\mathbf{I}}_-, \hat{\mathbf{I}}_z \right] = \hat{\mathbf{I}}_-$$

$$\left[\hat{\mathbf{I}}_+, \hat{\mathbf{I}}_- \right] = 2\hat{\mathbf{I}}_z$$

*6.2   Use the first commutator relationship in Problem 1 to show that $|\beta\rangle = \hat{\mathbf{I}}_-|\alpha\rangle$.

*6.3   Show that the approximation that the diagonal elements of the Hamiltonian for the first-order two-spin system are good estimates of the energies of the stationary states is reasonable in light of the more exact solution given for nonfirst-order spectra (see Chapter 4).

*6.4   The rotation operator for an **x** pulse on a single spin is given by Equation 6.26, while the inverse of the rotation operator, $\hat{R}_{\varphi x}^{-1}$ is given by:

$$\begin{pmatrix} \cos(\varphi/2) & -i\sin(\varphi/2) \\ -i\sin(\varphi/2) & \cos(\varphi/2) \end{pmatrix} = \hat{R}_{\varphi x}^{-1}$$

Using matrix multiplication, show that the effect of a 90° +**x** pulse on the operator $\hat{\mathbf{I}}_{1z}$ is the same as is shown in Figure 6.3.

*6.5   Using the product operator formalism, describe the evolution of $\hat{\mathbf{I}}_{1x}$ under the influence of the chemical shift operator Ω_1 and the two coupling operators $\pi J_{12}\mathbf{I}_{1z}\mathbf{I}_{2z}$ and $\pi J_{13}\mathbf{I}_{1z}\mathbf{I}_{3z}$ during a time t.

6.6   Consider any operator $\hat{B}$ that you can think of as a matrix having elements $kk' = \langle k|\hat{B}|k'\rangle$. If $\hat{B}$ is Hermitian, $\langle k|\hat{B}|k'\rangle = \langle k'|\hat{B}|k\rangle*$. Let the operator $\hat{A} = \hat{B}_2$ where $\hat{B}^2 = \hat{B}\cdot\hat{B}$ according to the rules of matrix multiplication. Show that the $Trace(\hat{A}) = \sum_k \langle k|A|k'\rangle = \sum_{kk'} |\langle k'|B|k\rangle|^2$ is equal to the sum of the square of all matrix elements.

6.7   (a)   Given the expansion of the density operator shown in Equations 6.14 and 6.15, show how σ* differs from σ, i.e. why is σ Hermitian?

       (b)   Given that:

$$\hat{\mathbf{I}}_y = \frac{\hbar}{2}\begin{pmatrix} 0 & i \\ i & 0 \end{pmatrix}$$

and that $\langle\langle\hat{\mathbf{I}}_y\rangle\rangle = Trace(\hat{\mathbf{I}}_y\sigma)$, show that coefficient id from the fourth term of Equation 6.14 is equal to $\langle\langle\hat{\mathbf{I}}_y\rangle\rangle$.

6.8   Using the matrix representations of the Cartesian operators $\hat{\mathbf{I}}_x$, $\hat{\mathbf{I}}_y$ and $\hat{\mathbf{I}}_z$, calculate matrix representations of the corresponding population and coherence operators $\hat{\mathbf{I}}_\alpha$, $\hat{\mathbf{I}}_\beta$, $\hat{\mathbf{I}}_-$ and $\hat{\mathbf{I}}_+$. Expand the density matrix in terms of these operators, starting from the definition in Equation 6.21.

*6.9   Calculate the expectation values for the following operators and wave functions:

       (a)   The operator $\hat{\mathbf{I}}_z = 1/2\begin{pmatrix} 1 & 0 \\ 0 & -1 \end{pmatrix}$ for the wave functions $\begin{pmatrix} 1 \\ 0 \end{pmatrix}, \begin{pmatrix} 0 \\ 1 \end{pmatrix}, \begin{pmatrix} i \\ 0 \end{pmatrix}, \begin{pmatrix} \alpha \\ \beta \end{pmatrix}$

       where α and β are any complex numbers of magnitude $1/\sqrt{2}$ such that $\alpha*\alpha = \beta*\beta = 1/2$.

       (b)   The operator $\hat{\mathbf{I}}_x = 1/2\begin{pmatrix} 0 & 1 \\ 1 & 0 \end{pmatrix}$ for the wave functions

$$\begin{pmatrix} 1 \\ 0 \end{pmatrix}, \begin{pmatrix} 0 \\ i \end{pmatrix}, \frac{1}{\sqrt{2}}\begin{pmatrix} 1 \\ 1 \end{pmatrix}, \frac{1}{\sqrt{2}}\begin{pmatrix} 1 \\ -1 \end{pmatrix}, \frac{1}{\sqrt{2}}\begin{pmatrix} 1 \\ i \end{pmatrix}.$$

(c)   The operator $\hat{\mathbf{I}}_y = 1/2\begin{pmatrix} 0 & i \\ -i & 0 \end{pmatrix}$ for the wave functions

$$\frac{1}{\sqrt{2}}\begin{pmatrix} 1 \\ -1 \end{pmatrix},\ \frac{1}{\sqrt{2}}\begin{pmatrix} 1 \\ i \end{pmatrix},\ \frac{1}{\sqrt{2}}\begin{pmatrix} e^{i\omega t/2} \\ e^{-i\omega t/2} \end{pmatrix}.$$

References

1. G. Mateescu and A. Valeriu, *2D NMR Density Matrix and Product Operator Treatment*. Prentice Hall Englewood Cliffs, New Jersey (1993).
2. H. Kessler *et al.*, *Angewandte Chemie Int. Ed. Eng.* 27, 490–536 (1988).
3. R. Ernst, G. Bodenhausen, and A. Wokaun, *Principles of Nuclear Magnetic Resonance in One and Two Dimensions*. Oxford Science Publications, Clarendon Press, Oxford (1992).

7

MULTIDIMENSIONAL NMR: HOMONUCLEAR EXPERIMENTS AND COHERENCE SELECTION

From the earliest days of the technique, a primary advantage of NMR over other more sensitive spectroscopic methods has been the ability to correlate spins via through-bond (scalar coupling) or through-space (dipolar coupling) effects. Prior to the introduction of multidimensional NMR methods (Reference 1), such correlations were typically established using double-resonance techniques. However, the introduction of multiple dimensions to NMR allowed multiple correlations to be established in the same experiment, and led to the possibility of characterizing very large molecules (e.g. proteins, polymers and polynucleic acids). There are literally hundreds of multidimensional NMR experiments published, and this text is not intended as a comprehensive index. Rather, we hope to provide the student with the building blocks and logic by which such experiments are constructed. Even the most complicated symphony is built from simple motifs, and the most sophisticated multidimensional NMR experiments are composed of relatively simple elements. We will now introduce concepts of **homonuclear** multidimensional NMR. The formalisms introduced in Chapter 6 allow us to consider how these experiments are put together. A homonuclear experiment is one in which all of the spins of interest have the same gyromagnetic ratio (e.g. all ^{1}H or all ^{13}C). This implies that a single rotating frame of reference can be used to describe the experiment, and nonselective pulses are generated using the same carrier frequency. While homonuclear experiments are most commonly used for ^{1}H–^{1}H correlations, the logic remains the same regardless of the type of nucleus being observed.

A simple two-dimensional NMR experiment

Consider the experiment, shown in Figure 7.1, in which a single spin with a Larmor frequency ω_A is observed. The spectrometer carrier frequency (i.e. the rotating frame frequency) is ω_0. If a $\pi/2_x$ pulse is applied, the population operator $\hat{\mathbf{I}}_{Az}$ is transformed into $-\hat{\mathbf{I}}_{Ay}$. The $-\hat{\mathbf{I}}_{Ay}$ term evolves as described in the previous chapter under the influence of the chemical shift operator $\Omega_A t \hat{\mathbf{I}}_{Az}$, where $\Omega_A = \omega_A - \omega_0$. After some time, the evolution has progressed such that the magnitude of $-\hat{\mathbf{I}}_{Ay}$ remaining on the $-\mathbf{y}$ axis is given by:

$$-\hat{\mathbf{I}}_{Ay}\left(t_1\right) = -\hat{\mathbf{I}}_{Ay}\left(0\right)\cos\Omega_A t_1 \tag{7.1}$$

where $-\hat{\mathbf{I}}_{Ay}(0)$ represents the amplitude of $-\hat{\mathbf{I}}_{Ay}$ immediately after the pulse and the time delay after the pulse is t_1, often called the **evolution time**. The magnitude of the component along the $+\mathbf{x}$ axis is given by:

$$\hat{\mathbf{I}}_{Ax}\left(t_1\right) = \hat{\mathbf{I}}_{Ay}\left(0\right)\sin\Omega_A t_1 \tag{7.2}$$

As t_1 increases, the amplitude of the cosine-modulated term in Equation 7.1, a maximum at $t_1 = 0$, will decrease. On the other hand, the term in Equation 7.2 has

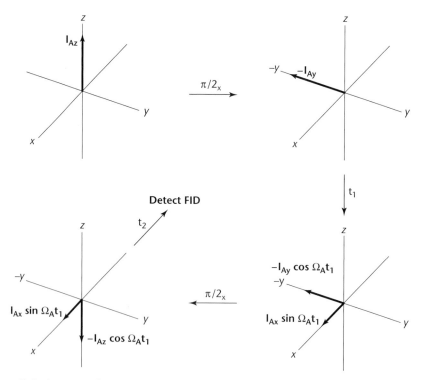

Figure 7.1 A two-pulse experiment with two independent time variables, t_1 and t_2. Two $\pi/2_x$ pulses are separated by the variable delay t_1. After the second pulse, the FID is detected during a second time t_2. Two independent FTs, one with respect to each time variable, yields a two-dimensional NMR experiment.

0 amplitude at $t_1 = 0$ and increases in magnitude as t_1 approaches $\pi/(2\Omega_A)$. At $t_1 = \pi/(2\Omega_A)$, the term in Equation 7.2 will be at a maximum and that in Equation 7.1 will be zero. Now consider what happens if a second $\pi/2_x$ pulse is applied after the evolution time t_1. This pulse converts the term shown in Equation 7.1 into $\hat{I}_{Az} \cos\Omega_A t_1$. As this term is not in the $\mathbf{x}$, $\mathbf{y}$ plane, it is not detectable. The only observable coherence is the component that was on the $+\mathbf{x}$ axis when the second pulse was applied (the term in Equation 7.2), and was therefore unaffected by the pulse. An FID detected immediately after the second pulse would show a signal oscillating at $\Omega_A = \omega_A - \omega_0$ during the acquisition time (usually called t_2) that gives rise to a peak at Ω_A after FT. However, the *amplitude* of the signal detected during t_2 depends upon the duration of t_1, i.e. the amplitude of the detected signal is modulated by $\sin\Omega_A t_1$. When $t_1 = 0$, no signal is observed. When $t_1 = \pi/(2\Omega_A)$, the maximum positive signal is observed (the term in Equation 7.2 is at a maximum). The amplitude again passes through zero when $t_1 = \pi/(\Omega_A)$, reaches a negative maximum when $t_1 = 3\pi/(2\Omega_A)$, and finally returns to zero when the cycle is completed, at $t_1 = 2\pi/(\Omega_A)$. Figure 7.2 shows a series of FIDs acquired as a function of systematically increasing t_1. We have ignored relaxation, which would also attenuate the amplitude of the signal as a function of increasing t_1.

After FT of each of the FIDs with respect to t_2 (as the experimental acquisition time is usually called) one can plot the series of transformed frequency domain spectra $S(\omega_2)$ where the amplitude of signal A is modulated by the factor $\sin\Omega_A t_1$ as shown in Figure 7.3.

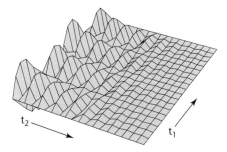

Figure 7.2 Plot of the series of FIDs obtained as a function of changing the value of t_1 in the experiment shown in Figure 7.1. The modulation occurs at the frequency of the single resonance in the spectrum, Ω_A.

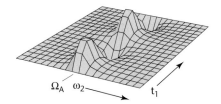

Figure 7.3 Plot of the series of spectra obtained after FT with respect to t_2 (the "real" FID dimension), giving rise to a series of spectra $S(\omega_2)$ that are amplitude-modulated with respect to t_1 at a frequency corresponding to the only signal in the spectrum, Ω_A.

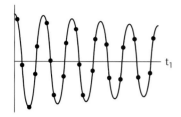

Figure 7.4 Plot of the series of points as a function of t_1 at the column corresponding to the shift Ω_A. This built-up time domain signal is called an interferogram. Unlike Figure 7.3, this representation includes the effects of relaxation.

If one extracts the column of points that represent the frequency Ω_A in all of the $S(\omega_2)$ spectra, and plots those points as a function of t_1, one obtains a time-domain signal that oscillates at a frequency $\sin\Omega_A$ with respect to t_1. The amplitude of this signal will be also reduced by T_2 relaxation as t_1 increases (not shown in Figure 7.3), so the plotted amplitude (called an **interferogram**) will look like Figure 7.4.

Figure 7.4 looks like an FID, and an interferogram can be Fourier-transformed, giving rise to a frequency domain spectrum $S(\omega_1)$ that has only one peak, at frequency $\Omega_A = (\omega_A - \omega_0)$. We ignore the problem of line shape for now, since the transformed spectrum will contain both absorptive and dispersive line shape elements, and both real and imaginary components are needed in order to phase it. For now, we get around this requirement by displaying the spectrum in **magnitude mode** (displaying the square root of the square of the value at each point) so that only positive points are displayed, and in the early days of two-dimensional NMR, that is precisely what was done. After FT of the interferogram with respect to t_1, one obtains a two-dimensional spectrum, a function of two independent frequency variables ω_1 and ω_2 derived from two independent time variables t_1 and t_2 (see Figure 7.5).

The information content of the two-dimensional spectrum in Figure 7.5 is no greater than that of a one-dimensional NMR spectrum, since the single peak is at the same frequency in both dimensions. However, the experiment contains the basic elements of all multidimensional NMR experiments, i.e. the evolution of coherence with respect to multiple independent time variables. An easy way to visualize a two-dimensional data set is as a matrix consisting of rows and columns of data. Each discrete value of t_1 gives rise to one row of the matrix, and each column is composed of points corresponding to the same t_2 value in each of the digitized FIDs. The data is Fourier-transformed first with respect to rows ($t_2 \rightarrow \omega_2$), and then with respect to columns ($t_1 \rightarrow \omega_1$). Only those columns that contain real signal amplitude will transform into a spectrum, i.e. noise is noise, whether it is in the time domain or the frequency domain. However, typically all rows and all columns are transformed, since it is computationally simple to do.

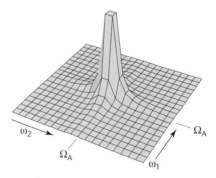

Figure 7.5 Plot of the two-dimensional spectrum obtained after Fourier transformation of the data shown in Figure 7.3 with respect to t_1. This spectrum is a function of two frequency variables, $S(\omega_1)$ and $S(\omega_2)$. The single peak in the spectrum occurs at the same frequency in both dimensions, Ω_A.

Coherence transfer in multidimensional NMR

Using the product operator formalism, we can show that transverse coherence evolves under the influence of **z**-directed operators such as chemical shift or *J*-coupling according to the expressions in Table 7.1. Of particular interest for multi-dimensional NMR is the evolution of a transverse operator under the influence of coupling:

$$\hat{\mathbf{I}}_{1x} \xrightarrow{\pi J_{12}\hat{\mathbf{I}}_{1z}\hat{\mathbf{I}}_{2z}} \hat{\mathbf{I}}_{1x} \cos \pi J_{12} t + 2\hat{\mathbf{I}}_{1y}\hat{\mathbf{I}}_{2z} \sin \pi J_{12} t \qquad (7.3)$$

Recalling that an **x** pulse leaves coherence on the *x* axis unaffected, i.e. $\hat{\mathbf{I}}_{1x} \xrightarrow{\beta x} \hat{\mathbf{I}}_{1x}$, the application of a $\pi/2_x$ pulse results in transformation only of the second term on the right-hand side of Expression 7.3:

$$2\hat{\mathbf{I}}_{1y}\hat{\mathbf{I}}_{2z} \sin \pi J_{12} t \xrightarrow{\pi/2_x} -2\hat{\mathbf{I}}_{1z}\hat{\mathbf{I}}_{2y} \sin \pi J_{12} t \qquad (7.4)$$

Expression 7.4 shows that the net result of this (pulse–t_1–pulse) sequence is to take coherence that is transverse with respect to spin 1 and convert it into a term that is transverse with respect to spin 2. After the second pulse, evolution of the term on the right-hand side of 7.4 due to a **z**-directed operator such as chemical shift or *J*-coupling is now with respect to spin 2 alone. This conversion is critical for multidimensional NMR: it allows coherence to evolve on one spin during t_1 and to then be transferred to another for further evolution during t_2, thus "painting" the coherence with a history of evolution on both spins. This history can then be read by FT with respect to both independent time variables.

Table 7.1 Operator transformations in the Cartesian basis due to pulses, *J*-coupling and chemical shift.

y-pulses

$$\pm\hat{\mathbf{I}}_x \xrightarrow{+\beta\hat{\mathbf{I}}_y} \pm\hat{\mathbf{I}}_x \cos\beta \mp \hat{\mathbf{I}}_z \sin\beta$$

$$\pm\hat{\mathbf{I}}_x \xrightarrow{-\beta\hat{\mathbf{I}}_y} \pm\hat{\mathbf{I}}_x \cos\beta \pm \hat{\mathbf{I}}_z \sin\beta$$

$$\pm\hat{\mathbf{I}}_y \xrightarrow{+\beta\hat{\mathbf{I}}_y} \pm\hat{\mathbf{I}}_y$$

$$\pm\hat{\mathbf{I}}_z \xrightarrow{+\beta\hat{\mathbf{I}}_y} \pm\hat{\mathbf{I}}_z \cos\beta \pm \hat{\mathbf{I}}_x \sin\beta$$

$$\pm\hat{\mathbf{I}}_z \xrightarrow{+\beta\hat{\mathbf{I}}_y} \pm\hat{\mathbf{I}}_z \cos\beta \mp \hat{\mathbf{I}}_x \sin\beta$$

Chemical shift (z-directed operator)

$$\pm\hat{\mathbf{I}}_x \xrightarrow{+\Omega t\hat{\mathbf{I}}_z} \pm\hat{\mathbf{I}}_x \cos\Omega t \pm \hat{\mathbf{I}}_y \sin\Omega t$$

$$\pm\hat{\mathbf{I}}_x \xrightarrow{-\Omega t\hat{\mathbf{I}}_z} \pm\hat{\mathbf{I}}_x \cos\Omega t \mp \hat{\mathbf{I}}_y \sin\Omega t$$

$$\pm\hat{\mathbf{I}}_y \xrightarrow{+\Omega t\hat{\mathbf{I}}_z} \pm\hat{\mathbf{I}}_y \cos\Omega t \mp \hat{\mathbf{I}}_x \sin\Omega t$$

$$\pm\hat{\mathbf{I}}_y \xrightarrow{-\Omega t\hat{\mathbf{I}}_z} \pm\hat{\mathbf{I}}_y \cos\Omega t \pm \hat{\mathbf{I}}_x \sin\Omega t$$

$$\pm\hat{\mathbf{I}}_z \xrightarrow{\pm\Omega t\hat{\mathbf{I}}_z} \pm\hat{\mathbf{I}}_z$$

x-pulses

$$\pm\hat{\mathbf{I}}_x \xrightarrow{\pm\beta\hat{\mathbf{I}}_x} \pm\hat{\mathbf{I}}_x$$

$$\pm\hat{\mathbf{I}}_y \xrightarrow{+\beta\hat{\mathbf{I}}_x} \pm\hat{\mathbf{I}}_y \cos\beta \pm \hat{\mathbf{I}}_z \sin\beta$$

$$\pm\hat{\mathbf{I}}_y \xrightarrow{-\beta\hat{\mathbf{I}}_x} \pm\hat{\mathbf{I}}_y \cos\beta \mp \hat{\mathbf{I}}_z \sin\beta$$

$$\pm\hat{\mathbf{I}}_z \xrightarrow{+\beta\hat{\mathbf{I}}_x} \pm\hat{\mathbf{I}}_z \cos\beta \mp \hat{\mathbf{I}}_y \sin\beta$$

$$\pm\hat{\mathbf{I}}_z \xrightarrow{-\beta\hat{\mathbf{I}}_x} \pm\hat{\mathbf{I}}_z \cos\beta \pm \hat{\mathbf{I}}_y \sin\beta$$

Weak J-coupling (positive coupling constant)

$$\hat{\mathbf{I}}_{1z} \xrightarrow{\pi J_{12}\hat{\mathbf{I}}_{1z}\hat{\mathbf{I}}_{2z}} \hat{\mathbf{I}}_{1z}$$

$$\pm\hat{\mathbf{I}}_{1x} \xrightarrow{\pi J_{12}\hat{\mathbf{I}}_{1z}\hat{\mathbf{I}}_{2z}} \pm\hat{\mathbf{I}}_{1x} \cos\pi J_{12}t \pm 2\hat{\mathbf{I}}_{1y}\hat{\mathbf{I}}_{2z} \sin\pi J_{12}t$$

$$\pm\hat{\mathbf{I}}_{1y} \xrightarrow{\pi J_{12}\hat{\mathbf{I}}_{1z}\hat{\mathbf{I}}_{2z}} \pm\hat{\mathbf{I}}_{1y} \cos\pi J_{12}t \mp 2\hat{\mathbf{I}}_{1x}\hat{\mathbf{I}}_{2z} \sin\pi J_{12}t$$

$$\pm 2\hat{\mathbf{I}}_{1x}\hat{\mathbf{I}}_{2z} \xrightarrow{\pi J_{12}\hat{\mathbf{I}}_{1z}\hat{\mathbf{I}}_{2z}} \pm 2\hat{\mathbf{I}}_{1x}\hat{\mathbf{I}}_{2z} \cos\pi J_{12}t \pm \hat{\mathbf{I}}_{1y} \sin\pi J_{12}t$$

$$\pm 2\hat{\mathbf{I}}_{1y}\hat{\mathbf{I}}_{2z} \xrightarrow{\pi J_{12}\hat{\mathbf{I}}_{1z}\hat{\mathbf{I}}_{2z}} \pm 2\hat{\mathbf{I}}_{1y}\hat{\mathbf{I}}_{2z} \cos\pi J_{12}t \mp \hat{\mathbf{I}}_{1x} \sin\pi J_{12}t$$

The COSY experiment

The simplest two-dimensional NMR experiment is called the COSY (*correlated spectroscopy*) experiment (Reference 2). COSY consists of a two-pulse sequence:

$$\pi/2_y - t_1 - \pi/2_y - t_2 \left(acquire \right) \tag{7.5}$$

Consider what happens in a three-spin system, with coupling constants J_{12}, J_{13} and J_{23}, in the course of a COSY experiment. We will follow the evolution of spin 1 after the first pulse, and assume that similar considerations also apply to the other two spins. During t_1, the $\hat{\mathbf{I}}_{1x}$ operator created from the equilibrium $\hat{\mathbf{I}}_{1z}$ by the first $\pi/2_y$ pulse evolves under the influence of chemical shift and J-coupling operators, $\Omega_1\hat{\mathbf{I}}_{1z}$, $\pi J_{12}\hat{\mathbf{I}}_{1z}\hat{\mathbf{I}}_{2z}$ and $\pi J_{13}\hat{\mathbf{I}}_{1z}\hat{\mathbf{I}}_{3z}$. Assuming weak coupling, we consider each evolution separately even though they are occurring simultaneously. Chemical shift evolution gives:

$$\hat{\mathbf{I}}_{1x} \xrightarrow{\Omega_1 t_1 \hat{\mathbf{I}}_{1z}} \hat{\mathbf{I}}_{1x} \cos\Omega_1 t_1 + \hat{\mathbf{I}}_{1y} \sin\Omega_1 t_1 \tag{7.6}$$

The coherence evolves under the influence of $\Omega_1\hat{\mathbf{I}}_{1z}$ during t_1 and is said to be *frequency-labeled* by that evolution. Thus the time t_1 is often called the **frequency labeling** period. The second pulse will move any coherence left on the **x** axis (represented by the term $\hat{\mathbf{I}}_{1x} \cos\Omega_1 t_1$) to the −**z** axis, so we can ignore further evolution of this term. The $\hat{\mathbf{I}}_{1y} \sin\Omega_1 t_1$ term evolves under the influence of coupling operators as follows:

$$\hat{\mathbf{I}}_{1y} \sin\Omega_1 t_1 \xrightarrow{\pi J_{12} t_1 \hat{\mathbf{I}}_{1z}\hat{\mathbf{I}}_{2z}} \hat{\mathbf{I}}_{1y} \sin\Omega_1 t_1 \cos\pi J_{12} t_1 - 2\hat{\mathbf{I}}_{1x}\hat{\mathbf{I}}_{2z} \sin\Omega_1 t_1 \sin\pi J_{12} t_1 \tag{7.7}$$

The first term on the right-hand side of expression 7.7 will give rise to a $2\hat{\mathbf{I}}_{1x}\hat{\mathbf{I}}_{3z}$ term under the influence of $\pi J_{13} t_1$ and results in coherence transfer to spin 3, among others. We will follow the second term, $-2\hat{\mathbf{I}}_{1x}\hat{\mathbf{I}}_{2z} \sin\Omega_1 t_1 \sin\pi J_{12} t_1$, that gives rise to an operator with transverse $\hat{\mathbf{I}}_2$ terms (observable coherence on spin 2) after the second pulse. This is the term we follow, after allowing it to evolve with the coupling $\pi J_{13}\hat{\mathbf{I}}_{1z}\hat{\mathbf{I}}_{3z}$:

$$-2\hat{\mathbf{I}}_{1x}\hat{\mathbf{I}}_{2z} \sin\Omega_1 t_1 \sin\pi J_{12} t_1 \xrightarrow{\pi J_{13} t_1 \hat{\mathbf{I}}_{1z}\hat{\mathbf{I}}_{3z}} -2\hat{\mathbf{I}}_{1x}\hat{\mathbf{I}}_{2z} \sin\Omega_1 t_1 \sin\pi J_{12} t_1 \cos\pi J_{13} t_1$$
$$+4\hat{\mathbf{I}}_{1y}\hat{\mathbf{I}}_{2z}\hat{\mathbf{I}}_{3z} \sin\Omega_1 t_1 \sin\pi J_{12} t_1 \sin\pi J_{13} t_1 \tag{7.8}$$

After the second pulse, the second term ($4\hat{\mathbf{I}}_{1y}\hat{\mathbf{I}}_{2z}\hat{\mathbf{I}}_{3z}$) is converted into multiple-quantum coherence (more than one transverse operator) and will not be observed. The first term on the right-hand side of Expression 7.8 becomes:

$$-2\hat{\mathbf{I}}_{1x}\hat{\mathbf{I}}_{2z} \sin\Omega_1 t_1 \sin\pi J_{12} t_1 \cos\pi J_{13} t_1 \xrightarrow{\pi/2_y} 2\hat{\mathbf{I}}_{1z}\hat{\mathbf{I}}_{2x} \sin\Omega_1 t_1 \sin\pi J_{12} t_1 \cos\pi J_{13} t_1 \tag{7.9}$$

During the FID acquisition time t_2, the term on the right-hand side of expression 7.9 will evolve under the influence of spin-2 operators $\Omega_2\hat{\mathbf{I}}_{2z}$, $\pi J_{12}\hat{\mathbf{I}}_{1z}\hat{\mathbf{I}}_{2z}$ and $\pi J_{23}\hat{\mathbf{I}}_{2z}\hat{\mathbf{I}}_{3z}$:

$$2\hat{\mathbf{I}}_{1z}\hat{\mathbf{I}}_{2x}\sin\Omega_1 t_1 \sin\pi J_{12}t_1 \cos\pi J_{13}t_1 \xrightarrow{\Omega_2 t_2 \hat{\mathbf{I}}_{2z}}$$

$$2\hat{\mathbf{I}}_{1z}\hat{\mathbf{I}}_{2x}\sin\Omega_1 t_1 \sin\pi J_{12}t_1 \cos\pi J_{13}t_1 \cos\Omega_2 t_2$$

$$+2\hat{\mathbf{I}}_{1z}\hat{\mathbf{I}}_{2y}\sin\Omega_1 t_1 \sin\pi J_{12}t_1 \cos\pi J_{13}t_1 \sin\Omega_2 t_2 \qquad (7.10)$$

Replacing the t_1 evolution terms $\sin\Omega_1 t_1 \sin\pi J_{12}t_1 \cos\pi J_{13}t_1$ with A, we continue the evolution with:

$$2\hat{\mathbf{I}}_{1z}\hat{\mathbf{I}}_{2x}A\cos\Omega_2 t_2 + 2\hat{\mathbf{I}}_{1z}\hat{\mathbf{I}}_{2y}A\sin\Omega_2 t_2$$

$$\xrightarrow{\pi J_{12}t_2 \hat{\mathbf{I}}_{1z}\hat{\mathbf{I}}_{2z}} \hat{\mathbf{I}}_{2y}A\cos\Omega_2 t_2 \sin\pi J_{12}t_2$$

$$-\hat{\mathbf{I}}_{2x}A\sin\Omega_2 t_2 \sin\pi J_{12}t_2 + \dots \qquad (7.11)$$

Although four terms result from the evolution described in Expression 7.11 (see Problem 7.1), only the two nonproduct terms shown will result in observable coherence under the influence of the $\pi J_{23}\hat{\mathbf{I}}_{2z}\hat{\mathbf{I}}_{3z}$ operator, and so it is their further evolution that we consider:

$$\hat{\mathbf{I}}_{2y}A\cos\Omega_2 t_2 \sin\pi J_{12}t_2 - \hat{\mathbf{I}}_{2x}A\sin\Omega_2 t_2 \sin\pi J_{12}t_2 \xrightarrow{\pi J_{23}t_2 \hat{\mathbf{I}}_{2z}\hat{\mathbf{I}}_{3z}}$$

$$\hat{\mathbf{I}}_{2y}A\cos\Omega_2 t_2 \sin\pi J_{12}t_2 \cos\pi J_{23}t_2 \underline{-\hat{\mathbf{I}}_{2x}A\sin\Omega_2 t_2 \sin\pi J_{12}t_2 \cos\pi J_{23}t_2} + \dots \quad (7.12)$$

Again, four terms result from this evolution, but we only consider the two that represent observable coherence. The underlined term in Expression 7.12 is the **transfer term** of interest, representing amplitude modulation of the coherence evolving during t_2 at the frequency of spin 2 by the frequency of spin 1 during t_1. If the spectral component resulting from this term along the x axis is phased absorptive, the other term $\hat{\mathbf{I}}_{2y}A\cos\Omega_2 t_2 \sin\pi J_{12}t_2 \cos\pi J_{23}t_2$ represents the dispersive portion of the signal. Normally, the two components would both contain dispersive and absorptive components, and phasing would be required in the usual fashion. If we strip the modulation terms from the expressions that we have generated, we see that the net result of the COSY sequence is the conversion $\hat{\mathbf{I}}_{1x}(t_1) \rightarrow -\hat{\mathbf{I}}_{2x}(t_2)$.

The spectral intensity resulting from these terms will be represented as a **cross-peak** between the resonances of spins 1 and 2 in the two-dimensional COSY spectrum obtained by FT with respect to the two time variables, t_1 and t_2. (The modulation by the Larmor frequency of spin 1 during t_1 is contained in A.) The cross-peak will show up at the Larmor frequency of spin 1 in ω_1 and at the Larmor frequency of spin 2 in ω_2. This peak will be split by J_{12} and J_{13} in both dimensions, since both couplings modulate the signal during t_1 and t_2. This peak is the goal of the COSY experiment, because it unambiguously identifies a J-coupling between spins 1 and 2.

We should also consider what happens to coherence that remains on the same spin during both evolution periods, and so will be represented by a **diagonal peak** (a peak with the same chemical shift in both frequency dimensions). This results from the evolution of the first term on the right-hand side of Expression 7.7, i.e.

coherence not transferred to spin 2 by the second pulse. After evolving during t_2 under the influence of the appropriate operators, two observable terms develop:

$$\hat{\mathbf{I}}_{1y}B \rightarrow$$

$$\hat{\mathbf{I}}_{1y}B \cos\Omega_1 t_2 \cos\pi J_{12}t_2 \cos\pi J_{13}t_2$$

$$-\hat{\mathbf{I}}_{1x}B \sin\Omega_1 t_2 \cos\pi J_{12}t_2 \cos\pi J_{13}t_2 \qquad (7.13)$$

where $B = \sin\Omega_1 t_1 \cos\pi J_{12}t_1 \cos\pi J_{13}t_1$. These are the dispersive and absorptive components of the signal leading to the diagonal peak for spin 1.

We have now described two of the three observable terms that develop from $\hat{\mathbf{I}}_{1x}$ in the course of the experiment. The third term is similar in form to the underlined term in Expression 7.12, but reflects coherence transfer from spin 1 to spin 3. In order to predict the appearance of each peak, we need to consider how they are generated. Two independent frequency-domain signals, one in ω_1 and the other in ω_2, result from Fourier transformation of each modulated operator term with respect to t_1 and t_2, respectively. We have already seen at the end of Chapter 6 how the sine and cosine factors representing chemical shift and coupling evolution that modulate $\hat{\mathbf{I}}_x$ determine the relative signs and phase characteristics of the transformed spectra. In the cross-term (underlined term in 7.12), the $\hat{\mathbf{I}}_{2x}$ coherence is modulated by one cosine and two sine factors that evolve during t_2. One sine factor is due to the chemical shift of spin 2 and the other due to the active coupling (i.e. the one responsible for the coherence transfer), J_{12}. The cosine factor is due to the passive coupling constant J_{23}. The term A (that represents evolution during t_1) also contains two sine factors and one cosine factor. Again, the two sine factors are due to the active coupling J_{12} and chemical shift (except during t_1 it was the chemical shift of spin 1), while the passive coupling constant J_{13} determines the cosine modulation. Since $\hat{\mathbf{I}}_{2x}$ is modulated in both time domains by the same types of terms, they will have similar phase characteristics after transformation. In fact, their appearance in either dimension is precisely the situation described at the end of the last chapter (Figure 6.5D). As such, they will have the appearance shown in Figure 7.6, with all absorptive peaks, antiphase with respect to active coupling and in phase with respect to the passive coupling.

What about diagonal peaks? The second term in Expression 7.13, $-\hat{\mathbf{I}}_{1x}$, describes coherence that evolves on spin 1 during both t_1 and t_2 (no coherence transfer). This term is modulated by one sine and two cosine factors. The sine factor is due to chemical shift, and the cosine factors are both due to couplings. Term B, that represents evolution during t_2, is similar, meaning that the phase characteristics of the diagonal peak will be the same in both dimensions. Regrouping terms, we find:

$$-\hat{\mathbf{I}}_{1x}A \sin\Omega_1 t_2 \cos\pi J_{12}t_2 \cos\pi J_{13}t_2$$

$$-\hat{\mathbf{I}}_{1x}A\Big/4 \left[\begin{array}{l} \sin\left(\Omega_1 - \pi J_{12} - \pi J_{13}\right)t_2 + \sin\left(\Omega_1 + \pi J_{12} - \pi J_{13}\right)t_2 \\ \sin\left(\Omega_1 - \pi J_{12} + \pi J_{13}\right)t_2 + \sin\left(\Omega_1 + \pi J_{12} + \pi J_{13}\right)t_2 \end{array} \right] \qquad (7.14)$$

FT of the four terms on the right-hand side of Equation 7.14 with respect to t_2 will result in four in-phase dispersive signals (all terms having the same sign). Hence,

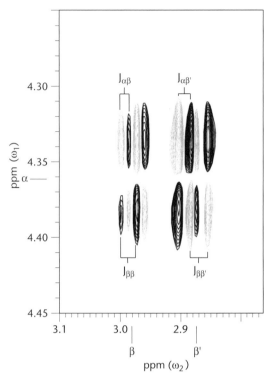

Figure 7.6 Absorptive anti-phase cross-peak between the $C_\alpha H$ and diastereotopic $C_\beta H_2$ protons of a phenylalanine side chain in the COSY spectrum of gramicidin, a peptide-based antibiotic. The active couplings are distinguished by antiphase signals, while the passive coupling (the one not responsible for the cross-peak) gives rise to in-phase splitting. Black peaks are negative, gray are positive. Note that the values of identical couplings are the same in ω_1 and ω_2, but the scales are different as shown. The digital resolution of the ω_1 dimension is less than that of the ω_2 dimension, resulting in a broader line in the ω_1 dimension than in ω_2.

the diagonal peaks in COSY are 90° out of phase with respect to the cross-peaks, that are usually selected to be antiphase absorptive.

Quadrature detection in multidimensional NMR

Although we assumed pure-phase characteristics of spectral features described for COSY, this is unlikely; even a one-dimensional spectrum usually needs phase correction. A more glaring omission is any discussion of how quadrature detection is obtained in the ω_1 dimension. All of the problems of differentiating positive and negative frequencies relative to the carrier frequency in the rotating frame that were discussed in Chapter 3 have counterparts in the indirectly detected dimensions of multidimensional NMR. The first two-dimensional NMR experiments were not performed using quadrature detection in t_1, nor could they be phased, but had to be displayed in magnitude mode. This permits one to display peaks that have both dispersive and absorptive ("phase-twist") character without too much distraction (see the double-quantum filtered COSY experiment in Figure 7.7). However,

Figure 7.7 (A) DQF-COSY spectrum of gramicidin in d$_6$-DMSO (dimethylsulfoxide) acquired at 500 MHz without a pure-phase acquisition scheme such as TPPI or States. The line shape is mixed absorptive–dispersive (phase-twist). (B) Same spectrum, but processed with a strong sine bell window function and presented in magnitude mode (all same sign). (C) Same sample, but the DQF-COSY spectrum was acquired using States-TPPI scheme for obtaining pure antiphase absorptive cross-peaks and diagonal peaks.

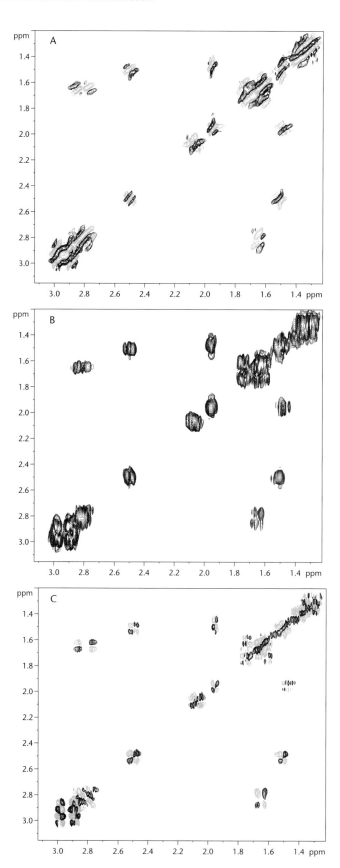

dispersive lines are broader than absorptive lines (they die off more slowly as one moves away from the center of the peak), so magnitude mode peaks have "wings" that reduce resolution considerably, and information concerning fine structure due to coupling is also lost.

Two schemes have been derived that permit quadrature detection in indirectly detected dimensions in multidimensional NMR. The first is often called "**States**" (Reference 3), which requires that two experiments be performed at each increment of t_1 in order to generate the imaginary part of the data for phasing. The first pulse of the experiment generating the imaginary data is shifted by $\pi/2$ rad relative to the first pulse in the "real" experiment performed at the same value of t_1. This changes the relative phase of the coherence generated correspondingly. For example, if the first pulse (P1) in COSY is a **y** pulse, $\hat{\mathbf{I}}_x$ coherence is generated, which evolves into $\hat{\mathbf{I}}_x \cos\Omega t_1 + \hat{\mathbf{I}}_y \sin\Omega t_1$. If the second pulse (P2) is also a **y** pulse, then after P2 the observable term $\hat{\mathbf{I}}_y \sin\Omega t$ evolves during acquisition (t_2). A shift in phase for P1 by $-\pi/2$ rad to **x** results in $-\hat{\mathbf{I}}_y$ coherence after P1, which evolves into $-\hat{\mathbf{I}}_y \cos\Omega t_1 + \hat{\mathbf{I}}_x \sin\Omega t_1$. P2 is still a **y** pulse, and so will generate $-\hat{\mathbf{I}}_y \cos\Omega t_1$ as the observable **y** term. In the first case, the $\hat{\mathbf{I}}_y$ term is sine-modulated, while in the second case it is cosine-modulated, a situation analogous to the outputs of the two mixers used for quadrature detection (Chapter 3). The two data sets generated in this way are stored separately, and are called **hyper-real** (cosine-modulated) and **hyper-imaginary** (sine-modulated) data sets. Each data set is processed separately for the first (t_2) Fourier transformation, with the $\hat{\mathbf{I}}_x$ observable providing the imaginary part of the signal, while the $\hat{\mathbf{I}}_y$ term provides the real part. The spectra are phased in the ω_2 dimension in the usual way, and the imaginary portions of the spectra (dispersive after phase correction) are discarded. For the second (t_1) transform, corresponding columns from the two data sets (i.e. at the same values of ω_2) are combined to give a set of complete complex interferograms of the form:

$$\left(\text{hyper-real real FT},\, \omega_2, t_1\right) + i\left(\text{hyper-imaginary real FT},\, \omega_2, t_1\right)$$
$$= \cos\left(\Omega_i t_1\right) + i\sin\left(\Omega_i t_1\right) = \exp\left(i\Omega_i t_1\right) \qquad (7.15)$$

for a signal at Ω_i. The sign of the frequency Ω_i at which the coherence evolved during t_1 is now uniquely determined. After Fourier transformation with respect to t_1, phasing is accomplished in the usual way. In this way, a pure-phase spectrum is obtained with quadrature detection in both dimensions. Characteristic data obtained using the States method is shown in Figure 7.7C.

An alternative scheme for obtaining quadrature detection in multidimensional NMR is called **TPPI**, or **time-proportional phase incrementation**. For TPPI quadrature detection of a single-quantum coherence, a $\pi/2$ rad shift is applied to the phase of the pulse preceding (or immediately following) the t_1 evolution period. This phase shift is applied each time t_1 is incremented. The phase shift results in a net retardation of frequencies lower than the carrier frequency and a net increase in frequencies higher than the carrier. To be precise, the signal amplitude with frequency Ω is modulated by the frequency $(\omega_{\max} - \omega_0)$, where $(\omega_{\max} - \omega_0) = \pi/(2\Delta t_1)$, Δt_1 being the increment in t_1 (i.e. the dwell time for the t_1 dimension). ω_0 is the carrier frequency and ω_{max} is the frequency at the edge of the spectral window. If Ω is negative (lower than the carrier) the shift is regressive. If Ω is positive, the shift is progressive. Δt_1 is determined by the spectral width, with two times the spectral width $(\omega_{max} - \omega_0)$ equal to $1/\Delta t_1$ for TPPI. The result is that the apparent frequencies of the signals are shifted, with the lowest frequency that is sampled shifted to zero frequency while the highest frequency is shifted to twice the carrier frequency. Since all frequencies are positive, the data are treated with a cosine Fourier transform ("real" transform). The TPPI scheme is diagrammed in Figure 7.8. Note that the TPPI experiment requires twice the number of t_1 increments as a States experiment with equivalent digital resolution.

Axial peaks

One other characteristic of the COSY experiment that should be considered is the phenomenon of axial peaks (see Figure 7.9). These are a series of peaks that occur at zero frequency (in the center of the spectrum) in ω_1. These peaks result from magnetization that returns to the +**z** axis during t_1 owing to T_1 relaxation. Frequency labeling is thereby lost, and results in zero frequency modulation in the t_1 dimension. After the second 90° pulse, the magnetization returns to the **x**–**y** plane, where it evolves at the characteristic frequency for the particular spin during t_2. Fourier transform of this term yields a series of peaks that are at their characteristic shifts Ω in the ω_2 domain, but at zero frequency in the ω_1 domain. Since TPPI moves zero frequency to one end of the spectrum, the axial peaks show up at the edge in TPPI experiments, but are found in the center of States experiments, which is usually undesirable. **States-TPPI** is a modification of the States scheme that moves axial peaks to the edge of the spectrum, and has come into general use in recent years for quadrature detection of indirectly detected dimensions in multidimensional NMR experiments. As with TPPI, States-TPPI moves axial peaks to the edge of the spectrum. The data sets with even values of t_1 (0, $2\Delta t_1$, $4\Delta t_1$, etc.) are obtained with the first pulse and receiver phases π rad out of phase with the data obtained at odd

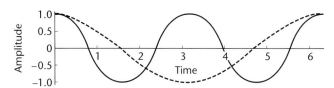

Figure 7.8 Change in frequency of a signal as a result of phase shift, as used in TPPI. The dashed line represents the "real" frequency that is sampled [$(\omega - \omega_0) = 1$ rad/time]. The solid line shows the apparent frequency after TPPI incrementation. The "real" frequency is sampled at the Nyquist frequency (2 rad/time), and the phase is incremented progressively $\pi/2$ radians at each step. A $\pi/2$ phase shift added to $\pi/2$ is π, so when the "real" wave has advanced in phase by $\pi/2$ radians (zero amplitude), the detected signal has advanced in phase by π (maximum negative amplitude). Thus the apparent frequency is double that of the real frequency. If the real frequency were $(\omega - \omega_0) = -1$ rad/time, every $\pi/2$ increment in phase is offset by a $\pi/2$ decrement, giving a net zero phase displacement, or zero frequency.

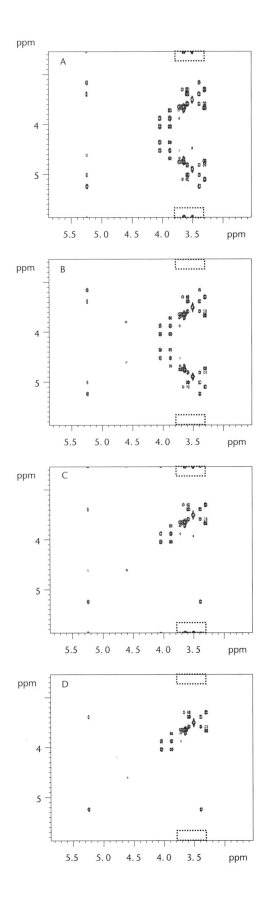

Figure 7.9. Effects of phase cycling in a magnitude-mode two-pulse [P1–t_1–P2–t_2(acquire)] COSY spectrum of sucrose in D_2O obtained at 14 T (600 MHz ^{1}H). (A) No phase cycling of pulses or receiver (both pulses are phased 0, receiver phase = 0). Axial peaks (indicated by dotted line boxes on edges of the spectrum) are present and there is no quadrature detection in ω_1. (B) Even–odd selection (P2 is cycled between 0 and π radians on alternating scans, P1 constant at 0 radians). Receiver phase is constant. Axial peaks are now suppressed, but there is still no quadrature detection in ω_1. (C) Quadrature detection in ω_1, obtained by phase cycling P2 between 0 and $\pi/2$, resulting in separate cosine- and sine-modulated signals, which combine to give quadrature detection in ω_1. Receiver phase is cycled between 0 and π radians. There is no axial peak suppression. (D) Combined axial peak suppression (even–odd) and quadrature detection in ω_1. P1 phase is constant at 0, P2 is cycled 0, π, $\pi/2$, 3 $\pi/2$. The receiver is cycled 0, 0, π, π.

values of t_1. Since the desired coherence follows the phase of the first pulse, it follows the receiver phase. However, axial peaks are modulated only by the receiver phase shift, so they gain an apparent frequency $1/(2\,\Delta t_1)$, moving their signals to the edge of the spectrum.

Phase cycling and coherence order selection: the DQF-COSY experiment

Even if a COSY experiment is acquired in phase-sensitive mode, the dispersive nature of the diagonal peaks is fairly distracting, in that the long "tails" of dispersive signals can bury cross-peaks near the diagonal. A modification of the COSY experiment called the double-quantum-filtered COSY (DQF-COSY) experiment (Figure 7.10, Reference 4) provides a way around this problem. This experiment is designed to select only for coherence that has passed through a two-quantum state. Such **coherence selection** can be accomplished using phase cycling of pulse and receiver phases.

DQF-COSY is a three-pulse experiment, $\pi/2x - t_1 -\pi/2_x - \Delta - \pi/2_y - t_2\ (acquire)$ where Δ is a very short delay time (usually just the time needed to switch pulse phase). In order to appreciate the DQF-COSY experiment, we convert from the Cartesian to the single-element basis. Now see what happens after the second pulse:

$$\hat{\mathbf{I}}_{1z} \xrightarrow{\pi/2_x} \hat{\mathbf{I}}_{1y} \xrightarrow{t_1} \left(\sin\right)2\hat{\mathbf{I}}_{1x}\hat{\mathbf{I}}_{2z} \xrightarrow{\pi/2_x} 2\hat{\mathbf{I}}_{1x}\hat{\mathbf{I}}_{2y} = \overset{\text{double-quantum}}{\tfrac{1}{2}\left(\hat{\mathbf{I}}_{1+}\hat{\mathbf{I}}_{2+} - \hat{\mathbf{I}}_{1-}\hat{\mathbf{I}}_{2-}\right)} - \overset{\text{zero-quantum}}{\tfrac{1}{2i}\left(\hat{\mathbf{I}}_{1+}\hat{\mathbf{I}}_{2-} - \hat{\mathbf{I}}_{1-}\hat{\mathbf{I}}_{2+}\right)} \Rightarrow$$

$$\tfrac{1}{2}\left(2\hat{\mathbf{I}}_{1x}\hat{\mathbf{I}}_{2y} + 2\hat{\mathbf{I}}_{1y}\hat{\mathbf{I}}_{2x}\right) \xrightarrow{\pi/2_y} \tfrac{1}{2}\left(2\hat{\mathbf{I}}_{1z}\hat{\mathbf{I}}_{2y} + 2\hat{\mathbf{I}}_{1y}\hat{\mathbf{I}}_{2z}\right)$$

$$(7.16)$$

In the above expression, the $2\hat{\mathbf{I}}_{1x}\hat{\mathbf{I}}_{2y}$ term has been expanded in terms of raising and lowering operators to show that the term consists of both double- and zero-quantum coherences. The Cartesian operators, although useful for describing the effect of pulses and time evolution on a spin system, are not as helpful for understanding coherence order and selection. Expansion of the Cartesian operators in terms of the raising and lowering operators helps clarify how phase cycling works to select a particular coherence transfer pathway. The phase cycling for the DQF-

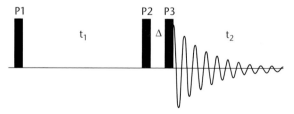

Figure 7.10 DQF-COSY pulse sequence. In the pulse sequence described here, P1 and P2 are both $\pi/2_x$ pulses, Δ is very short (typically just the spectrometer phase switching time), and P3 is a $\pi/2_y$ pulse.

COSY experiment is designed to select only the double-quantum terms of $2\hat{\mathbf{I}}_{1x}\hat{\mathbf{I}}_{2y}$ that are generated from the sine-modulated term from J-coupling evolution during t_1. Returning the double-quantum terms to the Cartesian basis, we obtain $\frac{1}{2}(2\hat{\mathbf{I}}_{1x}\hat{\mathbf{I}}_{2y} + 2\hat{\mathbf{I}}_{1y}\hat{\mathbf{I}}_{2x})$, that after the third pulse becomes $\frac{1}{2}(2\hat{\mathbf{I}}_{1z}\hat{\mathbf{I}}_{2y} + 2\hat{\mathbf{I}}_{1y}\hat{\mathbf{I}}_{2z})$. Note that the two terms in this expression are evenly divided between coherence that is transverse on spin 1 (and will therefore be on the diagonal after transformation, since it evolved on spin 1 during both t_1 and t_2), and coherence that is transverse on spin 2, and will generate the cross-peak between spin 1 and spin 2. Because they have the same history and both evolve from the same double-quantum term, the cross-peaks and diagonal peaks in the DQF-COSY experiment have the same phase characteristics, and can be phased identically (usually antiphase absorptive).

How does one select the double-quantum terms using phase cycling? First, look at the evolution of coherence under the influence any arbitrary **z**-directed evolution operator (such as chemical shift or J-coupling):

$$\hat{\mathbf{I}}_{1z} \xrightarrow{\;\zeta_z\;} \hat{\mathbf{I}}_{1z}$$

$$\hat{\mathbf{I}}_x \xrightarrow{\;\zeta_z\;} \hat{\mathbf{I}}_x \cos\zeta_z t + \hat{\mathbf{I}}_y \sin\zeta_z t$$

$$\hat{\mathbf{I}}_y \xrightarrow{\;\zeta_z\;} \hat{\mathbf{I}}_y \cos\zeta_z t - \hat{\mathbf{I}}_x \sin\zeta_z t \tag{7.17}$$

In order to understand how particular orders of coherence are selected, one must know how the $\hat{\mathbf{I}}_+$ and $\hat{\mathbf{I}}_-$ operators evolve with time. Remembering that $\hat{\mathbf{I}}_+ = \hat{\mathbf{I}}_x + i\hat{\mathbf{I}}_y$, the evolution of $\hat{\mathbf{I}}_+$ due to a **z**-directed operator ζ_z is given by:

$$\hat{\mathbf{I}}_x + i\hat{\mathbf{I}}_y \xrightarrow{\;\zeta_z\;} \hat{\mathbf{I}}_x \cos\zeta_z t + \hat{\mathbf{I}}_y \sin\zeta_z t + i\left(\hat{\mathbf{I}}_y \cos\zeta_z t - \hat{\mathbf{I}}_x \sin\zeta_z t\right)$$

$$= \cos\zeta_z t\left(\hat{\mathbf{I}}_x + i\hat{\mathbf{I}}_y\right) + \sin\zeta_z t\left(-i\hat{\mathbf{I}}_x + \hat{\mathbf{I}}_y\right)$$

$$= \cos\zeta_z t\left(\hat{\mathbf{I}}_x + i\hat{\mathbf{I}}_y\right) + i^3 \sin\zeta_z t\left(\hat{\mathbf{I}}_x + i\hat{\mathbf{I}}_y\right)$$

$$= \hat{\mathbf{I}}_+ \cos\zeta_z t - i\hat{\mathbf{I}}_+ \sin\zeta_z t$$

$$= \hat{\mathbf{I}}_+\left(\cos\zeta_z t - i\sin\zeta_z t\right)$$

$$= \hat{\mathbf{I}}_+ e^{-i\zeta_z t} \tag{7.18}$$

This is a simple result. By the same token, evolution of the lowering operator gives:

$$\hat{\mathbf{I}}_- = \hat{\mathbf{I}}_x - i\hat{\mathbf{I}}_y \Rightarrow \hat{\mathbf{I}}_- e^{i\zeta_z t} \tag{7.19}$$

Note that the results for $\hat{\mathbf{I}}_+$ and $\hat{\mathbf{I}}_-$ differ in the sign of the argument for the exponent. For zero- and double-quantum coherence, we find, respectively:

$$\hat{\mathbf{I}}_{1+}\hat{\mathbf{I}}_{2-} \xrightarrow{\;\zeta_z t\;} \hat{\mathbf{I}}_{1+} e^{-i\zeta_z t}\hat{\mathbf{I}}_{2-} e^{i\zeta_z t} = \hat{\mathbf{I}}_{1+}\hat{\mathbf{I}}_{2-}$$

$$\hat{\mathbf{I}}_{1+}\hat{\mathbf{I}}_{2+} \xrightarrow{\;\zeta_z t\;} \hat{\mathbf{I}}_{1+} e^{-i\zeta_z t}\hat{\mathbf{I}}_{2+} e^{-i\zeta_z t} = \hat{\mathbf{I}}_{1+}\hat{\mathbf{I}}_{2+} e^{-2i\zeta_z t} \tag{7.20}$$

In other words, zero-quantum coherence does not evolve with the **z**-directed operator, and double-quantum terms evolve at the sum of the evolution frequencies of the single-quantum operators. Putting an additional factor ϕ into the expression, $e^{-i\zeta_z t} = e^{-i\phi\zeta_z t}$, $\phi = 1$ in the exponent for the $\hat{\mathbf{I}}_+$ term and -1 for the $\hat{\mathbf{I}}_-$ term. The sum of the all of the ϕ factors for the multiplied operators gives the coherence order of the product term. For the flip-flop operator, the coherence order is 0 (zero-quantum coherence) and for the $\hat{\mathbf{I}}_{1+}\hat{\mathbf{I}}_{2+}$ operator, the coherence order is 2. Similarly, for the $\hat{\mathbf{I}}_{1-}\hat{\mathbf{I}}_{2-}$ operator, the coherence order is -2.

The phase advance of a given order of coherence under the influence of **z**-directed evolution ζ_z is given by $\exp(-i\phi\zeta_z)$ This observation leads to the basic logic of phase cycling, that the receiver phase is set to follow the phase advance of a desired change in coherence order according to the expression:

$$\text{receiver phase} = -\Delta\phi\Delta p \qquad (7.21)$$

for a change in coherence order of $\Delta\phi$ and a change in pulse phase of Δp.

Armed with this information, we re-examine the simplest case of phase cycling, CYCLOPS, used to select against quadrature images. As shown from the expansion of a single transverse Cartesian operator in terms of raising and lowering operators, both $+1$ and -1 coherences are present in the signal detected. The system started at a coherence order of zero (i.e. no coherence), so these coherences represent $\Delta\phi$ values of $+1$ and -1 respectively (see Figure 7.11). If one of the two is not selected, peaks at both $-\Omega$ and $+\Omega$ will be observed in the transformed spectrum. This is just another way of stating the need for quadrature detection. In general, we select only the -1 coherence (that will be detected at frequency $+\Omega$). We can phase cycle the pulse and receiver phases in order to select for one of the two frequencies around the carrier frequency. One phase cycle that can accomplish this is:

Pulse phase	Receiver phase
0	0
1	1
2	2
3	3

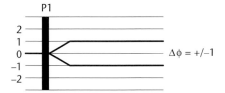

Figure 7.11 Generation of $+1$ and -1 quantum coherence by a single pulse P1. The starting state prior to the first pulse (P1) is a population operator that has no coherence (order zero). After the pulse, $+1$ and -1 quantum coherences develop under the influence of chemical shift and coupling.

Multiplying the number representing the phase step by $\pi/2$ yields the phase in radians. The logic of this phase cycle can be followed in Figure 7.12. In Case 1 shown in Figure 7.12, pulse phase (which is cycled in steps of $\Delta p = \pi/2$) is tracked by receiver phase, and all signals, when co-added, will reinforce. According to Equation 7.21, the change in phase of -1 coherence follows the pulse phase, and so also follows the receiver phase. The change in phase of $+1$ coherence is opposite (Case 2), and so receiver phase does not follow the change in phase of $+1$ coherence. In this case, co-addition of all four spectra results in cancellation.

The phase cycle shown in Figure 7.12 successfully discriminates $\Delta\phi = +1$ and $\Delta\phi = -1$. Do we actually need a four-step cycle in order to obtain such discrimination? The answer is no. The number of steps required in a phase cycle is determined by how many adjacent levels must be discriminated when selecting for a particular $\Delta\phi$. If one particular order of coherence is desired out of n adjacent levels, then an n-step phase cycle consisting of $2\pi/n$ radian increments in pulse phase $\Delta p = 2\pi/n$ is required. Based on Figure 7.8, we are selecting against $\Delta\phi = +1$ and for $\Delta\phi = -1$ (three adjacent levels). A two-step phase cycle ($\Delta p = \pi$) would be insufficient for this purpose, since the signals obtained at $p = 0$ and $p = \pi$ are identical in both the $\Delta\phi = +1$ and $\Delta\phi = -1$ cases: the $p = \pi/2$ and $p = 3\pi/2$ signals are required to cancel

Case 1

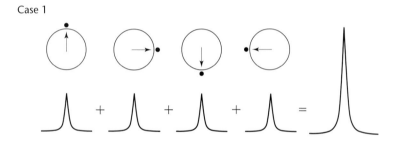

Case 2

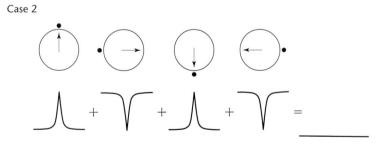

Figure 7.12 Phase cycling for selection of -1 versus $+1$ quantum coherence. In case 1, the evolution of -1 quantum coherence upon pulse phase changes $\Delta p = \pi/2$ is followed by the appropriate receiver phase shifts (represented by the black dot). The appearance of the transformed spectrum expected is shown underneath, as well as the resulting constructive addition spectrum. In case 2, the same pulse phase cycle and receiver phase cycle is used, but the evolution of $+1$ coherence is shown. The resulting spectra will cancel upon addition.

the signals due to the $\Delta\phi = +1$ coherence. However, a three-step phase cycle is sufficient to select for $\Delta\phi = -1$ in the presence of $\Delta\phi = +1$.

The cancellation of the terms due to $+1$ coherence is less obvious in Figure 7.13 than in Figure 7.12, but if we draw vectors reflecting the *difference* between the receiver phase and coherence phase, we find that those vectors cancel out upon addition, as shown in Figure 7.14. Now, the cancellation is evident from the equal and opposite vectors.

Now consider the phase cycle required for a DQF-COSY experiment. The desired coherence transfer pathways for DQF-COSY are shown in Figure 7.15. The first pulse of the DQF-COSY generates, as always, $+1$ and -1 coherence. After time t_1, a second pulse is applied that generates all possible orders of coherence. However, the phase cycle should select only those coherences that exist during the delay Δ as double-quantum coherence. This means selecting for changes in coherence orders of $\Delta\phi = -3$ and $\Delta\phi = +1$ after the third pulse, since we only detect -1 coherence during t_2.

If we select for $\Delta\phi = -3$ and against $\Delta\phi = -2, -1, 0$, we need a four-step phase cycle (one out of four adjacent levels selected). We can arbitrarily choose any of four paths to our final destination through double-quantum coherence:

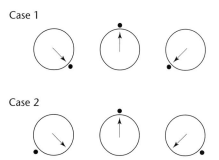

Figure 7.13 Differentiation between $+1$ and -1 quantum coherence by a three-step phase cycle with steps of $\Delta p = 2\pi/3$.

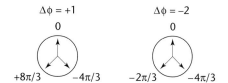

Figure 7.14 Vectors representing the difference between the receiver phase and the coherence phase in the three-step phase cycle shown in Figure 7.15 for the $\Delta\phi = +1$ and $\Delta\phi = -2$ pathways. Cancellation of these vectors upon addition indicates that the phase cycle selects against these pathways.

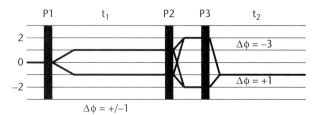

Figure 7.15 Changes in coherence order during the DQF-COSY experiment. The starting state prior to the first pulse (P1) is a population operator which has no coherence (order zero). After the first pulse, +1 and −1 coherences develop under the influence of chemical shift and coupling. After the second pulse (P2), the single quantum terms give rise to all possible orders of coherence (from −2 to +2 quantum), with different evolution operators giving cross-overs to different orders. Phase cycling must select for $\Delta\phi = -3$ and $\Delta\phi = +1$ after the last pulse.

$$\Delta\phi_1 = +1, \Delta\phi_2 = +1, \Delta\phi_3 = -3,$$
$$\Delta\phi_1 = +1, \Delta\phi_2 = -3, \Delta\phi_3 = +1,$$
$$\Delta\phi_1 = -1, \Delta\phi_2 = -1, \Delta\phi_3 = +1,$$
$$\Delta\phi_1 = -1, \Delta\phi_2 = +3, \Delta\phi_3 = -3$$

We need both +1 and −1 coherence generated during t_1 for quadrature detection in the ω_1 dimension, so the first and second pulses are cycled together. Choosing the first pathway, the receiver phase is set according to the expression:

$$\text{receiver phase} = -\left[\left(+2\right)p_2 + \left(-3\right)p_3\right] \tag{7.22}$$

However, the same receiver phase will also suffice for any of the other three possible pathways (Problem 7.3). A four-step phase cycle of the first two pulses with the receiver phase properly set follows, where the numbers shown are multipliers for $\pi/2$ rad:

Step #	P1	P2	P3	Receiver
1	0	0	0	0
2	1	1	0	2
3	2	2	0	0
4	3	3	0	2

In order to eliminate axial peaks that develop during t_1, there is an additional phase cycling of the first two pulses called **odd–even** selection. Odd–even selection involves shifting the phase of the first pulse by π rad at each step. If the receiver phase follows the pulse phase, odd orders of $\Delta\phi$ are selected. This will select for $\Delta\phi = -1$ and +1 (odd) and against the **z** magnetization, that behaves as $\Delta\phi = 0$ (even).

This completes the phase cycle for the DQF-COSY experiment, with the numbers showing phase being multipliers for $\pi/2$ rad:

Step #	P1	P2	P3	Receiver	Step #	P1	P2	P3	Receiver
1	0	0	0	0	5	2	0	0	2
2	1	1	0	2	6	3	1	0	0
3	2	2	0	0	7	0	2	0	2
4	3	3	0	2	8	1	3	0	0

Other multiple-quantum filters in COSY

Although DQF-COSY is the most commonly used *multiple-quantum-filtered* (MQF) experiment, it is sometimes useful to select for a particular **spin topology** in order to simplify a complex spectrum. For this purpose, a higher-order quantum filter can be used. For example, the alkyl methyl resonance region of a protein ^{1}H NMR spectrum is often quite complicated, containing the CH–CH$_3$ cross-peaks of valine and leucine as well as the CH–C$_{\gamma2}$H$_3$ and C$_{\gamma1}$H$_2$–C$_{\delta1}$H$_3$ cross-peaks of isoleucines. The coupling patterns of the CH–CH$_3$ methyls differ from those of the isoleucine C$_{\gamma1}$H$_2$–C$_{\delta1}$H$_3$ in that the (equivalent) protons of CH–CH$_3$ methyls have a single observable coupling, to the methine proton (CH). At most, then, CH–CH$_3$ groups can develop double-quantum coherence. The isoleucine C$_{\gamma1}$H$_2$–C$_{\delta1}$H$_3$ group has three mutually coupled resonances, the two diastereotopic (nonequivalent) methylene CH$_2$ protons and the equivalent protons of the CH$_3$ group. These three couplings can be used to generate triple-quantum coherence. A *triple-quantum filter* in a COSY experiment (TQF-COSY, Reference 5) can thus be used to distinguish these different spin-coupling topologies. In a TQF-COSY, only the isoleucine CH$_2$–CH$_3$ couplings will be observed.

The simplified coherence transfer pathway for an MQF-COSY is shown in Expression 7.23:

$$\hat{\mathbf{I}}_{1z} \xrightarrow{\pi/2_x} \hat{\mathbf{I}}_{1y} \xrightarrow{t_1} 2^{n-1}\hat{\mathbf{I}}_{1x}\hat{\mathbf{I}}_{2z}\hat{\mathbf{I}}_{3z}\ldots\hat{\mathbf{I}}_{nz} \xrightarrow{\pi/2_x} 2^{n-1}\hat{\mathbf{I}}_{1x}\hat{\mathbf{I}}_{2y}\hat{\mathbf{I}}_{3y}\ldots\hat{\mathbf{I}}_{ny}$$

$$\xrightarrow{MQF} \hat{\mathbf{I}}_{1x}\hat{\mathbf{I}}_{2y}\hat{\mathbf{I}}_{3y}\ldots\hat{\mathbf{I}}_{ny} + \hat{\mathbf{I}}_{1y}\hat{\mathbf{I}}_{2x}\hat{\mathbf{I}}_{3y}\ldots\hat{\mathbf{I}}_{ny} \xrightarrow{\pi/2_x}$$

$$\hat{\mathbf{I}}_{1x}\hat{\mathbf{I}}_{2z}\hat{\mathbf{I}}_{3z}\ldots\hat{\mathbf{I}}_{nz} + \hat{\mathbf{I}}_{1z}\hat{\mathbf{I}}_{2x}\hat{\mathbf{I}}_{3z}\ldots\hat{\mathbf{I}}_{nz} \xrightarrow{t_2} 2^{n-1}\hat{\mathbf{I}}_{2x} + 2^{n-1}\hat{\mathbf{I}}_{1x} \qquad (7.23)$$

The coherence transfer diagram for a TQF-COSY is shown in Figure 7.16. Note that the correct $\Delta\phi = -4$ change in coherence must be selected from six adjacent levels ($\Delta\phi = -4$ from $\Delta\phi = -3 \rightarrow +1$), so a six-step phase cycle is required. Note that each phase increment is $\pi/3$ rad for this phase cycle (see Problem 7.4).

Multiple-quantum spectroscopy

Now consider the MQF experiment in a different way. Instead of allowing coherences to evolve during the time between the first two pulses, the t_1 evolution time is placed *after* MQ coherence is created (i.e. after the second pulse). Now, the indi-

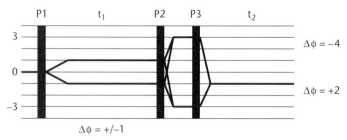

Figure 7.16 Changes in coherence order during the TQF-COSY experiment. As for DQF-COSY, the starting state prior to the first pulse (P1) is a population operator which has no coherence (order zero). After the first pulse, +1 and −1 quantum coherences develop under the influence of chemical shift and coupling. Both pathways are required for quadrature detection in the indirectly detected dimension. After the second pulse (P2), the single quantum terms give rise to all possible orders of coherence. Phase cycling must select for $\Delta\phi = -4$ and $\Delta\phi = +2$ after the last pulse.

rectly detected dimension will represent evolution of the observed coherences as MQ coherences (recall that MQ coherence cannot be directly detected!). Since MQ coherences are invariant to the active coupling, the coherences do not evolve with respect to the active coupling (i.e. the coupling that generated the coherence) during t_1. However, chemical shift evolution will occur at the sum of the chemical shifts of the coherences involved in the evolution. This is shown for +DQ coherence:

$$2\left(\hat{\mathbf{I}}_{1x}\hat{\mathbf{I}}_{2y} + \hat{\mathbf{I}}_{1y}\hat{\mathbf{I}}_{2x}\right) \xrightarrow{\left(\Omega_1 \hat{\mathbf{I}}_{1z} + \Omega_2 \hat{\mathbf{I}}_{2z}\right)t_1} 2\left(\hat{\mathbf{I}}_{1x}\hat{\mathbf{I}}_{2y} + \hat{\mathbf{I}}_{1y}\hat{\mathbf{I}}_{2x}\right)\cos\left(\Omega_1 + \Omega_2\right)t_1$$
$$+2\left(\hat{\mathbf{I}}_{1y}\hat{\mathbf{I}}_{2y} - \hat{\mathbf{I}}_{1x}\hat{\mathbf{I}}_{2x}\right)\sin\left(\Omega_1 + \Omega_2\right)t_1$$
$$2\left(\hat{\mathbf{I}}_{1x}\hat{\mathbf{I}}_{2y} - \hat{\mathbf{I}}_{1y}\hat{\mathbf{I}}_{2x}\right) \xrightarrow{\left(\Omega_1 \hat{\mathbf{I}}_{1z} + \Omega_2 \hat{\mathbf{I}}_{2z}\right)t_1} 2\left(\hat{\mathbf{I}}_{1y}\hat{\mathbf{I}}_{2y} - \hat{\mathbf{I}}_{1x}\hat{\mathbf{I}}_{2x}\right)\cos\left(\Omega_1 + \Omega_2\right)t_1$$
$$-2\left(\hat{\mathbf{I}}_{1x}\hat{\mathbf{I}}_{2y} + \hat{\mathbf{I}}_{1y}\hat{\mathbf{I}}_{2x}\right)\sin\left(\Omega_1 + \Omega_2\right)t_1 \tag{7.24}$$

The generalized pulse sequence shown in Figure 7.17 provides for evolution of MQ coherence during t_1 (Reference 6). In order to generate maximum MQ coherence, a delay time (2Δ, the **mixing time**) is inserted between the first and second $\pi/2$

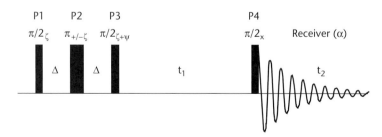

Figure 7.17 MQ experiment for the selection of *p*-quantum coherence. Phase cycling of pulses and receiver are as shown (Reference 6):

$$\Delta = \frac{1}{4J}, \zeta = \frac{k\pi}{p}; k = 0, 1, 2, \ldots 2p - 1; \psi = 0(even), \frac{\pi}{2}(odd); \alpha = n\zeta.$$

pulses in order to allow the development of $I_{1x}I_{2z}I_{3z}...I_{(n-1)z}$ terms. For generation of DQ coherence between spins 1 and 2, this time is set equal to $1/2J_{12}$. For higher orders of coherence the precise delay times are compromises between different couplings involved in generating the coherences. Note that a π pulse is inserted in the middle of the mixing time, generating an internal spin–echo sequence. As discussed earlier, a spin echo refocuses chemical shift effects at the height of the echo; hence, at the time of the second $\pi/2$ pulse, only coupling evolution will remain and chemical shift effects on the efficiency of multiple-quantum coherence generation will be removed. The phase cycling of pulses and receiver determines the order of MQ coherence selected (Figure 7.17).

The MQ experiment has some interesting features. Following the evolution of coherence due to a single coupling, we find:

$$\hat{\mathbf{I}}_{1x} + \hat{\mathbf{I}}_{2x} \xrightarrow{2\Delta} 2\left(\hat{\mathbf{I}}_{1y}\hat{\mathbf{I}}_{2z} + \hat{\mathbf{I}}_{1z}\hat{\mathbf{I}}_{2y}\right) \xrightarrow{\pi/2_y} 2\left(\hat{\mathbf{I}}_{1y}\hat{\mathbf{I}}_{2x} + \hat{\mathbf{I}}_{1x}\hat{\mathbf{I}}_{2y}\right) \xrightarrow{\pi/2_x} 2\left(\hat{\mathbf{I}}_{1z}\hat{\mathbf{I}}_{2x} + \hat{\mathbf{I}}_{1x}\hat{\mathbf{I}}_{2z}\right)$$

First, the coherence that develops during the evolution time t_1 ends up evenly distributed between the two spins during t_2, so cross-peaks show up at the ω_2 frequencies of *both* of the nuclei involved in the double-quantum (DQ) coherence. This differs from COSY and DQF-COSY, in which symmetrically related cross-peaks result from two separate coherence transfers. The cross-peaks in the DQ spectrum will be arranged symmetrically with respect to the DQ diagonal ($\omega_1 = 2\omega_2$, see Figure 7.18), and result from the direct transfer terms: $2\hat{\mathbf{I}}_{1x}\hat{\mathbf{I}}_{2y} + 2\hat{\mathbf{I}}_{1y}\hat{\mathbf{I}}_{2x} \xrightarrow{\pi/2x} 2\hat{\mathbf{I}}_{1x}\hat{\mathbf{I}}_{2z}$ $+ 2\hat{\mathbf{I}}_{1z}\hat{\mathbf{I}}_{2x}$. Another difference between DQ and DQF-COSY is that all detected coherences evolve as DQ coherence during t_1 (i.e. they will be labeled with the sum of the frequencies of all spins contributing to the transverse coherence during t_1), and so *there will be no diagonal peaks in the final transformed spectrum*. This can be valuable in crowded spectra. All of the direct transfer cross-peaks are aligned in ω_1 at the frequency that represents the *sum* of the chemical shifts of the two spins giving rise to the DQ coherence, $\omega_1 = \Omega_1 + \Omega_2$ (see Equation 7.24). Because active couplings do not evolve during t_1, the cross-peaks do not show the symmetric multiplet structure in two dimensions that DQF-COSY and COSY cross-peaks do; they are pure-phase absorptive without splitting by the active coupling in ω_1 (see Figure 7.18). However, MQ coherence does evolve with passive couplings, as shown in Expression 7.25 for the evolution of multiple-quantum coherence between spins 1 and 2 under the influence of a third mutual coupling partner, spin 3:

$$2\left(\hat{\mathbf{I}}_{1y}\hat{\mathbf{I}}_{2x} + \hat{\mathbf{I}}_{1x}\hat{\mathbf{I}}_{2y}\right) \xrightarrow{(\pi J_{13}2\hat{\mathbf{I}}_{1z}\hat{\mathbf{I}}_{3z} + \pi J_{23}\hat{\mathbf{I}}_{2z}\hat{\mathbf{I}}_{3z})t_1} 2\left(\hat{\mathbf{I}}_{1y}\hat{\mathbf{I}}_{2x} + \hat{\mathbf{I}}_{1x}\hat{\mathbf{I}}_{2y}\right)\cos\pi\left(J_{13}+J_{23}\right)t_1$$
$$-4\left(\hat{\mathbf{I}}_{1y}\hat{\mathbf{I}}_{2y} + \hat{\mathbf{I}}_{1x}\hat{\mathbf{I}}_{2x}\right)\hat{\mathbf{I}}_{3z}\sin\pi\left(J_{13}+J_{23}\right)t_1$$
$$2\left(\hat{\mathbf{I}}_{1x}\hat{\mathbf{I}}_{2x} + \hat{\mathbf{I}}_{1y}\hat{\mathbf{I}}_{2y}\right) \xrightarrow{(\pi J_{13}2\hat{\mathbf{I}}_{1z}\hat{\mathbf{I}}_{3z} + \pi J_{23}2\hat{\mathbf{I}}_{2z}\hat{\mathbf{I}}_{3z})t_1} 2\left(\hat{\mathbf{I}}_{1x}\hat{\mathbf{I}}_{2x} + \hat{\mathbf{I}}_{1y}\hat{\mathbf{I}}_{2y}\right)\cos\pi\left(J_{13}+J_{23}\right)t_1$$
$$-4\left(\hat{\mathbf{I}}_{1x}\hat{\mathbf{I}}_{2y} + \hat{\mathbf{I}}_{1y}\hat{\mathbf{I}}_{2x}\right)\hat{\mathbf{I}}_{3z}\sin\pi\left(J_{13}+J_{23}\right)t_1 \tag{7.25}$$

In-phase splitting will therefore be observed in ω_1 of MQ experiments because of passive couplings. Another interesting result of the evolution due to passive

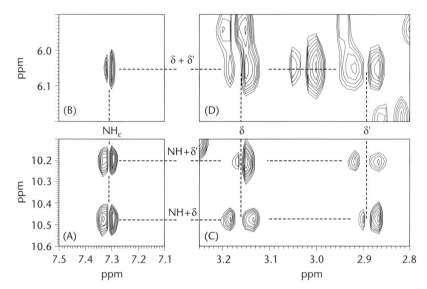

Figure 7.18 Correlations for the NεH-CδH and NεH-CδH' spins of an arginine residue in the DQ spectrum of putidaredoxin, a ferredoxin, in 90% H_2O/D_2O obtained at 500 MHz. (A) Direct correlations for NεH-CδH and NεH-CδH' at the NεH shift; (B) remote correlation to the NεH at the sum of the CδH and CδH' shifts. Remote correlations are opposite in phase to the direct correlations. (C) A mix of direct and remote correlations at the shifts of the diastereotopic CδH and CδH' protons. (D) The direct correlations between the CδH and CδH' spins.

coupling during t_1 is the observation of peaks that arise from the transformation of the sine-modulated terms of Expression 7.25:

$$4\hat{\mathbf{I}}_{1y}\hat{\mathbf{I}}_{2y}\hat{\mathbf{I}}_{3z} \xrightarrow{\pi/2x} 4\hat{\mathbf{I}}_{1z}\hat{\mathbf{I}}_{2z}\hat{\mathbf{I}}_{3y} \xrightarrow{t_2} \hat{\mathbf{I}}_{3x}\left(\Omega_1+\Omega_2,\Omega_3\right) \tag{7.26}$$

Note that although these coherences evolved as DQ terms on spins 1 and 2 during t_1, they are relayed via the passive coupling to spin 3 and evolve on that spin during t_2. The peaks resulting from these terms in the MQ spectrum are called relayed or **remote peaks**, and appear at the chemical shift of the passively coupled spin in ω_2 (see Figure 7.18). The remote peaks thus represent MQ coherence that evolves between two spins, but this is transferred to a third spin by the passive coupling evolution and evolves on that spin during t_2.

Effect of π pulses on coherence

We noted in passing that the MQ experiment incorporates a π pulse in the middle of the mixing time 2Δ in order to remove chemical shift modulation in the development of the terms that result in MQ coherence. This is a commonly used device in multiple-pulse experiments where one wishes to separate the effects of chemical shift from J-coupling. In terms of coherence operators, the effect of a π pulse is to change the sign of the affected coherences. That is, $\hat{\mathbf{I}}_-$ coherence becomes $\hat{\mathbf{I}}_+$, $\hat{\mathbf{I}}_{1-}\hat{\mathbf{I}}_{2-}$ becomes $\hat{\mathbf{I}}_{1+}\hat{\mathbf{I}}_{2+}$, and so on. Based on our earlier discussion concerning the time evolution of such coherences, the direction of evolution is switched, and so a π pulse can be thought of as performing a **time reversal**. This is the opposite of the intuitive picture used in Chapter 4: our classical explanation of the spin–echo experiment had coupling evolution reversing direction while the chemical shift evolution was unaffected (however, you were also warned not to fall in love with this picture!). Using the operator approach, we can see that coupling evolution of an operator is unaffected by homonuclear nonselective π pulses while the sense of chemical shift evolution is reversed.

$$\mathbf{I}_x \cos \Omega t + \mathbf{I}_y \sin \Omega t \xrightarrow{\;\pi_x\;} \mathbf{I}_x \cos \Omega t - \mathbf{I}_y \sin \Omega t$$
$$\mathbf{I}_x \cos \Omega t - \mathbf{I}_y \sin \Omega t = \mathbf{I}_x \cos \Omega \left(-t\right) + \mathbf{I}_y \sin \Omega \left(-t\right) \tag{7.27}$$

For coupling, on the other hand, the same pulse does not affect evolution, as long as the pulse affects both coupling partners:

$$\hat{\mathbf{I}}_{1x} \cos \pi J_{12} t + 2 \hat{\mathbf{I}}_{1y} \hat{\mathbf{I}}_{2z} \sin \pi J_{12} t \xrightarrow{\;\pi_x\;} \hat{\mathbf{I}}_{1x} \cos \pi J_{12} t - 2 \hat{\mathbf{I}}_{1y} \left(-\hat{\mathbf{I}}_{2z}\right) \sin \pi J_{12} t$$
$$\hat{\mathbf{I}}_{1x} \cos \pi J_{12} t - 2 \hat{\mathbf{I}}_{1y} \left(-\hat{\mathbf{I}}_{2z}\right) \sin \pi J_{12} t = \hat{\mathbf{I}}_{1x} \cos \pi J_{12} t + 2 \hat{\mathbf{I}}_{1y} \hat{\mathbf{I}}_{2z} \sin \pi J_{12} t \tag{7.28}$$

Pulsed-field gradients for coherence selection

In recent years, advances in NMR hardware have made it possible to select particular orders of coherence without phase cycling via the application of **pulsed-field gradients** (**PFGs**). Solution NMR spectroscopists have traditionally been obsessed with keeping the applied magnetic field $\vec{B}_0$ as homogeneous as possible over the entire sample volume. This is accomplished by shimming the magnetic field. However, a $\vec{B}_0$ field upon which a linear **field gradient** has been superimposed has some interesting and useful properties. If the magnetic field depends in a linear fashion upon the **z** coordinate of the sample volume in the magnet (with the magnet bore being along the **z** axis) the resonant frequency ω_{obs} of a nucleus is determined by:

$$\omega_{obs} = \gamma \left(B_0 + B_g\right) = \gamma B_0 + \gamma B_g$$
$$\omega_{obs} = \omega_0 + \gamma B_g = \omega_0 + \gamma G_z z = \omega_0 + \Delta \omega$$
$$\Delta \omega = \gamma G_z z \tag{7.29}$$

where G_z is the field gradient given in units of change in field strength change per unit displacement (usually G/cm) along the **z** axis and z is the displacement along the **z** axis. A sample that in a normal homogeneous magnetic field gives a sharp resonance line would ideally have the appearance of a "mesa" in the presence of a linear field gradient along the long axis of the sample (usually the laboratory **z** axis). The raised features at the signal edges in a real sample (Figure 7.19) are due to distortions in the magnetic field as it passes through the boundary between the tube inserts and the samples that have slightly different magnetic susceptibilities. If one knows the field dependence of displacement along the **z** axis of the sample, one can calculate the length of the sample along the direction of the gradient. This is the basis of NMR imaging (MRI), and will be discussed in more detail in Chapter 12.

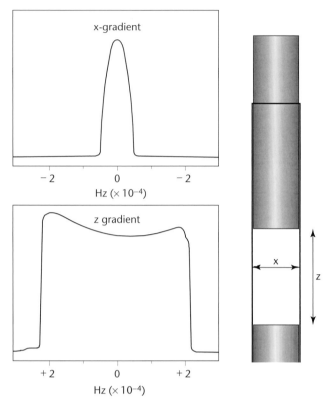

Figure 7.19 (Top) ^{1}H NMR image profile of a water sample in a 5 mm susceptibility-matched NMR tube (shown at right) in the presence of an **x**-direction magnetic field gradient. The profile corresponds to signal intensity in the horizontal **x** direction (perpendicular to the tube axis). The center of the sample is at 0 Hz, the walls of the tube are at ~0.6 × 10^4 Hz. (Bottom) Same sample as in the top spectrum, but in the presence of a **z** gradient (along the tube long axis). The sample is centered in the receiver coil, so the middle of the sample is at 0 Hz. At the top and bottom of the sample are susceptibility-matched plugs located at 150 mm above and below the center of the sample, and are at ~2.2 × 10^4 Hz in the spectrum. This is the basis of magnetic resonance imaging.

For now, the relevant question is how a field gradient affects coherence evolution. We know that the time evolution of the raising operator $\hat{\mathbf{I}}_+$ (representing +1 coherence) under the influence of a **z**-directed operator is $\omega\hat{\mathbf{I}}_z$ given by:

$$\hat{\mathbf{I}}_+ \xrightarrow{\omega\hat{\mathbf{I}}_z} \hat{\mathbf{I}}_+ \exp\left(-i\omega t\right) \tag{7.30}$$

In the presence of a field gradient, an extra term is present:

$$\hat{\mathbf{I}}_+ \xrightarrow{w\hat{\mathbf{I}}_z} \hat{\mathbf{I}}_+ \exp\left[-i\left(\omega t + \Delta\omega t\right)\right] \tag{7.31}$$

Likewise, for the lowering operator that represents –1 coherence:

$$\hat{\mathbf{I}}_- \xrightarrow{\ \omega \hat{\mathbf{I}}_z\ } \hat{\mathbf{I}}_- \exp\left[i\left(\omega t + \Delta\omega t\right)\right] \tag{7.32}$$

For DQ coherence:

$$\hat{\mathbf{I}}_{1+}\hat{\mathbf{I}}_{2+} \xrightarrow{\ \omega \hat{\mathbf{I}}_z\ } \hat{\mathbf{I}}_{1+}\hat{\mathbf{I}}_{2+} \exp\left[-2i\omega t\right] \times \exp\left[-2i\Delta\omega t\right] \tag{7.33}$$

The term that determines evolution for double-quantum coherence due to the field gradient has a factor of two relative to the corresponding term for single-quantum coherence. Thus, DQ coherence in a small volume of the sample goes out of phase with other **isochromats** in the direction of the field gradient twice as fast as single-quantum coherences. An isochromat is a set of coherences in a small sample volume δV that all exhibit the same gradient-dependent time evolution. The loss of phase between isochromats is called **defocusing**. On the other hand, for zero-quantum coherence (ZQ), evolution in the presence of a field gradient is given by:

$$\hat{\mathbf{I}}_{1+}\hat{\mathbf{I}}_{2-} \xrightarrow{\ \omega \hat{\mathbf{I}}_z\ } \hat{\mathbf{I}}_{1+}\hat{\mathbf{I}}_{2-} \exp\left[-i\left(\omega_1 - \omega_2\right)t\right] \times \exp\left[-i\left(\Delta\omega_1 - \Delta\omega_2\right)t\right] \tag{7.34}$$

Assuming that the shifts of the two spins are similar (as we do implicitly to get the factor of two for the DQ case), then both exponents are essentially zero, and the ZQ coherence is to a first approximation not affected by field gradients. This is generally true: *ZQ coherence (as well as any coherence stored along the **z** axis, where it is invariant to chemical shift) is unaffected by field gradients.* This includes the longitudinal terms that contribute to the NOE.

It is important to clarify a common misconception: the **x**, **y** and **z** coordinates used to describe field gradients are laboratory frame coordinates, similar to those used to describe the shim coils. They do not refer to the coordinate axes of the rotating frame in which the operator formalisms are applied. To state this in another way, all coherences of the same order are affected the same way by a given gradient. Only *spatial distributions* of spins in the sample are differentially affected by a field gradient.

Clearly, a field gradient during acquisition is a bad thing (unless one is doing an imaging experiment), since we want spectral lines to be as narrow as possible. However, it is potentially useful to be able to cause different orders of coherence to do different things prior to acquisition. For this reason, PFGs have been developed that can be applied for a short time during an experiment in order to affect coherences in a desired fashion without causing broadening during acquisition.

Unlike RF pulses, there is no change in coherence order in the presence of a PFG, since a gradient does not cause RF frequency modulation, only field changes. Only the relative phase of coherences within different isochromats are affected by PFGs. This phase change is given by:

$$\hat{\mathbf{I}}_+ \xrightarrow{\ \omega \hat{\mathbf{I}}_z\ } \hat{\mathbf{I}}_+ \exp\left[-i\left(\omega t + \phi\right)\right] \tag{7.35}$$

where $\phi = \gamma p G_z z t$, p is the order of coherence, G_z is the gradient amplitude (usually given in units of G/cm), z is the position of the spin in the sample and t is the duration of the gradient pulse.

By taking advantage of these phase shifts, gradients can be used to select a certain order of coherence by applying a series of gradients of a particular intensity and duration. In order to do this selection, one needs a series of gradients that defocus all orders of coherence but refocus only the desired order. This means that the sum of all gradient-generated phase factors for the desired order of coherence must equal zero over the pulse sequence:

$$\sum_i \gamma p_i G_i z t_{gi} = 0 \qquad (7.36)$$

A simple example of such selection is given by the **gradient echo** experiment, consisting of a $\pi/2$ pulse followed by two gradients, one positive and one negative (Figure 7.20).

In order to observe maximum signal (-1 quantum coherence), $\gamma p_1 G_1 z t_{g1} + \gamma p_2 G_2 z t_{g2} = 0$, and the simplest way to accomplish that is to set $t_{g1} = t_{g2}$ and $G_1 = -G_2$, although other combinations will yield the same results.

The gradient COSY experiment

PFGs can be used to select for the desired coherence order in a COSY experiment (Figure 7.21). If G_1 and G_2 (Figure 7.21) are of equal strength and duration, then $\gamma(+1)G_1 z t_{g1} + \gamma(-1)G_2 z t_{g2} = 0$, and coherence orders $\phi = +1$ during t_1 and $\phi = -1$ during t_2 are selected.

There is an important difference between coherence selection using phase cycling and coherence selection using gradients. Phase cycles select for *changes* in coherence order $\Delta\phi$, rather than particular orders of coherence ϕ. Gradients produce different results depending on the order of the coherence, and as such can be used to select for one particular coherence pathway. Furthermore, selection of only

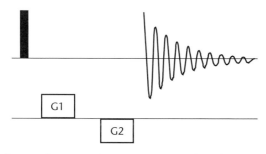

Figure 7.20 Gradient echo experiment. Two equal and opposite gradients are applied to coherence generated by a single pulse. The sum of the gradient-induced phase shifts ϕ will then be equal to zero, and the coherence will be refocused.

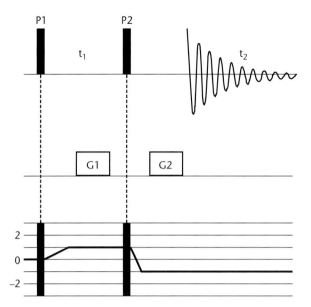

Figure 7.21 Gradient COSY experiment. Experiment is in magnitude mode, as only +1 coherence is selected during t_1. Both gradients are of equal strength and duration.

one of the two possible coherence orders ($\phi = +1$ versus $\phi = -1$) during t_1 of the gradient COSY experiment makes it unnecessary to use a quadrature detection scheme for the indirectly detected dimension (see Figure 7.22). However, this selection (called **p-selection** for +1 and **n-selection** for –1 coherence selection) comes at the price of not being able to phase the two-dimensional experiment, requiring the data to be displayed in magnitude mode. For concentrated samples with well-dispersed signals, such experiments are still desirable. Gradient COSY acquisitions can be very short (sometimes a minute or less), since no phase cycling is required, and only a single FID needs to be acquired at each t_1 value.

The DQF-COSY experiment can also be performed using gradients for coherence selection. If a magnitude mode experiment is acceptable, the simplest gradient-selected DQF-COSY sequence is shown in Figure 7.23.

In order to achieve the desired coherence selection in the DQF-COSY experiment (p-selection in t_1, n-selection in t_2), the following must be true:

$$\gamma\left(+1\right)G_1 z t_{g1} + \gamma\left(+2\right)G_2 z t_{g2} + \gamma\left(-1\right)G_3 z t_{g3} = 0$$
$$G_1 + 2G_2 - G_3 = 0 \tag{7.37}$$

One possibility would be to use equal durations for all three gradients, with the gradient strengths set proportionally as $G_1 = G_2 = 1$, $G_3 = 3$. The spectrum would be displayed in magnitude mode, since only +1 coherence is selected during t_1 (see Figure 7.24). In order to obtain a DQF-COSY spectrum with phase-sensitive quadrature detection in ω_1, gradients are not be applied during t_1. A scheme for a

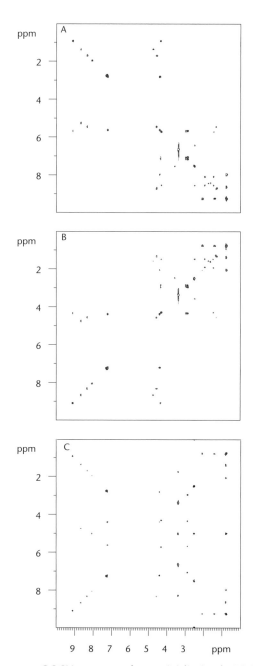

Figure 7.22 Single-scan COSY spectra of gramicidin in d_6-DMSO obtained at 11.7 T (500 MHz ^{1}H) using the pulse sequence shown in Figure 7.21. (A) G1 = (−G2) and −1 coherence is selected in t_1. (B) G1 = G2 and +1 coherence is selected in t_1. (C) No gradients are applied.

phase-sensitive DQF-COSY experiment using PFG coherence selection is shown in Figure 7.25.

Gradients satisfying the following relationship would perform the coherence selection shown in Figure 7.25:

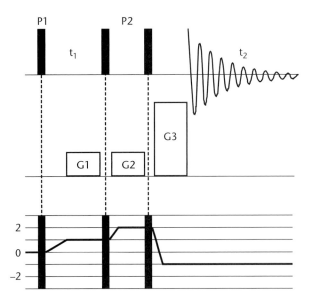

Figure 7.23 Gradient DQF-COSY pulse sequence. Experiment is in magnitude mode. One possible gradient strength combination is G1 = G2 = 1, G3 = 3, if the gradient times are all equal.

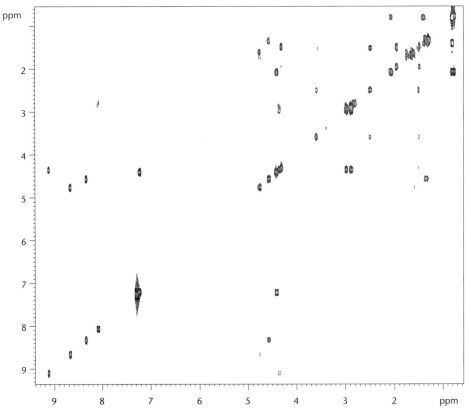

Figure 7.24 DQF-COSY spectrum of gramicidin in d_6-DMSO obtained at 11.7 T (500 MHz ^{1}H) using the pulse sequence shown in Figure 7.23. One scan was acquired per t_1 value.

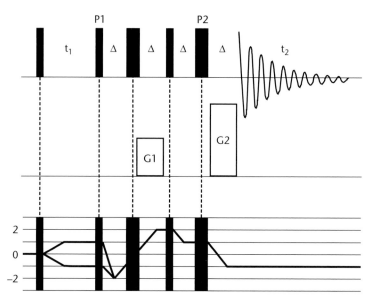

Figure 7.25 Pulse scheme and coherence selection diagram for a gradient DQF-COSY experiment with quadrature detection in t_1. Thicker vertical lines represent π pulses, thinner vertical lines are $\pi/2$ pulses. One possible gradient strength combination is G1 = 1, G2 = 2, if the gradient times are equal. The delay times Δ are all of the same duration, and are for refocussing of chemical shift evolution that occurs during the application of the gradients.

$$\gamma\left(+2\right)G_1 z t_{G1} + \gamma\left(-1\right)G_2 z t_{G2} = 0$$
$$2G_1 + \left(-1\right)G_2 = 0 \tag{7.38}$$

Two spin echos $(\pi/2 - \Delta - \pi - \Delta)$ have been inserted into the sequence shown in Figure 7.25 in order to refocus chemical shift evolution that occurs during the selection gradients. The length of Δ is determined by the duration of the gradient pulse.

It is possible to use PFG coherence selection rather than phase cycling and still obtain pure-phase multidimensional NMR spectra using rather more complicated spectral subtraction schemes (see Figure 7.26). Such schemes are commonly used in heteronuclear multidimensional NMR experiments, and will be discussed in detail in Chapter 8.

"Zero-quantum filtered COSY": NOESY and incoherent transfer

We now return to the DQF-COSY sequence (Figure 7.8). In the abbreviated product operator treatment of DQF-COSY (Expression 7.16), only the sine-modulated term generated after the first $\pi/2_x$ pulse $(2\hat{\mathbf{I}}_{1x}\hat{\mathbf{I}}_{2z})$ was of interest, since the cosine term $(\hat{\mathbf{I}}_{1y})$ would be selected against by the phase cycling. However, this term is also frequency labeled during t_1 and we can just as easily design a phase cycle or a gradient selection to select for this term and against the sine-modulated term. The second $\pi/2_x$ pulse converts the frequency-labeled $\hat{\mathbf{I}}_{1y}$ term as follows:

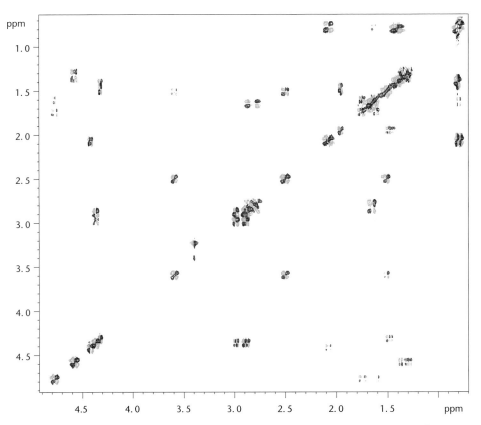

Figure 7.26 Single-scan phase-sensitive DQF-COSY spectrum of gramicidin in d_6-DMSO obtained at 11.7 T (500 MHz ^{1}H) using gradients for coherence selection.

$$\hat{\mathbf{I}}_{1y} \xrightarrow{\Omega_1 t_1 \hat{\mathbf{I}}_{1z}} \hat{\mathbf{I}}_{1y} \cos \Omega_1 t_1 \xrightarrow{\pi/2_x} -\hat{\mathbf{I}}_{1z} \cos \Omega_1 t_1 \qquad (7.39)$$

In Expression 7.39, $-\hat{\mathbf{I}}_{1z}$ represents a nonequilibrium population distribution for spin 1, and although it will not evolve during the delay between the second and third pulse under the influence of chemical shift and coupling, it *does* undergo T_1 relaxation processes that tend to return spin 1 to equilibrium. These processes include dipole–dipole interactions responsible for the NOE. The magnitude of the $-\hat{\mathbf{I}}_{1z}$ term is modulated by chemical shift and J-coupling during t_1. Since the magnitude of the NOE depends on the extent to which populations of cross-relaxing spins are perturbed from equilibrium, the $-\hat{\mathbf{I}}_{1z}$ term will give rise to variable NOEs throughout the experiment. In turn, this means the extent of cross-relaxation will be modulated by the chemical shifts of cross-relaxing spins. If we replace the $\pi/2_y$ pulse of DQF-COSY with a $\pi/2_x$ pulse after the mixing time Δ, the amplitudes of the signal of spin 2 will be modulated by the cross-relaxation with spin 1:

$$-\hat{\mathbf{I}}_{1z} \cos \Omega_1 t_1 \xrightarrow[\hat{\mathbf{I}}_{1z}, \hat{\mathbf{I}}_{2z}]{\text{cross-relaxation,}} \pm\hat{\mathbf{I}}_{2z} \cos \Omega_1 t_1 \xrightarrow{\pi/2_x} \pm\hat{\mathbf{I}}_{2y} \cos \Omega_1 t_1$$

$$\pm\hat{\mathbf{I}}_{2y} \cos \Omega_1 t_1 \xrightarrow{t_2} \pm\hat{\mathbf{I}}_{2y} \cos \Omega_1 t_1 \cos \Omega_2 t_2 \qquad (7.40)$$

The sign of the transfer depends upon whether double- or zero-quantum processes predominate, as discussed in Chapter 4. In either case, FT of the cross-term with respect to t_1 and t_2 results in peaks representing the dipolar cross-relaxation between spins. This is the NOESY experiment (*nuclear Overhauser effect spectroscopy*) (Reference 7). In order for significant NOE to be observed, the delay between the second and third pulses Δ must be considerably longer than in the DQF-COSY experiment, on the order of T_1 for the spins involved, and is usually called the mixing time, or τ_m. The phase cycling for a NOESY experiment selects for zero-order coherence during τ_m. If the mixing time is very short (as it is in DQF-COSY), the experiment would simply be a zero-quantum-filtered COSY. In fact, at short mixing times, coherent transfer (zero-quantum coherence due to J-coupling between spins) becomes a consideration in the interpretation of NOESY spectra. The NOESY pulse sequence and coherence transfer diagram is shown in Figure 7.27.

A usable phase cycle for NOESY is shown below (a phase step of 1 corresponding to $\pi/2$ rad):

Step #	P1	P2	P3	Receiver	Step #	P1	P2	P3	Receiver
1	0	0	0	0	5	2	0	0	2
2	1	1	0	0	6	3	1	0	2
3	2	2	0	0	7	0	2	0	2
4	3	3	0	0	8	1	3	0	2

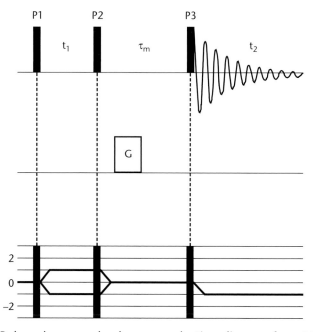

Figure 7.27 Pulse scheme and coherence selection diagram for a NOESY experiment with quadrature detection in t_1. The delay time τ_m is less than (but on the same order of magnitude as) T_1 for the spins of interest. An optional "crusher" gradient G may be included during the mixing time to destroy any transverse coherence that may be present.

Even–odd cycling of the first two pulses is used to remove axial peaks due to longitudinal relaxation during t_1.

The NOESY experiment is extremely important, particularly for the determination of molecular and macromolecular structure by NMR. Entire books have been devoted to the proper analysis of data obtained from NOE experiments (Reference 8). However, as a lead-in to the next section, we simply recall from Chapter 4 that the sign and intensity of the NOE has a strong dependence on the molecular correlation time τ_c. If $\omega\tau_c \gg 1.12$, (true for large molecules such as proteins or small molecules in viscous solvents), NOEs are negative (signal intensity is decreased by the NOE), and have a maximum value of -100%, i.e. complete loss of signal intensity. In the two-dimensional experiment, this results in decreased diagonal intensity and increased cross-peak intensity, with cross-peaks and diagonal peaks having the same sign and phase (i.e. they can both be phased pure-phase absorptive). If $\omega\tau_c \gg 1.12$ (small molecules in nonviscous solvents) NOEs are small and positive, i.e. they represent signal enhancements, and can have a maximum value of 50%. In a NOESY spectrum of a small molecule, cross-peaks and diagonal peaks are opposite in sign for this reason. As we noted earlier, there is an uncomfortable region for molecules with $\omega\tau_c \approx 1$, in which case no NOE is observed because the double- and zero-quantum processes cancel each other out. This is often the case for molecules with molecular weights in the low thousands. In order to observe NOEs for such molecules, either the correlation time τ_c must be changed (by changing temperature or solvent viscosity) or the frequency ω at which the nuclei resonate must be changed. It would seem that the only way to accomplish this is to obtain a larger (or smaller) magnet. Or is it?

Rotating frame NOEs: CAMELSPIN and ROESY

During t_1 of a two-dimensional NMR experiment, transverse coherence develops that contains both sine- and cosine-modulated terms of the form:

$$\hat{\mathbf{I}}_{1x} \xrightarrow{\ \Omega_1 t_1 \hat{\mathbf{I}}_{1z}\ } \hat{\mathbf{I}}_{1x} \cos \Omega_1 t_1 + \hat{\mathbf{I}}_{1y} \sin \Omega_1 t_1 \tag{7.41}$$

If more than one type of spin is present, the **x** and **y** components of all of these coherences are present, each with magnitudes modulated by the chemical shift and J-coupling evolution with which we are now familiar. In fact, cross-relaxation takes place between parallel components of coherences on spins that interact via dipolar coupling (i.e. they would normally give rise to NOEs with each other) during t_1. These interactions are completely analogous to cross-relaxation between the $\hat{\mathbf{I}}_{iz}$ components during the mixing time of NOESY. However, because parallel components of coherences due to different spins evolve with different chemical shifts (e.g. $\hat{\mathbf{I}}_{1x} \cos\Omega_1 t_1$ and $\hat{\mathbf{I}}_{2x} \cos\Omega_1 t_1$), they are constantly changing phase with respect to each other, so the net result of this cross-relaxation is *not* magnetization transfer but simply line broadening. For this reason, no amplitude modulation (required for cross-peaks to develop in two-dimensional NMR) results from the cross-relaxation.

However, if one could force parallel components of different coherences to stay in phase with each other, net magnetization transfer would take place, and it would be possible to obtain cross-peaks due to dipolar cross-relaxation of these transverse components. This is accomplished by preventing free precession of the coherences in the rotating frame, a result that can be obtained using a **spin-locking pulse**. A spin-locking pulse can be simply a long, relatively low power pulse $\vec{B}_1$, usually applied at the carrier frequency. This appears (like any pulse at the carrier frequency) as a vector fixed along one transverse axis of the rotating frame. During the application of the pulse, transverse coherences are forced to precess around the effective field $\vec{B}_{eff}$ that results from the vector addition of $\vec{B}_1$ and the offset due to chemical shift, $\Delta \vec{B}$ (see Problem 3.11 and Figure 7.28). As with any precession, the component of the coherence parallel to the field is not affected, but the component perpendicular to the field will precess at a frequency γB_{eff}. If the spin-lock pulse is applied for a long time relative to the precession frequency, inhomogeneities in $\vec{B}_1$ will dephase the perpendicular components of the spin-locked coherences so that they no longer contribute to signal after the spin-locking pulse is removed.

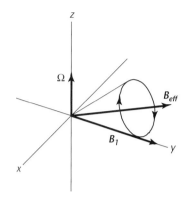

Figure 7.28 Effective field $\vec{B}_{eff}$ generated by a spin-locking pulse. The vector addition of the chemical shift offset Ω and the applied RF field $\vec{B}_1$ generates $\vec{B}_{eff}$. The conic represents the path of precession of spin-locked coherences around $\vec{B}_{eff}$.

The net effect of the spin-locking field is to prevent coherences from evolving in the **x, y** plane as they normally would. As soon as a coherence moves away from the axis along which $\vec{B}_1$ lies, it will be forced to precess around $\vec{B}_{eff}$ and will end up back on the opposite side of $\vec{B}_1$. The net result is no free precession; the coherences are "locked" along one transverse axis by $\vec{B}_{eff}$.

A number of interesting phenomena occur during spin-locking. The most relevant at the moment is that dipolar cross-relaxation takes place between the parallel components of the coherences of dipolar-coupled spins. Because the relative phases of the coherences are held constant by the lack of free precession, mutual amplitude modulation between dipolar-coupled spins occurs. After the spin-locking field is removed, the resonances evolve as usual, but their amplitudes have been modulated by the cross-relaxation that occurred during the spin-lock. After transformation, peak integrations will reflect that modulation, i.e. they exhibit NOE. The first observation of such **transverse NOEs** obtained by spin-locking was by Bothner-By *et al.* (Reference 9), who called the experiment CAMELSPIN, after the figure-skating routine that the locked spins appear to emulate (Figure 7.29).

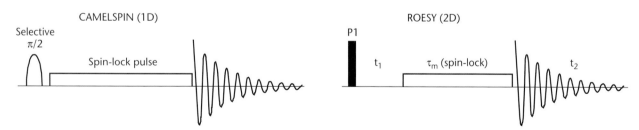

Figure 7.29 Rotating-frame NOE experiments. The original one-dimensional experiment, CAMELSPIN, involves a selective pulse at the resonance with which NOEs are desired, followed by spin-locking. Rotating frame NOEs (ROEs) develop during the spin-lock on spins coupled by dipolar interactions to the selectively excited spin. The two-dimensional version, called ROESY, replaces the selective pulse with a nonselective $\pi/2$ pulse followed by the frequency-labeling period t_1, which in turn is followed by the mixing time τ_m, during which the spin-lock pulse is applied. After the spin-lock is complete, the spins are allowed to precess freely for detection during t_2.

If coherences are frequency-labeled during a t_1 period prior to spin-locking, the amplitude modulation will result in cross-peaks in a two-dimensional experiment. These cross-peaks reflect NOE-type cross-relaxation transfers, but ones that take place in the transverse dimensions (**x** and **y**) rather than along the **z** axis. For this reason, the two-dimensional experiment in which a t_1 period is followed by a spin-lock mixing period is called ROESY (*rotating frame nuclear Overhauser effect spectroscopy*) (Reference 10; see Figure 7.29). A point of particular importance is that the spectral density functions that govern relaxation in the rotating frame of reference are quite different from those that determine relaxation for population operators. Since $\vec{B}_{eff}$ is invariably much weaker than the static magnetic field, transition frequencies (determined by $\gamma\vec{B}_{eff}$) are much lower, and the spectral densities at all of the relevant transition frequencies (W_0, W_2 or W_1) will be equally well represented. This means that the extreme narrowing condition applies to molecules of all sizes and virtually any correlation time in liquids. Thus, rotating frame NOEs can be observed over a wide range of molecular sizes, even for molecules that are in the uncomfortable region for longitudinal NOE, where $\omega\tau_c \approx 1$. However, there are complications that sometimes make ROESY a difficult experiment to interpret.

The most important complication in ROESY is the result of **coherent energy transfer** (or **polarization transfer**) between J-coupled spins. In all of the homonuclear correlation experiments discussed so far (except for NOESY and ROESY), the transfer of coherence can be thought of as a transfer of information between coupled spins. Loss of that information means a loss of coherence (the result of entropic, or T_2, processes). The transfer of information in COSY, MQ and related experiments does not involve a change in spin populations for coupled spins, but rather, a sorting of spins according to the states of their coupling partners, leading to a decrease in entropy. On the other hand, both NOESY and ROESY result in **incoherent polarization transfer**, i.e. changes in the relative populations of the dipolar-coupled spins. Such transfers are affected by T_1 processes (or the transverse equivalent, known as $T_{1\rho}$, that will be discussed later) and are enthalpic in nature. However, coherent energy transfer can occur between two coupled oscillators that have the same natural frequency. When one of the oscillators begins to vibrate, in time the vibration will be transferred from one oscillator to the other. Figure 7.30 provides a conceptual example of this involving two pendula. Two coupled spins that fulfill the following expression:

$$\gamma_1 B_1 = \gamma_2 B_2 \tag{7.42}$$

are said to fulfill the **Hartmann–Hahn condition**, and coherent energy transfer can take place between them. In the simplest (homonuclear) case, $\gamma_1 = \gamma_2$, and B is B_{eff}. Therefore, two coupled spins that detect the same B_{eff} during a spin-lock (such as two coupled spins resonating at $\pm\Delta\nu$ from the carrier) can transfer spin polarization via coherent pathways as well as incoherent (dipolar) ones. This can give rise to cross-peaks in ROESY spectra that are due to J-coupling rather than through-space interactions (Figure 7.31). For phase-sensitive ROESY experiments, the sign of a coherence transfer peak is opposite to that of a ROESY peak, and the two effects can be distinguished. However, coherent transfer from spin 1 to spin 2 via J-coupling can be relayed via dipolar interactions to a spin 3. This gives rise to a ROESY-type

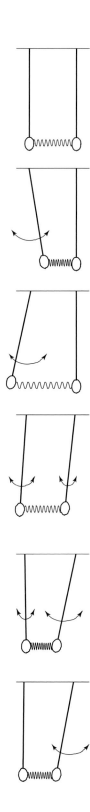

Figure 7.30 Transfer of energy between weakly coupled oscillators with the same natural frequency. Two pendula that have the same mass and length (and so oscillate at the same frequency) are connected by a loose spring, which provides weak coupling between the pendula. After the first pendulum is pushed (top), it begins to oscillate. With time, some of the energy is transferred to the second pendulum (middle). Still later it is completely transferred to the second pendulum (bottom). The process reverses (bottom to top) and the energy returns to the first pendulum.

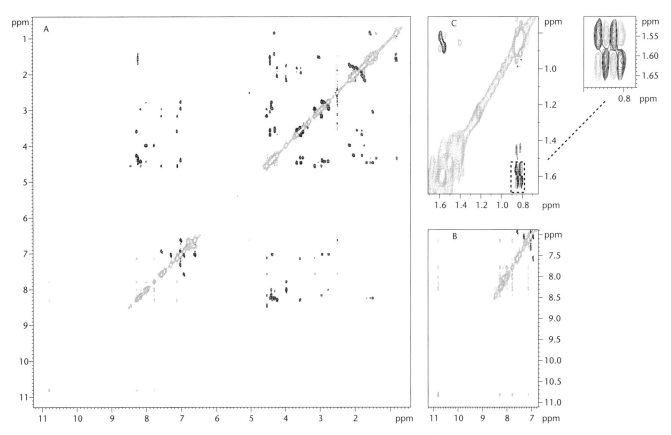

Figure 7.31 Different types of transfer in ROESY experiments. Shown is a ^{1}H ROESY spectrum of LHRH (luteinizing hormone-releasing hormone), a peptide, 1 mM in d$_6$-DMSO. (A) Black peaks (positive, opposite in sign to diagonal peaks) are true NOE cross-peaks, indicating spatial proximity between connected protons. The series of cross-peaks in gray (negative) are due to chemical exchange. In particular, see cross-peaks between the tryptophan indole NH at 10.8 ppm and other amide protons, expanded in (B). (C) Expansion of the aliphatic region showing COSY-type cross-peaks between the methyl groups of a leucine residue and the directly coupled methine (CH) proton.

cross-peak between spin 1 and spin 3, regardless of whether there is any direct dipolar interaction between them. Furthermore, this "false" ROESY peak will be of the same sign as the "true" ROESY cross-peaks. Such relayed ROESY peaks are quite sensitive to the position of the carrier frequency, since it must be centered between the two peaks undergoing J-mediated coherence transfer, whereas true ROESY transfers are less sensitive to the power or placement of the spin-locking frequency. Generally, ROESY experiments are performed several times on the same sample with different carrier frequencies and/or B_1 power levels. This is usually sufficient to distinguish "true" and "false" ROESY peaks.

Spin-locking experiments for coherence transfer: TOCSY and composite pulse decoupling

The coherent transfer of spin polarization between J-coupled spins experiencing the same $\vec{B}_{eff}$ is a complication (or nuisance, if you prefer) in ROESY. However, this phenomenon is the basis for what is perhaps the single most useful homonuclear *J*-correlation experiment, TOCSY (*to*tally *c*orrelated *s*pectroscop*y*), also known as HOHAHA (*ho*monuclear *Ha*rtmann–*Ha*hn), depending upon how the mixing is accomplished (Figure 7.32) (References 11 and 12). In the TOCSY experiment, even spins that are not directly coupled to each other but are part of the same coupling network can transfer polarization to each other via stepwise transfers from one spin to the next (Figure 7.33 and 7.34). This is practically useful for proteins, in that ^{1}H spins on each amino acid represent separate spin systems. A TOCSY spectrum correlates the proton resonances of a single amino acid residue less ambiguously than a DQF-COSY experiment.

The requirement for efficient polarization transfer between two spins *a* and *b* in a spin-locking experiment is:

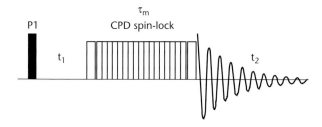

Figure 7.32 TOCSY pulse sequence.

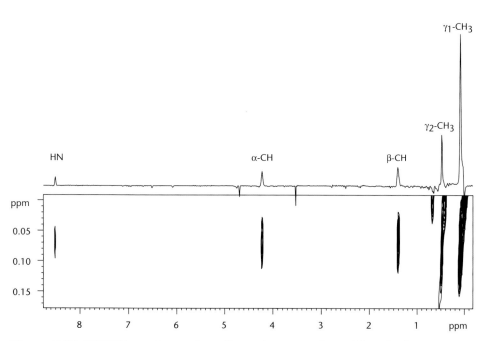

Figure 7.33 TOCSY spectrum of a valine spin system in putidaredoxin, a ferredoxin, in 90% H_2O/D_2O obtained at 500 MHz. The one-dimension slice at the γ_1-CH$_3$ frequency is shown above the two-dimensional spectrum. All members of the spin system are visible.

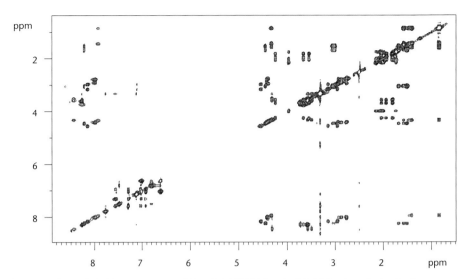

Figure 7.34 TOCSY spectrum of 1 mM LHRH in d_6-DMSO. A spin-lock time (mixing time) of 100 ms was used.

$$\frac{\left(\Delta B_a\right)^2 - \left(\Delta B_b\right)^2}{2B_1} << J_{ab} \qquad (7.43)$$

with all fields and offsets reported in frequency units. It is standard to report the strength of spin-locking fields in hertz. This is simply the number of times the spin-locked spins will precess around $\vec{B}_1$ per second, and is the inverse of the length of a 2π pulse at that RF field strength. For proton spin-locking experiments, with J-couplings between 2 and 10 Hz, a spin-locking field of 7500 Hz in an 11.74 T static field is usually sufficient to generate most of the desired coherence transfers.

In the ROESY experiment, the spin-locking field is generally a low power CW (continuous wave) pulse that is **phase coherent** with the high power pulse applied at the beginning of the sequence. Phase coherence means that the relative phases of the two pulses are defined. This is necessary so that spin-locking occurs with the same relative $\vec{B}_{eff}$ at each increment of the experiment. Phase coherence between pulses is best accomplished using the same frequency generator and amplifier to generate all pulses. Usually, a phase shift takes place between pulses with different power levels (i.e. different values of $\vec{B}_1$) even if the same carrier frequency is used to generate both pulses. However, these shifts can be calibrated or compensated in the phase adjustments of the two-dimensional spectrum after transformation.

In TOCSY, Equation 7.43 requires that in order to place spins over a large chemical shift range into the Hartmann–Hahn condition, $\vec{B}_1$ must be relatively large. This makes the process of spin-locking technically challenging. If one attempts to apply a CW pulse to spin-lock for TOCSY, the further away from the carrier frequency one gets, the more acute the angle of $\vec{B}_{eff}$ with $\vec{B}_1$, and the requirement for coherent transfer in TOCSY (that spins be experiencing essentially the same $\vec{B}_{eff}$) is lost.

What is needed is a way to ensure that all of the spins of interest experience the same $\bar{B}_{eff}$, or **broadband excitation** over the period of the spin-lock. This is accomplished via a series of phase-cycled **hard pulses** (i.e. pulses short enough that there is little variation in excitation over the bandwidth of interest) known as a **composite pulse decoupling** (**CPD**) sequence applied to the spins to be spin-locked. A CPD sequence puts the spins through a series of "spin gymnastics" that result in all spins of interest experiencing a net $\bar{B}_{eff}$ of zero during the sequence. Under such conditions, Equation 7.42 is true and the Hartmann–Hahn condition is obtained. CPD sequences are designed to provide the most efficient use of transmitter power to obtain the desired effect.

Another situation where excitation must be applied over a wide frequency range for a long time is **broadband decoupling**. Broadband decoupling is a requirement of many heteronuclear experiments, as we will see in Chapter 8. For example, when observing a ^{13}C spectrum, broadband decoupling is applied to ^{1}H in order to collapse multiplets and to give NOE at ^{13}C. This results in a dynamic averaging of ^{1}H spin states so the ^{13}C spins observe their attached protons half of the time in the α state, half in the β state, so there is no splitting. Furthermore, the population difference between the proton states approaches zero, since the states are being rapidly interchanged.

A variety of CPD sequences have been designed over the years, and many have been used for spin-locking in the TOCSY sequence. One of the most commonly used CPD sequences is MLEV-17. The acronym comes both from the multiple levels of RF power applied during the train, and the name of the inventor of the sequence (Reference 13). The MLEV-17 sequence is *ABBA–BBAA–BAAB–AABB*, where $A = \pi/2_y$, π_x, $\pi/2_y$ and $B = \pi/2_{-y}$, π_{-x}, $\pi/2_{-y}$. In recent years, computer-assisted design of CPD sequences has resulted in a variety of schemes that optimize the use of RF power so the most effective decoupling is obtained for the least amount of RF power. For the TOCSY experiment, one of the most efficient CPD schemes employed for spin-locking is the DIPSI sequence (*d*ecoupling *i*n the *p*resence of *s*calar *i*nteractions) (Reference 14). As with many optimized CPD sequences, DIPSI employs a variety of pulse lengths, phases and repetition patterns in order to accomplish the goal of the sequence, i.e. to have the spins experience an average $\bar{B}_{eff}$ of zero, the time-average environment of the spins is isotropic. For this reason, the TOCSY spin-lock is often called **isotropic mixing**. Since the Larmor frequencies of all the spins in an isotropic environment will be zero, then polarization transfer can occur according to Equation 7.42.

The efficiency of TOCSY coherence transfer is determined by the magnitude of the coupling between spins; the larger the coupling, the more efficient the transfer. For an isolated two-spin system, transfer from spin 1 to spin 2 (and vice versa) builds up as $\sin\pi Jt$ so that complete transfer is seen at mixing time $t = 2J$. In the presence of multiple coupled spins, it is usual to perform the experiment at multiple mixing times. At shorter mixing times, correlations with directly coupled partners are emphasized, while at longer mixing times, relayed correlations begin to dominate (see Figure 7.35). If still longer mixing times are possible (i.e. relaxation is sufficiently slow), the coherence returns via relayed transfer back to the source spin. A useful analogy is water in a bathtub. A child sliding in the tub causes waves that start

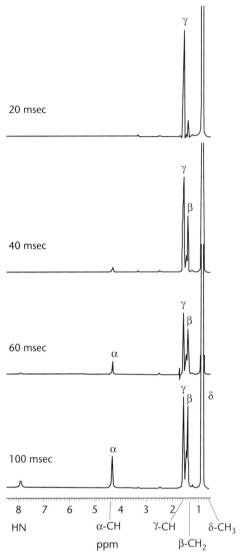

Figure 7.35 Corresponding one-dimensional slices from the TOCSY spectra of 1 mM LHRH in d_6-DMSO taken using different spin-lock mixing times. The slices are taken at the δ-CH$_3$ frequency of the leucine spin system.

at one end and move to the other end (energy transfer from one end of the tub to the other). Once the waves hit one end of the tub, they begin to move in the other direction, returning the energy to the source.

Another feature of TOCSY is that, because energy transfer occurs between coupled coherences with the same phase characteristics during spin-locking, the cross-peaks that result from the transfer have no antiphase characteristics, as in COSY or DQF-COSY, and so can be phased pure-phase absorptive in both dimensions (Figures 7.33–7.35). This removes the possibility of coincidental cancellation of overlapping cross-peaks in crowded regions of the spectrum, and makes TOCSY particularly adaptable for three-dimensional NMR experiments, as we will see in Chapter 9. Problem 8 deals in more detail with the transfer of coherence in the TOCSY experiment.

Problems

*7.1   What are the missing terms in Equation 7.11 dealing with the evolution of operators in the COSY experiment?

7.2   Show that Equation 7.14 dealing with the relative phases of cross-peaks in DQF-COSY is correct.

*7.3   Show that the receiver phase cycling shown in Equation 7.22 will suffice for any of the four possible coherence transfer pathways in the DQF-COSY experiment.

7.4   Including even–odd selection to remove axial peaks, develop a 12-step cycle (increment of one step = $\pi/3$ rad) for a triple-quantum filtered COSY experiment shown below.

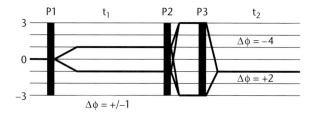

*7.5   Assuming that the following phase cycle selects for double quantum coherence, fill in the blanks on the phase cycle for the pulse sequence shown in Figure 7.17:

Step #	P1,3	P2	P4	Receiver
1	0	0		0
2	1	1		2
3	2	2		0
4	3	3		2
5	0		0	0
6	1		0	2
7	2		0	0
8	3		0	2
9		1	1	
10		2	1	
11		3	1	
12		0	1	
13	1	3	1	3
14	2	0	1	1
15	3	1	1	3
16	0	2	1	1

*7.6   Show that the effect of a spin echo is to remove evolution due to chemical shift at the end of the echo time by converting $\hat{\mathbf{I}}_x$ to single element operators.

*7.7   Two-dimensional NMR experiments are often classified as "sine-modulated" or "cosine-modulated" depending upon the nature of the interferogram produced in t_1. Based upon the product operators responsible for transfer between spins, what categories do you expect COSY, DQF-COSY and NOESY to fall into, respectively? At what value of t_1 do you expect to see the maximum signal in a sine-modulated experiment? What about a cosine-modulated experiment?

7.8   TOCSY is normally a cosine-modulated experiment. Evolution during t_1 following a $\pi/2_x$ pulse and prior to the spin-locking period results in the following terms for the evolution of coherence on spin 1 coupled to a second spin 2 by J_{12} (ignoring multiple-quantum evolution):

$$-\hat{\mathbf{I}}_{1y} \cos\Omega_1 t_1 \cos\pi J_{12} t_1 + 2\hat{\mathbf{I}}_{1x}\hat{\mathbf{I}}_{2z} \cos\Omega_1 t_1 \sin\pi J_{12} t_1$$
$$+\hat{\mathbf{I}}_{1x} \sin\Omega_1 t_1 \cos\pi J_{12} t + 2\hat{\mathbf{I}}_{1y}\hat{\mathbf{I}}_{2z} \sin\Omega_1 t_1 \sin\pi J_{12} t_1$$

Similar terms would also arise from coherences on spin 2. The two-spin terms (containing $2\hat{\mathbf{I}}_{1x}\hat{\mathbf{I}}_{2z}$ and $2\hat{\mathbf{I}}_{1y}\hat{\mathbf{I}}_{2z}$) would normally give rise to the antiphase multiplet structure associated with COSY cross-peaks, and during the spin-locking period these tend to destructively interfere with each other and can be ignored. This leaves:

$$-\hat{\mathbf{I}}_{1y} \cos\Omega_1 t \cos\pi J_{12} t_1 + \hat{\mathbf{I}}_{1x} \sin\Omega_1 t \cos\pi J_{12} t_1$$

During the spin-locking period τ_m, spin energy is transferred from spin 1 to spin 2 (and vice versa, using mirror-image operators), so that the terms shown above evolve into four terms:

$$-\hat{\mathbf{I}}_{1y} \tfrac{1}{4}\Big[\cos\big(\Omega_1 + \pi J_{12}\big)t_1 + \cos\big(\Omega_1 - \pi J_{12}\big)t_1\Big]\Big[1 + \cos\big(2\pi J_{12}\tau_m\big)\Big]$$
$$-\hat{\mathbf{I}}_{2y} \tfrac{1}{4}\Big[\cos\big(\Omega_1 + \pi J_{12}\big)t_1 + \cos\big(\Omega_1 - \pi J_{12}\big)t_1\Big]\Big[1 - \cos\big(2\pi J_{12}\tau_m\big)\Big]$$
$$+\hat{\mathbf{I}}_{1x} \tfrac{1}{4}\Big[\sin\big(\Omega_1 + \pi J_{12}\big)t_1 + \sin\big(\Omega_1 - \pi J_{12}\big)t_1\Big]\Big[1 + \cos\big(2\pi J_{12}\tau_m\big)\Big]$$
$$+\hat{\mathbf{I}}_{2x} \tfrac{1}{4}\Big[\sin\big(\Omega_1 + \pi J_{12}\big)t_1 + \sin\big(\Omega_1 - \pi J_{12}\big)t_1\Big]\Big[1 - \cos\big(2\pi J_{12}\tau_m\big)\Big] \quad \text{(P7.1)}$$

The second bracketed term in each line above represents the mixing that occurs during τ_m. The first and third lines represent magnetization that evolves on spin 1 during t_1 and remains there during the spin-lock period, giving rise to a diagonal peak. The second and fourth lines represent magnetization that evolves on spin 1 during t_1 but is transferred to spin 2 during τ_m, and would give rise to a cross-peak between spin 1 and spin 2 after FT. The terms in lines 1 and 2, corresponding to $-\hat{\mathbf{I}}_y$ coherence, are modulated differently than those in lines 3 and 4 ($\hat{\mathbf{I}}_x$). In order to obtain pure-phase TOCSY spectra, phase cycling is typically used to select for only the terms in lines 1 and 2. However, this leads to half of the signal being wasted (or a loss of $\sqrt{2}$ in S/N in the final transformed spectrum). Cavanaugh and Rance [*J. Magn. Reson.* 88, 72–85 (1990)], describe a trick whereby this wasted signal can be mostly recovered and still provide a pure-phase spectrum.

(a) Application of a π_y pulse at the end of τ_m would convert the terms in P7.1 to four new terms. What are those terms?

(b) Addition of the FIDs resulting after the π_y pulse to those resulting from P7.1 would cancel the $\hat{\mathbf{I}}_x$ components, while subtraction of the two would cancel only the $\hat{\mathbf{I}}_y$. Show that this is true.

(c) The resulting sum and difference FIDs would have the same modulation, and differ only by a 90° phase shift. What should be done to directly combine the two spectra to generate one new spectrum with improved S/N? (In fact, this "trick" has a good deal of similarity to the States method of obtaining pure-phase two-dimensional spectra discussed in this chapter. In Chapter 8, we will see that a variation on this theme is used to obtain pure phase spectra in combination with gradient coherence selection.)

*7.9   Assuming two coupled protons, an NH proton at 9.1 ppm and a CH proton at 4.3 ppm, what is the amplitude of the spin-locking field (in hertz) needed to simultaneously spin-lock these two spins at 11.74 T (500 Mz ^{1}H), 14 T (600 MHz ^{1}H) and 21 T (900 MHz)?

7.10   The phase-cycled homonuclear double-quantum experiment of Rance *et al.* (Reference 6) has the sequence and phase-cycling as shown below (with 1 = $\pi/2$ rad).

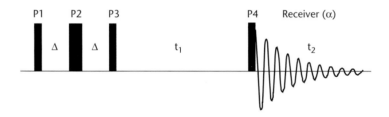

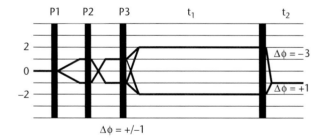

Step #	P1,3	P2	P4	Receiver
1	0	0	0	0
2	1	1	0	2
3	2	2	0	0
4	3	3	0	2
5	1	1	1	1
6	2	2	1	3
7	3	3	1	1
8	0	0	1	3

Describe all of the product operator evolutions that occur during t_1 and $\Delta = 1/4J$, and explicitly show that the phase cycling selects for coherences that evolve as 2Q coherence during t_1 and against those that do not.

*7.11   The phase cycling for DQF-COSY:

Step #	P1	P2	P3	Receiver
1	0	0	0	0
2	1	1	0	2
3	2	2	0	0
4	3	3	0	2

is equivalent to:

Step #	P1	P2	P3	Receiver
1	0	0	0	0
2	0	0	3	1
3	0	0	2	2
4	0	0	1	3

Why?

*7.12   What is the evolution of $\hat{I}_x$, $\hat{I}_y$ and $\hat{I}_z$ as a result of the sequence $\tau/2–\pi_x–\tau/2$ where $\tau/2 = 1/2J_{IS}$, and spin I interacts with a spin S with a scalar coupling of J_{IS}?

7.13   With the appropriate phase cycle, the NOESY sequence shown in Figure 7.27 selects for magnetization that is on the $-\mathbf{z}$ axis during t_{mix}. If a π pulse is applied in the middle of the mixing time for refocusing of chemical shift, and the phase cycling is modified to select for transverse rather than longitudinal terms, an experiment called **relayed coherence transfer (RCT)** is generated (see below). This experiment results in cross-peaks resembling COSY cross-peaks (antiphase absorptive) between spins that share a common coupling partner, but are not directly coupled. Directly coupled spins give rise to mixed phase multiplets.

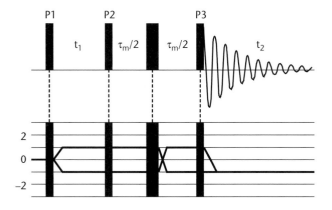

(a) Suggest a phase cycle for RCT based on the NOESY phase cycle discussed in the text (Figure 7.27).

(b) Describe the product operators that give rise to the RCT peaks, and show that the phase cycle will select for those terms.

(c) Show that the phase cycle selects against NOESY-type terms ($-\hat{\mathbf{I}}_z$).

(d) Why is the π pulse needed in the middle of τ_m for RCT, but not for NOESY?

*7.14 Show that the J-coupling term of the two-spin Hamiltonian $J_{12}\hat{\mathbf{I}}_1\hat{\mathbf{I}}_2$ expands to $J_{12}\hat{\mathbf{I}}_{1z}\hat{\mathbf{I}}_{2z} + J_{12}/2\hat{\mathbf{I}}_{1+}\hat{\mathbf{I}}_{2-} + \hat{\mathbf{I}}_{1-}\hat{\mathbf{I}}_{2+}$.

*7.15 The priniciple of phase incrementation used to shift the apparent frequency of a sampled signal in TPPI can be used to create new frequencies within NMR experiments. It is often necessary to apply a pulse in regions of a spectrum away from the carrier frequency in the course of an experiment. One way to do this without actually changing the spectrometer RF frequency (something that takes some time within the hardware to do) is to create phase-shifted pulses within the pulse sequence that result in a frequency that is higher (or lower) than the carrier frequency by a certain amount. Consider a ^{13}C carrier frequency of 100.00005 MHz. If an off-resonance pulse of 512 discrete increments of 2 μs each is desired to occur at a frequency 10 000 Hz (100 ppm) away from the carrier frequency, what phase shift would be required on each 2 μs increment in order to achieve that frequency shift? Assuming that this was a square pulse (all increments the same amplitude) what would the distance be in hertz from the center of the pulse to the first null?

References

1. J. Jeener, *Ampere International Summer School*, Basko Poljie, Yugoslavia (1971).
2. W. P. Aue, E. Bartholdi, and R. R. Ernst, *J. Chem. Phys.* 64, 2229 (1976).
3. D. J. States, R. A. Haberkorn, and D. J. Ruben, *J. Magn. Reson.* 48, 286 (1982).
4. U. Piantini, O. W. Sorensen, and R. R. Ernst, *J. Am. Chem. Soc.* 104, 6800 (1982).
5. N. Muller, R. R. Ernst, and K. Wuthrich, *J. Am. Chem. Soc.* 108, 6482 (1986).
6. M. Rance, W. J. Chazin, C. Dalvit, and P. E. Wright, *Meth. Enzymol.* 176, 114 (1989).
7. J. Jeener, B. H. Meier, P. Bachmann, and R. R. Ernst, *J. Chem. Phys.* 71, 4546 (1979).
8. D. Neuhaus and M. Williamson, *The Nuclear Overhauser Effect in Structural and Conformational Analysis*. VCH Publishers, New York (1989).
9. A. A. Bothner-By, R. L. Stephens, and J. Lee, *J. Am. Chem. Soc.* 106, 811 (1984).
10. A. Bax and D. G. Davis, *J. Magn. Reson.* 63, 207 (1985).
11. L. Braunschweiler and R. R. Ernst, *J. Magn. Reson.* 53, 521 (1983).
12. A. Bax and D. G. Davis, *J. Magn. Reson.* 65, 355 (1986).
13. M. H. Levitt, R. Freeman, and T. Frenkiel, *J. Magn. Reson.* 47, 328 (1982).
14. A. J. Shaka, C. J. Lee, and A. Pines, *J. Magn. Reson.* 77, 274 (1988).

8

HETERONUCLEAR
CORRELATIONS IN NMR

Virtually all of the two-dimensional NMR experiments described in Chapter 7 for detection of ^1H–^1H correlations can be used with little modification for other $\mathbf{I}$ = 1/2 nuclei. However, ^1H is the only $\mathbf{I}$ = 1/2 nucleus with high natural abundance that is often found in coupled networks, so application of such experiments to nuclei other than ^1H is relatively rare unless the molecules to be investigated are isotopically enriched. One exception is the application of the double-quantum correlation experiment to natural abundance ^{13}C in organic molecules, an experiment called INADEQUATE (the *i*ncredible *n*atural *a*bundance *d*ouble-*qua*ntum *t*ransfer experiment) (Reference 1). The INADEQUATE sequence (Figure 8.1) takes advantage of the large one-bond coupling ($^1J_{CC}$ ~ 40 Hz for sp^3 carbons) between covalently bonded ^{13}C atoms to generate and detect double-quantum coherence between the coupled ^{13}C spins (Figure 8.2). Considering that ^{13}C is only 1.1% natural abundance, the likelihood of finding two ^{13}C atoms adjacent to each other in a covalent bond is 1.1% of 1.1% or 0.0121%. As such, the sensitivity of the INADEQUATE experiment is low, and concentrated samples are required.

The low sensitivity of INADEQUATE can also be attributed to the lower gyromagnetic ratio of the ^{13}C relative to ^1H. At a given field strength, the sensitivity of a nucleus is proportional to the cube of the gyromagnetic ratio. If ^1H is assigned a sensitivity factor of 1, then ^{13}C, the next most commonly observed nucleus, has a sensitivity factor of 1.6×10^{-2}. The situation is even more dismal for ^{15}N (sensitivity factor of 1.04×10^{-3}). As such, even isotopic enrichment does not always improve the situation to the point where standard multidimensional NMR experiments are

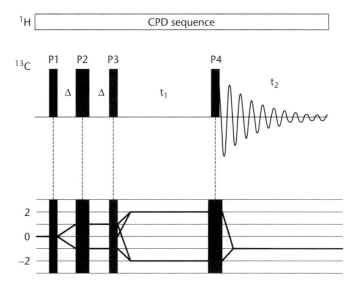

Figure 8.1 INADEQUATE experiment for detecting double-quantum coherence between adjacent ^{13}C nuclei. ^{1}H decoupling is applied throughout the sequence in order to remove coupling to protons. The phase cycling for INADEQUATE is discussed in detail in Problem 8.1.

practical for spins other than ^{1}H. How then can the situation be improved? One answer lies in a trick called **polarization transfer**. If the polarization (i.e. equilibrium population difference across the spin transition) of ^{1}H, the most sensitive common nucleus, is transferred to a heteronucleus one wishes to observe, this improves the S/N of the experiment significantly. This type of transfer is the basis for a number of NMR experiments. These experiments are also used as building blocks for multidimensional NMR experiments involving protons and other nuclei.

Heteronuclear polarization transfer and the INEPT experiment

Consider a single ^{1}H spin *J*-coupled to a single ^{13}C spin. The stationary states for that spin pair are shown in Figure 8.3. The population differences across the two W_{1S} transitions (S = sensitive nucleus, i.e. ^{1}H) are given by $2\delta_S$. The population differences across the S transitions are larger than the population differences across the two W_{1I} transitions (given by $2\delta_I$, where I is the insensitive nucleus, i.e. ^{13}C) by a factor of γ_S/γ_I. The reader should note that the reverse convention, $I = {}^1$H and S = insensitive nucleus, is also widely used.

Now, if one applies a selective π pulse at the frequency of one of the two ^{1}H transitions (we have chosen the lower-frequency transition), the populations across one of the two W_{1S} transitions is inverted and the differences between the bottom and top of the two ^{13}C transitions is as shown in Figure 8.4.

Examination of Figure 8.4 shows that the population difference across the top ^{13}C W_{1I} transition is now $2(\delta_S + \delta_I)$, with an inverted population difference $(-2(\delta_S + \delta_I))$ across the bottom two states. As signal intensity is proportional to the population

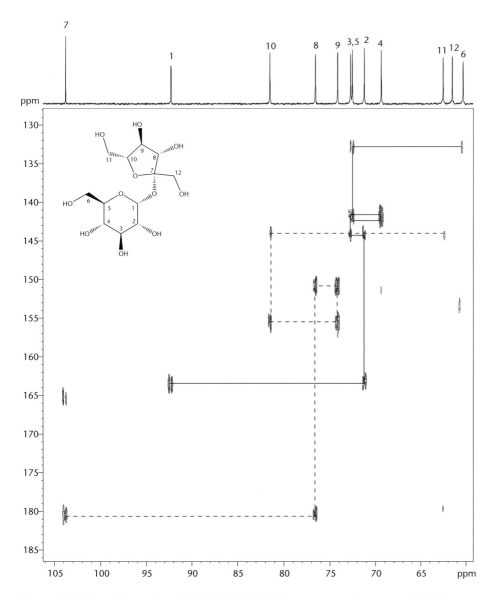

Figure 8.2 Two-dimensional ^{13}C INADEQUATE spectrum of sucrose in D$_2$O, acquired at 150.79 MHz. Proton-decoupled ^{13}C spectrum of sucrose is shown along the top of the two-dimensional spectrum, with the numbers indicating each resonance's assignment. The dashed lines indicate the sequential walk between atoms of the five-membered ring (7→8→9→10→11→12), while the solid lines indicate the sequential connectivites in the six-membered ring (1→2→3→4→5→6).

difference across a transition, the signals resulting from absorption across the two ^{13}C transitions will now be enhanced relative to the equilibrium experiment by a factor of $(1 \pm \gamma^1H/\gamma^{13}C)$, depending on the relative signs of δ_S or δ_I. If the ^{13}C signal is observed at this point, the signal intensity will be proportional to the population differences across the W_{1I} transitions. One would see inversion of one ^{13}C line relative to the other; compared with an intensity of 1 corresponding to $2\delta_I$, one line

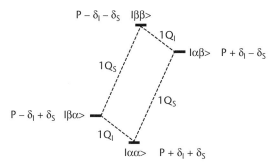

Figure 8.3 Heteronuclear *J*-coupled two-spin system, with the single-quantum transitions (1Q) for spin *I* (insensitive) and spin *S* (sensitive) indicated. The individual transitions have different energies and therefore can be excited selectively. Populations of the individual states at equilibrium are shown.

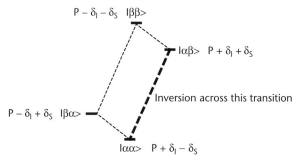

Figure 8.4 Same heteronuclear two-spin system as in Figure 8.3. Populations across the lower 1Q *S* transitions have been inverted by a selective π pulse, giving new populations as shown.

would have an intensity of 5 (higher than the ^{1}H intensity of 4) and the other an intensity of 3 (lower than ^{1}H intensity of 4). This is illustrated in Figure 8.5. What is accomplished by this simple experiment is a sorting of the carbon spins according to the spin state of the attached proton (spin up, α, or spin down, β). The net result is to enhance the signal of the less sensitive nucleus (^{13}C), by transferring to it the polarization of the more sensitive nucleus (^{1}H).

Now it would be useful if this polarization transfer could be achieved simultaneously for all C–H pairs in the spectrum regardless of chemical shift. This is accomplished, as one might guess, by incorporating a spin–echo sequence into the experiment. The pulse sequence for this experiment, called INEPT (*i*nsensitive *n*ucleus *e*nhancement by *p*olarization *t*ransfer), is shown in Figure 8.6 (Reference 2).

Product operator evolution of the interesting coherences in the INEPT experiment goes as:

$$\hat{\mathbf{S}}_z \xrightarrow{\pi/2_x(S)} -\hat{\mathbf{S}}_y \xrightarrow{1/4J_{SI} - \pi(S,I) - 1/4J_{SI}} -2\hat{\mathbf{S}}_x\hat{\mathbf{I}}_z \xrightarrow{\pi/2_y(S,I)} 2\hat{\mathbf{S}}_z\hat{\mathbf{I}}_x \quad (8.1)$$

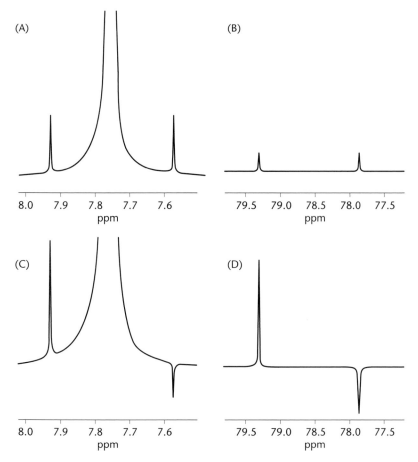

Figure 8.5 Selective inversion experiment demonstrating polarization transfer, using CHCl₃. (A) ¹H spectrum of CHCl₃ with the vertical scale expanded so the ¹³C satellites are visible. (B) Proton-coupled ¹³C spectrum. (C) ¹H spectrum after a selective π pulse has been applied to one of the satellites. (D) ¹³C spectrum after the selective ¹H pulse has been applied in (C).

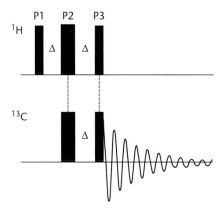

Figure 8.6 The INEPT experiment for transferring polarization from a sensitive nucleus (¹H) to an insensitive nucleus (¹³C). P1 = $\pi/2_x$ (¹H), P2 = simultaneous π_x (¹H), π_x (¹³C), P3 = simultaneous $\pi/2_y$, (¹H), $\pi/2_y$ (¹³C), $\Delta = 1/4J_{CH}$.

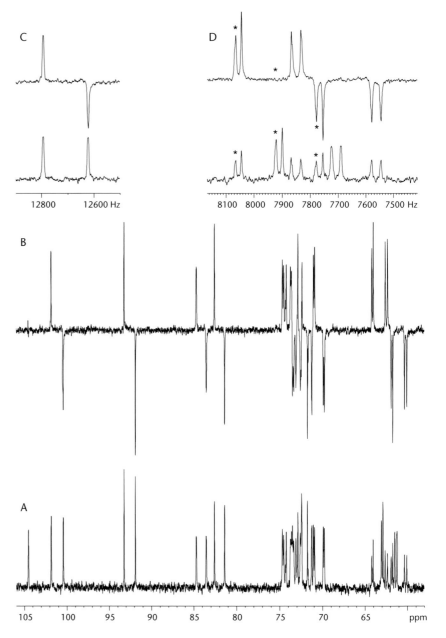

Figure 8.7 ^{13}C spectra of melezitose in D$_2$O acquired at 125.6 MHz. (A) Proton-coupled ^{13}C spectrum. (B) INEPT spectrum run under the same conditions. (C) Expansion of a region of (A) and (B) showing the doublet observed for a methine carbon. (D) Expansion showing the triplet (1:0:−1) observed for a methylene carbon. Asterisks indicate the corresponding triplet in the ^{13}C and INEPT spectra.

where $\hat{\mathbf{S}} = {}^1$H and $\hat{\mathbf{I}} = {}^{13}$C. Only the evolution due to J-coupling is considered, since the spin echo removes chemical shift evolution. Note also that the π_x pulse is not shown in the operator evolution in Expression 8.1, since it does not affect the coupling. The delay time ($2\Delta = 1/2J$) ensures that the evolution of $-\hat{\mathbf{S}}_y \rightarrow -2\hat{\mathbf{S}}_x\hat{\mathbf{I}}_z$ is complete prior to the second $\pi/2$ pulse in order to maximize the polarization transfer.

The observed ^{13}C spectrum would consist of a series of antiphase multiplets for methine (CH) carbons, each of which is enhanced by the ratio $\gamma^1H/\gamma^{13}C$ over what would be observed in a standard ^{13}C spectrum. For CH_2 groups, the center peak of the expected triplet is lost, since the π pulse interchanges the top and bottom transitions of the three-spin system. This leaves the middle transitions unaffected (see Problem 8.2) , while inverting one of the outer lines relative to another, resulting in a 1:0:–1 peak structure (Figure 8.7). However, the outer two lines will be enhanced by a factor of $2\gamma^1H/\gamma^{13}C$. For CH_3 groups, the multiplet structure will be 1:1:–1:–1, with enhancement of each line by $3\gamma^1H/\gamma^{13}C$.

The INEPT experiment is one of the few multipulse NMR experiments that lends itself to interpretation using macroscopic magnetization vectors, and it is worth thinking explicitly about how polarization transfer enhancements take place in a simple IS spin system using this model. Note that we need not one but two rotating frames of reference for the discussion, one that rotates at the I frequency and one that rotates at the S frequency (Figure 8.8). The equilibrium magnetization for S (^1H in the CH example) represents the equilibrium population difference between the α and β states of S. After the first $\pi/2_x$ pulse on S, during the $1/2J_{IS}$ delay (ignoring the π pulse in between the two shorter delays for chemical shift refocusing), the M_y^S term generated by the first pulse evolves into two components. One component corresponds to S spins attached to I spins in the α state ($I\alpha$), i.e. approximately one-half of the total I spins, and the other component composed of those spins attached to I spins in the β state ($I\beta$). In essence, the INEPT sequence sorts the S spins according to the spin state of their coupling partner I. After the second $\pi/2_y$ pulse on S, those S spins that are attached to $I\alpha$ are placed into the α state themselves (at least with the current arrangement of pulses), while those attached to $I\beta$ are placed in the β state themselves. Thus, instead of the random distribution of $S\alpha$ attached to both $I\alpha$ and $I\beta$ (and vice versa) that would be observed in a one-pulse I-observe experiment, the Boltzmann population difference for spin S is also present across the I transitions (see Figure 8.8).

Refocused INEPT

The antiphase structure of CH_n multiplets in INEPT is somewhat inconvenient for analysis. Therefore, it is common to allow the individual magnetization components that result from J-coupling to refocus by waiting for the appropriate length of time, and then decoupling in order to observe an enhanced singlet. This is called a **refocused INEPT** experiment (Figure 8.9). Note that this experiment is identical to the INEPT sequence until after the third pulse. Instead of detecting at this point, the $-2\hat{S}_z\hat{I}_x$ term is allowed to refocus into $-\hat{I}_y$. At this point, acquisition of the signal from I (^{13}C in the present case) is begun and the active coupling between I and S is prevented from re-evolving by the application of broadband S (^1H) decoupling during the acquisition period. This results in a simple singlet structure for the detected I signal.

The refocused INEPT sequence shown in Figure 8.9 with refocusing delays set to $\Delta = 1/4J_{CH}$ will only refocus CH doublets correctly. For CH_2 groups (a carbon triplet), the same delays would give no signal, since the vectors corresponding to the outer

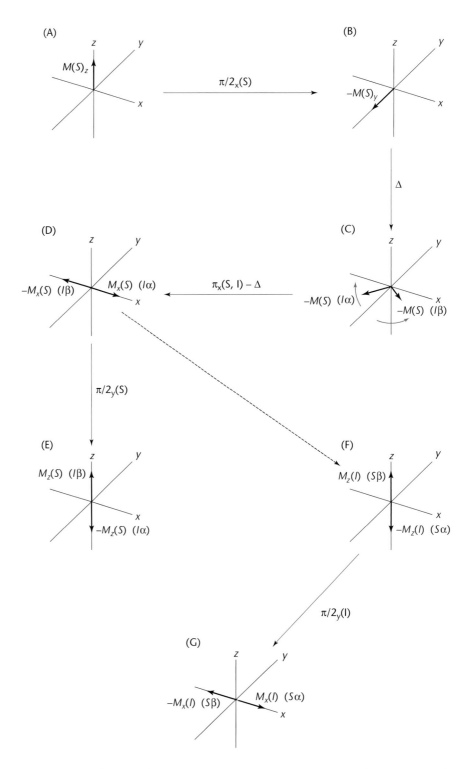

Figure 8.8 Evolution of macroscopic magnetization (*not* product operators) on *S* and *I* due to coupling between the two spins during the INEPT sequence. Although the first $\pi/2_y$ pulse on *I* is applied simultaneously with the second $\pi/2$ pulse on *S*, the effects are shown separately. Frames A–D show the rotating frame for nucleus *S*, frames F and G show the rotating frame for nucleus *I*.

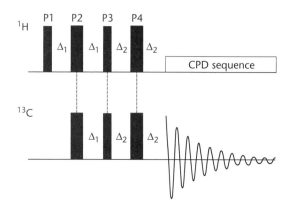

Figure 8.9 Refocused INEPT experiment for generating singlets with polarization transfer from ^1H to ^{13}C for signal enhancement. P1 = $\pi/2_x$ (^1H), P2 = simultaneous π_x (^1H and ^{13}C) followed by a refocusing delay. At this point, decoupling is applied at ^1H in order to prevent evolution of coupling in the ^{13}C signal. Complete phase cycling is not listed, but the phase of P3 is alternated between +**y** and –**y** with receiver phase cycled appropriately in order to remove signals from ^1H spins not attached to ^{13}C (see Problem 8.2).

lines of the multiplet would refocus in time $2\Delta = 1/2J_{CH}$, but in the opposite direction of the vector for the center line, resulting in signal cancellation. For multiplicities besides doublets, refocused INEPT requires different optimum refocusing delays. These delays can be calculated as follows: for CH groups, the refocused signal intensity is proportional to $\sin\theta$; for CH_2, it is proportional to $\sin 2\theta$; and for CH_3 groups, is proportional to $3/4(\sin\theta + \sin 3\theta)$, where $\theta = \pi J\Delta$, J is the C–H coupling, and Δ is the total time for the echo experiment ($\Delta/2$ on either side of the π pulse). A compromise can be reached to observe all signals with $\Delta = 1/7J_{CH}$ (see Figure 8.10).

Two-dimensional polarization transfer: HETCOR

The benefits of polarization transfer are readily extended to two-dimensional experiments. One of the most straightforward versions of two-dimensional polarization transfer is HETCOR (for *het*eronuclear *cor*relation) (see Figure 8.11) (Reference 3). The HETCOR sequence results in a $\gamma^1H/\gamma^{13}C$ signal enhancement to the detected ^{13}C–^1H correlations, as in the one-dimensional experiment. The delay times Δ_1 and Δ_2 are set independently. The first delay Δ_1 is optimized for the evolution of the multiplicity to be detected, and the second delay Δ_2 set so as to prevent any signal from being completely nulled after the refocusing. In practice, the values of Δ_1 and Δ_2 are often similar (Figure 8.12).

Sensitive nucleus (inverse) detection of an insensitive nucleus: the double INEPT or HSQC experiment

The sensitivity of HETCOR is limited by the fact that the insensitive nucleus is detected during acquisition (t_2). In recent years, the two-dimensional experiment of choice for detecting heteronuclear correlations is one in which the more sensitive

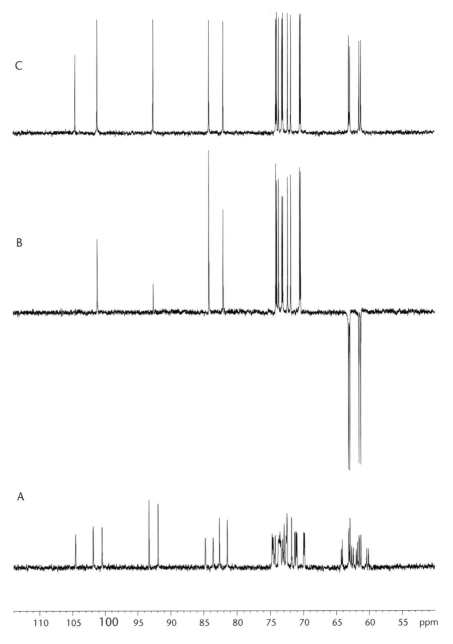

Figure 8.10 ^{13}C spectra of melezitose in D$_2$O acquired at 125.6 MHz. (A) Proton-coupled ^{13}C spectrum. (B) Refocused INEPT spectrum with $\Delta_2 = 1/3J_{CH}$ to observe all protonated carbons. (C) Proton-decoupled ^{13}C spectrum.

nucleus (usually ^{1}H) is detected: the double INEPT or HSQC (*heteronuclear single-quantum coherence*) experiment (Reference 4). Since the insensitive nucleus is detected indirectly, HSQC is always a two-dimensional experiment. The sequence for HSQC is shown in Figure 8.13. This experiment has gained wide acceptance both on its own and as a component in more complex three- and four-dimensional experiments. The HSQC (double INEPT) experiment returns a S/N enhanced over that of the INEPT sequence by a factor of $(\gamma_H/\gamma_X)^{3/2}$.

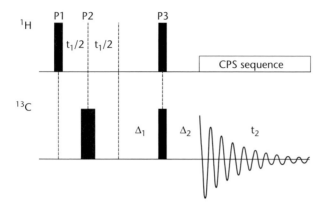

Figure 8.11 Two-dimensional HETCOR experiment for correlating ^{1}H to coupled ^{13}C nuclei. P1 = $\pi/2_x$ (^{1}H), P2 = π_x (^{13}C), P3 = simultaneous $\pi/2_{\phi 1}$ (^{1}H), $\pi/2_x$ (^{13}C), $\Delta_1 = 1/2J_{CH}$ for CH groups. $\Delta_2 = 1/3J_{CH}$ in order to avoid nulls for some terms. Phase ϕ and receiver phase are both cycled as 0 2 1 3 ($\pi/2$ increments).

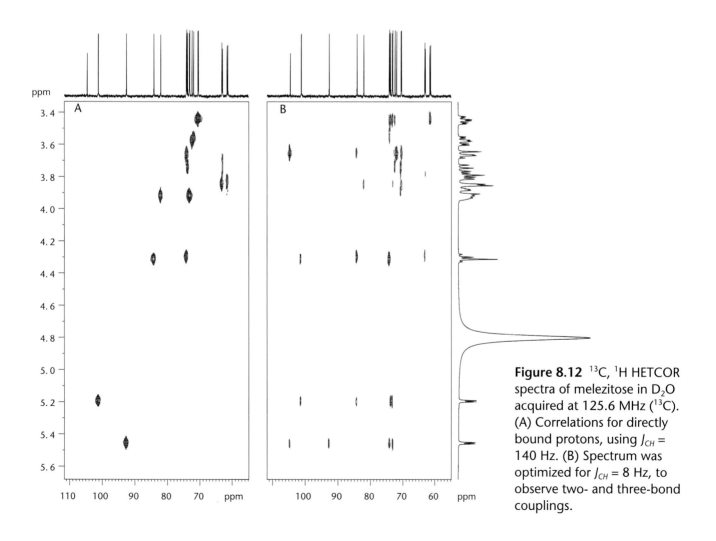

Figure 8.12 ^{13}C, ^{1}H HETCOR spectra of melezitose in D_2O acquired at 125.6 MHz (^{13}C). (A) Correlations for directly bound protons, using J_{CH} = 140 Hz. (B) Spectrum was optimized for J_{CH} = 8 Hz, to observe two- and three-bond couplings.

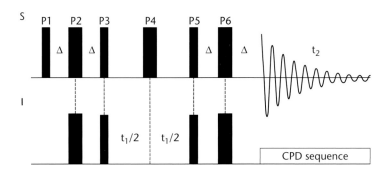

Figure 8.13 Double INEPT (HSQC) experiment for heteronuclear correlation with sensitive nucleus detection. First polarization transfer from S to I: P1 = $\pi/2_x$ (S), P2 = simultaneous π_x (S), π_x (I), P3 = simultaneous $\pi/2_y$ (S), $\pi/2_y$ (I), Δ = $1/4J_{CH}$, P4 = π_x (S) for refocusing of SI coupling during I chemical shift evolution period t_1. Reverse INEPT step transfers polarization back to S: P5 = simultaneous $\pi/2_y$ (S), $\pi/2_y$ (I), Δ = $1/4J_{CH}$, P6 = simultaneous π_x (S), π_x (I). During acquisition, decoupling is applied to spins I in order to prevent re-evolution of SI coupling terms on the S signal. Phase cycling is not included; plus/minus cycling of the third pulse with appropriate cycling of the receiver is required in order to remove S signals for spins not coupled to I; appropriate phase cycling for quadrature detection during t_1 is also required.

The HSQC experiment is an example of what we might call a "there and back again" sequence (recognizable by the symmetric arrangement of pulses in the sequence). The protons (S) are used to provide polarization to the insensitive nucleus of interest (I) to which the S spins are coupled. This polarization evolves during t_1 on I, and then is transferred back to S for detection. The detected proton (S) signals evolve at their own Larmor frequencies, and when the data are transformed with respect to t_1, the chemical shifts of the attached I spins are detected.

A simplified product operator analysis of the double INEPT experiment is as follows.

$$\hat{S}_z \xrightarrow{\pi/2_x(S)} -\hat{S}_y \xrightarrow{1/4J_{SI}-\pi(S,I)-1/4J_{SI}} -2\hat{S}_x\hat{I}_z \xrightarrow{\pi/2_y(S,I)} -2\hat{S}_z\hat{I}_x \xrightarrow{t_1}$$

$$-2\hat{S}_z\hat{I}_x \cos\Omega_I t_1 \xrightarrow{\pi/2_y(S,I)} -2\hat{S}_x\hat{I}_z \cos\Omega_I t_1 \xrightarrow{1/4J_{SI}\pi(S,I)-1/4J_{SI}} -\hat{S}_y \cos\Omega_I t_1 \quad (8.2)$$

By alternating the phase of the P3 pulse on I by 180° at each increment and doing the same with the receiver phase, coherences that pass through the heteronuclear couplings will be reinforced, whereas signals from protons (S) not attached to the heteronucleus of interest will be cancelled. As such, the phase-cycled HSQC sequence is really a difference experiment, and requires that at least the phase cycling of P3 be completed in order to select only the desired signals. A one-dimensional example of this sort of difference spectroscopy is called the **spin–echo difference** experiment (see Figure 8.14). The operator should be alert to the fact that in any experiment involving differences, it is the signal of the unsubtracted scan

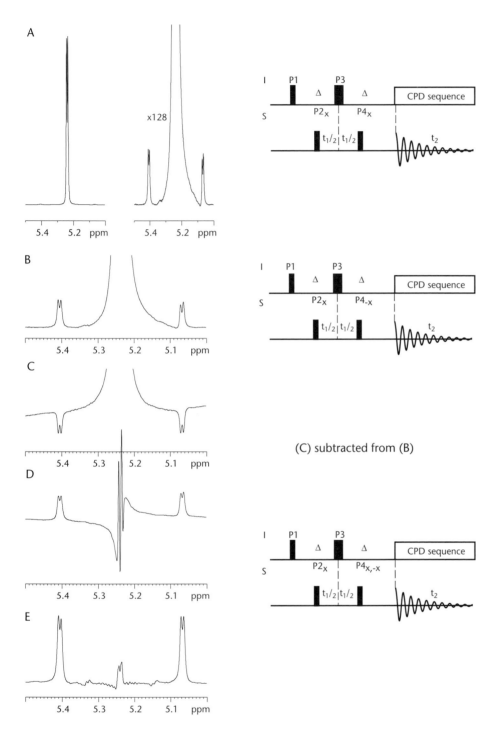

Figure 8.14 ^{1}H spectra of the anomeric proton of sucrose in D_2O acquired at 499.7 MHz. (A) The spectrum on the left shows the anomeric proton at full scale. This signal results from ^{12}C-bound protons. The spectrum on the right has the vertical scale increased so that the ^{13}C-bound protons are visible on each side of the center signal. (B) Proton spectrum obtained using the pulse sequence shown, with phase = **x** for P4. (C) Proton spectrum with phase = –**x** for P4. (D) Spectrum obtained in (C) is subtracted from that in (B). Result shows residual subtraction artifacts in the center from the cancellation of ^{12}C-bound protons and enhancement of the ^{13}C satellites. (E) Proton spectrum obtained when the subtraction is performed as part of the phase cycling.

that is important for determining the appropriate setting of the receiver gain. However, later in this chapter we will see that pulsed-field gradient coherence selection can be used to perform an HSQC experiment that is not a difference experiment, but returns only the desired signals in each scan.

Multiple-quantum approaches to heteronuclear correlation: DEPT and HMQC

An operator analysis of all of the INEPT-based experiments reveals that at no point is there more than one transverse (**x** or **y**) term in any of the product operators. Conversion of the Cartesian product operators to single element operators shows that only single-quantum coherences (one raising or lowering operator) are represented by such terms. However, higher-order coherences can be used to obtain polarization transfer and signal enhancement in heteronuclear NMR experiments. A one-dimensional NMR experiment providing such enhancement is called DEPT (*d*istortionless *e*nhancement by *p*olarization *t*ransfer) (see Figure 8.15) (Reference 5).

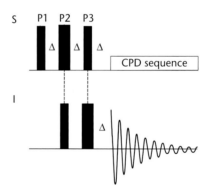

Figure 8.15 The DEPT experiment for identifying multiplicities of $^{13}CH_n$ nuclei. P1 = $\pi/2_x$ (S, usually ^1H), P2 = simultaneous π_y (S), $\pi/2_x$ (I, usually ^{13}C), P3 = simultaneous θ_y (S), π_y (I), $\Delta = 1/2J_{CH}$. The value of θ depends on the multiplicity of the I signal to be enhanced. For maximum enhancement of CH groups, $\theta = \pi/2$; for methylenes and methyls, $\theta = \pi/4$.

DEPT can be used to edit ^{13}C spectra according to the multiplicity of the carbon signals (i.e. whether they arise from CH, CH_2 or CH_3 groups). The first pulse generates transverse ^1H coherence that evolves under the influence of J_{CH} during the delay $\Delta = 1/2J_{CH}$: $\hat{S}_z \xrightarrow{\pi/2_x(S)} -\hat{S}_y \xrightarrow{1/2J_{SI}} 2\hat{S}_x\hat{I}_z$. This takes place the same for all protons no matter how many other protons are attached to the same ^{13}C atom. The second set of pulses converts the operator to a multiple quantum term: $2\hat{S}_x\hat{I}_z \xrightarrow{\pi/2_x(I),\pi_y(S)} 2\hat{S}_x\hat{I}_y$. Since this is multiple quantum coherence, the terms resulting from CH groups will not evolve during the subsequent evolution period with respect to coupling. However, those terms resulting from CH_2 or CH_3 groups will evolve with respect to passive couplings to the other protons on the same carbon:

$$2\hat{\mathbf{S}}_x\hat{\mathbf{I}}_y \xrightarrow{\ 1/2J_{SI}\ } -4\hat{\mathbf{S}}_x\hat{\mathbf{I}}_x\hat{\mathbf{S}}'_z \ (CH_2)$$

$$2\hat{\mathbf{S}}_x\hat{\mathbf{I}}_y \xrightarrow{\ 1/2J_{SI}\ } -8\hat{\mathbf{S}}_x\hat{\mathbf{I}}_x\hat{\mathbf{S}}'_z\hat{\mathbf{S}}''_z \ (CH_3) \qquad (8.3)$$

Application of the next set of pulses (θ_y (S), π_y (I),) converts the multiple-quantum terms back into single-quantum operators that are observable after the third evolution period. The relative intensities of the different terms will depend on the tip angle of the proton pulse θ_y according to the relationships:

$$2\hat{\mathbf{S}}_x\hat{\mathbf{I}}_y \to -2\hat{\mathbf{S}}_z\hat{\mathbf{I}}_y \sin\theta \ (CH)$$

$$-4\hat{\mathbf{S}}_x\hat{\mathbf{I}}_x\hat{\mathbf{S}}'_z \to -4\hat{\mathbf{S}}_z\hat{\mathbf{I}}_x\hat{\mathbf{S}}'_z \sin\theta\cos\theta \ (CH_2)$$

$$8\hat{\mathbf{S}}_x\hat{\mathbf{I}}_y\hat{\mathbf{S}}'_z\hat{\mathbf{S}}''_z \to 8\hat{\mathbf{S}}_z\hat{\mathbf{I}}_y\hat{\mathbf{S}}'_z\hat{\mathbf{S}}''_z \sin\theta\cos^2\theta \ (CH_3) \qquad (8.4)$$

By suitable linear combinations of DEPT spectra obtained using a variety of values of θ, it is possible to obtain edited subspectra that contain only carbons of one type (all CH, all CH_2 or all CH_3) (Figure 8.16). Note that chemical shift evolution is ignored in Expressions 8.3 and 8.4. This is due to the inclusion of the π pulses on each spin (at P2 for ^1H and P3 for ^{13}C). These pulses refocus ^1H shifts prior to the application of P3 and ^{13}C shifts prior to acquisition. Decoupling of ^1H (S) during acquisition prevents the evolution of J_{CH} during the acquisition time.

The inverse (^1H) detected two-dimensional version of heteronuclear multiple-quantum correlation spectroscopy, called HMQC (*h*eteronuclear *m*ultiple-*q*uantum *c*oherence) was one of the earliest methods used to obtain polarization transfer and inverse detection in isotopically labeled biological macromolecules (Reference 6). HMQC (also referred to in some earlier works as the "forbidden echo" experiment) is very simple in execution, consisting of only four pulses, two on each nucleus (Figure 8.17). This made it the sequence of choice on early inverse detection spectrometers, which lacked the versatile pulse programmers now available on standard NMR spectrometers. Transformed two-dimensional HMQC spectra are similar in appearance to HSQC spectra, except that line widths in the indirect dimension are determined by double-quantum relaxation rather than single-quantum effects, and so tend to be broader than those observed in HSQC (at least for large macromolecules). Furthermore, since ^1H coherence is transverse during t_1 in HMQC, passive ^1H–^1H couplings will evolve, and lines will be broadened in the indirect detection dimension by these couplings. Because the ^1H coherence is along the **z** axis in HSQC, the indirectly detected dimension is not broadened by passive couplings. Problem 8.4 details the evolution of coherences in the HMQC experiment.

Although, HSQC has supplanted HMQC as the standard two-dimensional heteronuclear correlation experiment on modern NMR spectrometers, the HMQC experiment remains an important building block for three-dimensional NMR experiments, as discussed in Chapter 9.

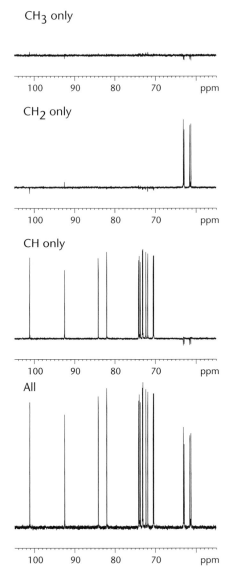

Figure 8.16 Linear combinations of DEPT spectra of melezitose in D_2O. Three spectra were obtained with the pulse sequence shown in Figure 8.15 with $\theta = \pi/4$ (spectrum **a**), $\theta = \pi/2$ (spectrum **b**), and $\theta = 3\pi/4$ (spectrum **c**). The following combinations were used to obtain the edited spectra: CH only = 2 * b; CH_2 only = a – c; CH_3 only = –0.77 * (2 * b) + a + c.

Gradient coherence selection in heteronuclear correlation NMR

Like HSQC, phase-cycled HMQC experiments are difference experiments, i.e. the phase cycling removes contributions to the accumulated FID from protons not attached to the heteronucleus of interest. In Chapter 7, we discussed how pulsed-field gradients (PFGs) are used to select for particular orders of coherence in NMR experiments. The requirement for selection is that the sum of the effects of all PFGs on the desired coherence order throughout the course of the experiment be

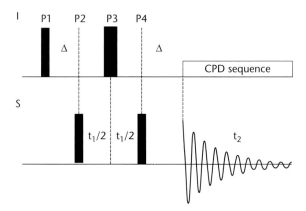

Figure 8.17 The HMQC experiment for correlating ^{1}H (S) to coupled I nuclei. P1 = $\pi/2_x$, P2 = $\pi/2_{x,-x}$ (I, often ^{15}N), P3 = π_x, P4 = $\pi/2_{x,x,-x,-x}$ (I), Δ = 1/2J_{CH}. Receiver phase is $x, -x, -x, x$. The I nucleus is decoupled using composite-pulse decoupling during acquisition of ^{1}H.

zero (Equation 7.36), $\sum_i \gamma_i p_i G_i z t_{gi} = 0$. Although the discussion in Chapter 7 focused on homonuclear coherence selection, Equation 7.36 is also valid for coherences resulting from heteronuclear couplings, and PFGs can be used to select for desired coherences in heteronuclear experiments. The restatement of Equation 7.36 for heteronuclear coherence selection is:

$$\sum_i (\gamma_I p_{Ii} + \gamma_S p_{Si}) G_i z t_{gi} = 0 \qquad (8.5)$$

where γ_I and γ_S are the gyromagnetic ratios of the spins I and S, respectively, the p_{Ii} and p_{Si} are the orders of coherence for the I and S spins, and the remaining terms are the same as in Equation 7.36. Assuming that diffusion is negligible during the experiment (that is, z remains constant), that term can be dropped and the resulting expression is used:

$$\sum_i (\gamma_I p_{Ii} + \gamma_S p_{Si}) G_i t_{gi} = 0 \qquad (8.6)$$

Consider the simplest heteronuclear correlation experiment, the HMQC sequence. A gradient version of this experiment used for ^{13}C–^{1}H correlations is shown in Figure 8.18. Recalling that the Cartesian operator can be represented as a linear combination of raising and lowering operators, after P1 on ^{1}H (S), the proton magnetization evolves to $\hat{S}_+$ and $\hat{S}_-$ (ignoring the population operators due to ^{13}C, that contribute to $\hat{I}_z$). These coherences are transformed at P2 on ^{13}C into four terms, $\hat{I}_+\hat{S}_+$, $\hat{I}_+\hat{S}_-$, $\hat{I}_-\hat{S}_+$ and $\hat{I}_-\hat{S}_-$. As in the phase-cycled HSQC, a π pulse (P3) is applied to the ^{1}H spins halfway through the t_1 evolution period, refocusing ^{1}H shift evolution. P3 interconverts $\hat{S}_+$ and $\hat{S}_-$ but leaves the I terms unaffected. The last $\pi/2$ pulse on ^{13}C, P4, converts the transverse I terms back into longitudinal terms, leaving just $\hat{S}_+$ and $\hat{S}_-$ terms for detection. In order to obtain P- or N-selection, three gradients are applied, $G1$,

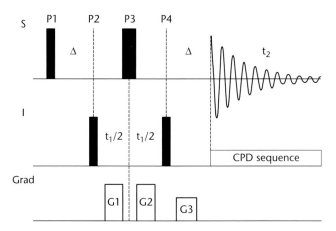

Figure 8.18 PFG coherence selection in HMQC. This experiment is not phase-sensitive in ω_1.

$G2$ and $G3$, at the points indicated in Figure 8.18. This can be diagrammed along with the various coherences that are present during the gradients as:

$$\hat{\mathbf{S}}_z \rightarrow \begin{matrix} \hat{\mathbf{S}}_+ \\ \hat{\mathbf{S}}_- \end{matrix} \rightarrow \begin{matrix} \hat{\mathbf{I}}_+\hat{\mathbf{S}}_+ \\ \hat{\mathbf{I}}_-\hat{\mathbf{S}}_+ \\ \hat{\mathbf{I}}_+\hat{\mathbf{S}}_- \\ \hat{\mathbf{I}}_-\hat{\mathbf{S}}_- \end{matrix} [G1] \rightarrow \begin{matrix} \hat{\mathbf{I}}_+\hat{\mathbf{S}}_- \\ \hat{\mathbf{I}}_-\hat{\mathbf{S}}_- \\ \hat{\mathbf{I}}_+\hat{\mathbf{S}}_+ \\ \hat{\mathbf{I}}_-\hat{\mathbf{S}}_+ \end{matrix} [G2] \rightarrow \begin{matrix} \hat{\mathbf{S}}_+ \\ \hat{\mathbf{S}}_- \end{matrix} [G3] \tag{8.7}$$

If we arbitrarily choose N-selection in the ^{13}C dimension, the three gradients should select the following pathway:

$$\hat{\mathbf{I}}_+\hat{\mathbf{S}}_+[G1] \rightarrow \hat{\mathbf{I}}_+\hat{\mathbf{S}}_-[G2] \rightarrow \hat{\mathbf{S}}_-[G3] \tag{8.8}$$

During the first half of the evolution period, gradient $G1$ is applied, dephasing all evolving coherences by $(\gamma_I p_{I1} + \gamma_S p_{S1})G1$. The desired coherence $(\hat{\mathbf{I}}_+\hat{\mathbf{S}}_+)$ will dephase by $(\gamma_I + \gamma_S)G1$. Following P3, the desired coherence is now of the form $\hat{\mathbf{I}}_+\hat{\mathbf{S}}_-$, and dephases by an amount $(\gamma_I + \gamma_S)G2$ under the influence of $G2$ (assuming for simplicity that all three gradients have equal durations). Remember that the gradient does not affect the order of coherence, so, other than the implied **z**-dependent phase factor, the product operator is not affected by the gradient. At the end of t_1, P4 returns the ^{13}C terms to the **z** axis, and the transverse proton terms are refocused prior to acquisition and detection. The third gradient ($G2$) is applied during the refocusing period. $G3$ must refocus the desired coherence $(\hat{\mathbf{S}}_-)$ for detection. Application of $G3$ will result in a phase shift $(-\gamma_S)G3$ for the desired coherence order. In order to select the appropriate coherence transfer pathway, $(\gamma_I + \gamma_S)G1 + (\gamma_I + \gamma_S)G2 + (-\gamma_S)G3 = 0$, according to Equation 8.6. Solving this for the case of ^{1}H and ^{13}C given that:

$$\frac{\gamma_{^1\text{H}}}{\gamma_{^{13}\text{C}}} \cong \frac{4}{1} \tag{8.9}$$

yields one possible solution of $G1 = 5$, $G2 = 3$, $G3 = 4$.

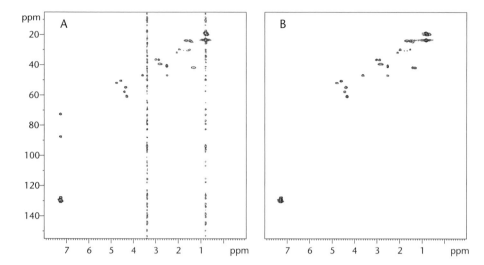

Figure 8.19 ^1H, ^{13}C HMQC spectra of 50 mM gramicidin S in d_6-DMSO acquired at 500.13 MHz (^1H). Gramicidin S is a cyclic peptide (-Val-Orn-Leu-(D-Phe)-Pro-)$_2$. Spectrum (A) was acquired with a receiver gain of 16 using phase-cycling, with a total of eight acquisitions per t_1 increment. Spectrum (B) was acquired using PFG coherence selection as shown in Figure 8.18, and one acquisition per t_1 increment. The signals due to protons bound to ^{12}C are ~100-fold greater than those due to protons bound to ^{13}C, so the absence of these signals in the PFG-selected FIDs results in a significant increase in the available dynamic range, i.e. the ADC is not filled by unused signal, as in the phase-cycled experiment. Since the experiment does not rely on difference spectroscopy, the receiver gain was increased to 4096. While increasing receiver gain does not automatically improve the signal-to-noise ratio, it does permit weaker signals to be digitized. Furthermore, below a threshold gain value, the noise from the electronic circuit will decrease signal-to-noise ratio.

Phase-sensitive gradient coherence selection experiments for heteronuclear correlations

As described in Chapter 7, PFGs select for only one coherence order, rather than the combination of positive and negative coherence orders that are used to obtain pure-phase quadrature detection in the indirect dimensions of phase-cycled experiments. Thus, the use of PFGs result in either P (positive frequency) or N (negative frequency) selection in indirectly detected dimensions of NMR experiments. This gives signals at the appropriate frequencies for correlations in PFG-selected experiments, but with mixed absorptive and dispersive line shapes that must be displayed in magnitude mode. This is the same situation shown in Figure 7.22.

As with homonuclear PFG-selected experiments, schemes have been devised that allow one to obtain pure-phase spectra with gradient selection (Figure 8.19). These schemes are based on the fact that the P- and N-type spectra contain both absorptive and dispersive line shape components, and by the appropriate linear combination of P- and N-type spectra, pure phase spectra can be obtained. Davis *et al.*

(Reference 7) described a series of pulse sequences for pure-phase heteronuclear correlation HSQC spectra using PFGs for coherence selection. One of these schemes is shown in Figure 8.20.

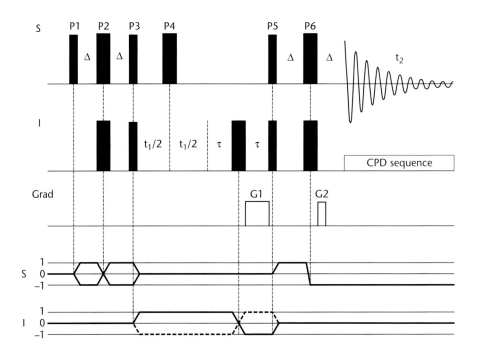

Figure 8.20 Phase-sensitive HSQC experiment for correlating ^{1}H (*S*) to coupled *I* nuclei, using pulsed-field gradients for coherence selection. The sign of *G1* determines P or N selection. Processing to obtain pure-phase spectra is performed as described in the text.

The HSQC experiment shown in Figure 8.20 proceeds as the phase-cycled version until the end of the t_1 evolution period, resulting in an *IS* term that is modulated by the chemical shift of *I* (^{13}C, in this example). At this point, a delay τ is added; this period is necessary for the eventual refocusing of chemical shifts evolving throughout the gradient pulse *G1* of length τ. *G1*, whose purpose is to dephase coherences at the end of the evolution period, is followed immediately by the back INEPT transfer to ^{1}H for detection. Following the last π pulses on *I* and *S*, a refocusing gradient *G2* is applied. The relative intensities of these gradients are determined by Equation 8.6 to be:

$$\frac{\gamma_S}{\gamma_I} = \frac{\pm G1}{G2} \tag{8.10}$$

where the positive *G1* selects for P-type coherence and the negative *G1* selects for N-type. The P- and N-selected spectra are stored separately. After FT with respect to t_2, a typical correlation signal in the P-type and N-type spectra can be represented respectively as:

$$S_P(t, \omega_2) = \exp(-i\Omega_I t_1)(A_2 + iD_2) \qquad (8.11)$$

$$S_N(t, \omega_2) = \exp(+i\Omega_I t_1)(A_2 + iD_2) \qquad (8.12)$$

ignoring the effects of relaxation. The exponential term on the right side of Equations 8.11 and 8.12 represents the modulation due to the I spin during t_1, at a positive shift (Ω_I) relative to the carrier in the P-type spectrum and at a negative shift relative to the carrier in the N-type spectrum. A_2 and D_2 are the absorptive and dispersive components of line shape at frequency ω_2. Taking the complex conjugate of 8.12 and adding it to 8.11 generates a new signal:

$$S_+(t_1, \omega_2) = 2\exp(-i\Omega_I t_1)A_2 \qquad (8.13)$$

Complex FT of this signal with respect to t_1 generates a complex spectrum of the form $(A_1 + iD_1)A_2$, the real portion of which is pure-phase absorptive in both ω_1 and ω_2.

Sensitivity enhancement in gradient coherence selection experiments

Besides the problem of obtaining pure-phase spectra, the other difficulty inherent in the use of PFGs for coherence selection is the decreased sensitivity of such experiments relative to phase-cycled experiments. Because only one coherence order is selected during t_1, rather than two, as in phase-cycled experiments, the net throughput of signal is smaller, resulting in a decrease of S/N by a factor of $\sqrt{2}$ relative to a phase-cycled experiment acquired in the same length of time. It is possible to recover some of this intensity in PFG-selected experiments using a data acquisition and processing scheme sometimes known as the "Rance trick." In the early 1990s, Rance *et al.* (Reference 8) introduced a modification to the acquisition and processing of two-dimensional experiments that significantly improved their sensitivity. They noted that many two-dimensional NMR experiments end up wasting a good deal of potential signal owing to the way in which frequency labeling is obtained. Although frequency information is contained in both the sine and cosine terms that develop throughout the frequency-labeling period of the two-dimensional experiment, generally only one of those terms is retained by the pulse sequence and converted into observable coherence. They proposed a scheme known as **sensitivity-enhanced spectroscopy** for recovering a portion of the frequency-labeled signal that is otherwise lost. We can follow the sensitivity enhancement scheme for an HSQC pulse sequence shown in Figure 8.21. Up to **A**, the sequence is the same as a standard HSQC experiment. For a single HX-coupled spin pair, the evolution up to that point is given by:

$$\hat{\mathbf{S}}_y \cos(\omega_I t_1) - 2\hat{\mathbf{S}}_x \hat{\mathbf{I}}_x \sin(\omega_I t_1) \qquad (8.14)$$

where $\hat{\mathbf{S}}$ is the proton and $\hat{\mathbf{I}}$ is the heteronucleus. In the standard HSQC experiment, detection begins at this point, and the second term in Equation 8.14, that

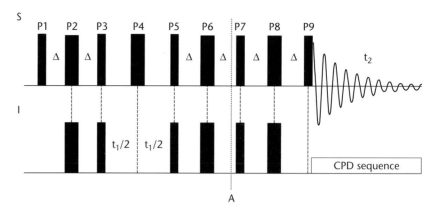

Figure 8.21 Pulse sequence for a sensitivity-enhanced HSQC experiment (Reference 9). Narrow lines represent $\pi/2$ pulses, broad lines represent π pulses. Delays are all $\Delta = 1/4J_{IS}$. Pulse phases for S (^{1}H) are P1 = x, P3 = $-y$, P5 = y, P7 = x and P9 = y. Pulse phases for I are P3: x, $-x$, x, $-x$; P5: x, x, $-x$, $-x$; P7: y, y, $-y$, $-y$; receiver: x, $-x$, $-x$, x. All other pulses on both I and S are x.

contains only zero- and double-quantum coherences, will not be detected. As such, only half of the initial proton polarization that was transferred to the heteronucleus in the first INEPT step is actually detected.

The additional pulses in the sensitivity-enhanced sequence result in the following evolution:

$$\hat{\mathbf{S}}_y \cos(\omega_I t_1) - 2\hat{\mathbf{S}}_x\hat{\mathbf{I}}_x \sin(\omega_I t_1) \xrightarrow{\pi/2_x(S),\pi/2_y(I)} \hat{\mathbf{S}}_z \cos(\omega_I t_1) + 2\hat{\mathbf{S}}_x\hat{\mathbf{I}}_x \sin(\omega_I t_1)$$

$$\xrightarrow{\tau-\pi_y(I,S)-\tau} -\hat{\mathbf{S}}_z \cos(\omega_I t_1) + \hat{\mathbf{S}}_y \sin(\omega_I t_1) \xrightarrow{\pi/2_y(S)} -\hat{\mathbf{S}}_x \cos(\omega_I t_1) + \hat{\mathbf{S}}_y \sin(\omega_I t_1) \quad (8.15)$$

Both terms at the end of this sequence contribute to observable signal, but the two terms are out of phase by 90°, resulting in a line shape that is a mixture of dispersive and absorptive components. However, a second experiment is performed in which the phase of the $\pi/2$ pulse on I immediately following the t_1 evolution period is inverted (ϕ_2 in Figure 8.18), resulting in coherence at the beginning of the detection period of the form $\hat{\mathbf{S}}_x \cos(\omega_I t_1) + \hat{\mathbf{S}}_y \sin(\omega_I t_1)$. Addition of the two FIDs resulting from the two experiments results in a time-domain signal $2\hat{\mathbf{S}}_y\sin(\omega_I t_1)$, while subtracting them gives a time-domain signal $-2\hat{\mathbf{S}}_x\cos(\omega_I t_1)$. It is computationally simple to phase shift one of these two deconvoluted signals so that the two signals are in phase. The signals can then be added to produce a spectrum that is pure phase in ω_1. The intensities of signals resulting from HX correlations (in which there is no second attached proton) is doubled with respect to the standard HSQC experiment performed with an equal number of transients. However, since the noise levels of the two experiments are statistically independent, the noise increases as $\sqrt{2}$, and so the net gain in S/N is $\sqrt{2}$ over a standard HSQC performed for the same length of time (Figure 8.23).

There are some practical limitations to sensitivity-enhanced experiments. First, as the sensitivity-enhanced sequence is longer than the standard experiment, there will be additional losses due to relaxation. Secondly, the multiplicity of coupling will affect how the sequence works for a particular signal. For example, in the HSQC experiment shown in Figure 8.21, the signals from heteronuclei with more than one attached proton are not enhanced, and as the noise is increased, there will be a loss in S/N for these resonances. Nevertheless, for experiments in which primarily one type of signal is detected (e.g. N–H amide correlations in ^{1}H, ^{15}N HSQC), the potential for recovering up to $\sqrt{2}$ in S/N makes the use of PFGs to select coherence orders more practical. Kay *et al.* described the combination of PFG coherence selection similar to that obtained using the pulse sequence shown in Figure 8.20 and the Rance sensitivity enhancement scheme to improve sensitivity (Figure 8.21) while still obtaining pure-phase two-dimensional spectra (Reference 9). The Kay sensitivity-enhanced gradient selection HSQC experiment is shown in Figure 8.22. As we will see in Chapter 9, the Kay experiment is the basis for a large number of what are called sensitivity-enhanced multidimensional NMR experiments.

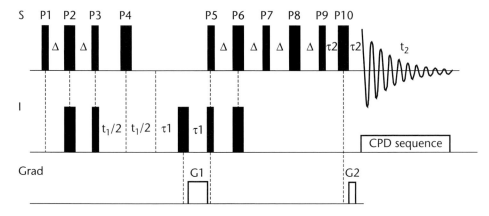

Figure 8.22 Sensitivity-enhanced gradient coherence selection HSQC. The use of signal recovery combinatorial processing (the Rance trick) makes the use of gradient coherence selection practical for heteronuclear correlations. All pulses are along *x*, except for P3 on *S* (*y*) and P7 on both *I* and *S* (both *y*). Evolution delays are all $\Delta = 1/4J_{IS}$. Delay times τ_1 and τ_2 correspond to the lengths of the gradient pulses G1 and G2, respectively. Gradient strengths are determined as in Equation 8.6. For each value of t_1, two FIDs are collected and stored separately, with the phase of P5 for spin I alternated between *x* and –*x* and the sign of G2 inverted between the two FIDs. Problem 8.11 discusses the processing of the data.

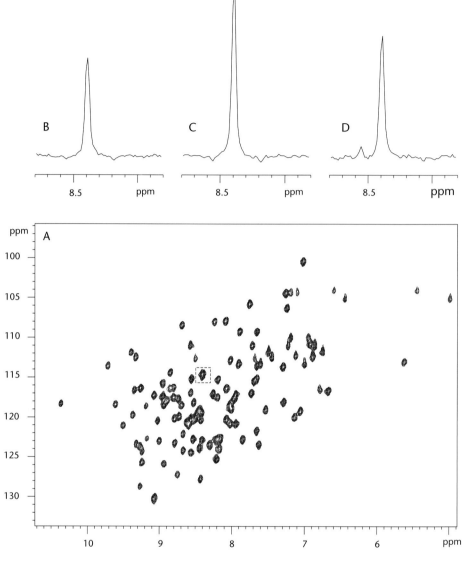

Figure 8.23 ^{1}H, ^{15}N HSQC spectra of 1 mM ^{13}C, ^{15}N-ribonuclease A in 90% H$_2$O/D$_2$O acquired at 500.13 MHz. (A) Two-dimensional spectrum using the sensitivity-enhanced pulse sequence in Figure 8.22; (B–D) rows through the boxed correlation peak, extracted from spectra obtained with different pulse sequences; (B) PFG-HSQC as shown in Figure 8.20; (C) SE-HSQC from spectrum shown in (A); (D) standard HSQC, as shown in Figure 8.13. The rows are scaled to the same noise amplitude to emphasize the different signal-to-noise ratios.

Problems

*8.1   The phase cycle for the INADEQUATE sequence shown in Figure 8.1 is given below. The phase cycling is eight steps ($1 = \pi / 4$) in order to achieve quadrature detection in ω_1 using TPPI.

(a) Why are eight steps necessary for quadrature detection in t_1 for INAD-EQUATE using TPPI when only four steps are required for a DQF-COSY obtained using TPPI?

(b) Demonstrate that the INADEQUATE phase cycle selects against +/–1 quantum coherences during t_1.

Step	P1	P2	P3	P4	Receiver
1	0	0	0	0	0
2	3	3	3	1	0
3	2	2	2	2	0
4	1	1	1	3	0
5	4	4	4	0	0
6	7	7	7	1	0
7	6	6	6	2	0
8	5	5	5	3	0

*8.2 (a) Using diagrams similar to Figure 8.2, show that the signal enhancement of the I spin (corresponding to intensities of ^{13}C lines), in which a selective inversion pulse is applied to one line of a ^1H J_{HC} doublet, is proportional to γ_S/γ_I in a 1:–1 doublet for CH, a $2\gamma_S/\gamma_I$ enhancement in a 1:0:–1 triplet for CH_2 and a 3 γ_S/γ_I in a 1:1:–1:–1 quartet for CH_3 groups, respectively. Assume equilibrium population differences proportional to the gyromagnetic ratios of the nuclei.

(b) These results ignore the contributions of the I nucleus (^{13}C in this case). This result is obtained experimentally if the last $\pi/2$ pulse on ^1H (S) is phase-shifted by π rad on alternating experiments and following this shift with a corresponding shift of the receiver phase. Show how this occurs.

*8.3 (a) The HMQC experiment is the simplest experiment for indirect detection of ^1H-coupled heteronuclear spins. Using the phase cycling described in the caption to Figure 8.9, explicitly determine all Cartesian product-operator terms that occur during in the evolution of a single ^{15}N–^1H spin pair coupled by $^1J_{NH}$ during both evolution times (t_1 and t_2) of an HMQC experiment. Show how phase cycling removes unwanted terms and gives rise to the desired terms after signal addition. (Note: this problem takes a while and requires careful book-keeping!).

(b) If you wished to obtain States-TPPI phase cycling for the HMQC experiment shown in Figure 8.9, which pulse(s) would you phase-cycle and how would that affect the receiver phase? Would you need to phase-cycle the π pulse on S (^1H)? If so, how?

*8.4 Calculate the relative strengths of gradients $G1$ and $G2$ needed for coherence selection of ^1H–^{15}N and ^1H–^{13}C correlations, respectively, in the HSQC experiment using Equation 8.6 and coherence transfer pathways shown in Figure 8.11. Do this for two cases, one in which the $G1$ and $G2$ gradient power levels are identical, and in the case where the gradient durations are identical.

8.5 Suggest a 4-step phase cycle for the refocused INEPT experiment shown in Figure 8.8.

*8.6   Describe explicitly the effect of each of the following pulses, π_x, π_y, π_{-x} and π_{-y} on the evolution of $\hat{I}_+$ in the course of a spin–echo sequence $\tau-\pi-\tau$. Repeat this for $\hat{I}_-$.

*8.7   What are the effects of the pulses in Problem 8.6 upon coherence order?

8.8   A π pulse directed between the x and y axes in the rotating frame (that is, $\pi/4$ rad or 45° from both axes) will interchange the x and y components of a given evolution operator. Show that this is equivalent to, $\hat{I}_+ \rightarrow i\hat{I}_-$. (Since $i=\exp[i(\pi/4)]$, this represents a $\pi/2$ shift in the result.)

8.9   The signal-to-noise ratio of an NMR experiment involving polarization transfer for a given sample at a fixed temperature is given by $S/N \sim \gamma_{ex}\gamma_{obs}^{3/2}(T_2)$, where γ_{ex} is the gyromagnetic ratio of the excited spin and γ_{obs} that of the observed spin, and T_2 is the transverse relaxation time of the observed coherence. For $^{15}N-^1H$ correlations in indirectly detected experiments such as HSQC and HMQC, this results in an enhancement factor of ~300, and the primary difference between HSQC and HMQC lies in the differences in the relaxation times that control line widths in the indirectly detected dimension. Given the individual relaxation times of the spins of interest, T_1^{1H}, T_1^{15N}, T_2^{1H} and T_2^{15N}, calculate the relevant relaxation rates for a given NH pair that determines the observed line width in the indirectly detected dimensions of HSQC and HMQC, respectively. Note that for large molecules, T_2 is usually shorter than T_1.

8.10   The PFG-selected sensitivity-enhanced HSQC experiment of Kay *et al.* (Figure 8.20, Reference 9) detects a signal $S(t_1)=\pm M_y \cos(\omega_I t_1)+M_y \sin(\omega_I t_1)$ assuming complete refocusing of the desired coherences by the PFGs. The sign of the first term is determined by the phase of pulse P5 on I, which is cycled between $+x$ and $-x$ for each value of t_1. The two spectra acquired with the different phase values of P5 are stored in separate locations. Specifically, what is required of the processing of such data in order to obtain a pure-phase two-dimensional spectrum in ω_1 after FT? (Hint: see section on the Rance trick in Chapter 7!).

References

1. A. Bax, R. Freeman, and S. P. Kempsell, *J. Am. Chem. Soc.* 102, 4849–4851 (1980).
2. G. A. Morris and R. Freeman, *J. Am. Chem. Soc.* 101, 760–762 (1979).
3. R. Freeman and G. A. Morris, *J. Chem. Soc. Chem. Commun.* 684 (1978).
4. G. Bodenhausen and D. Ruben, *J. Chem. Phys. Lett.* 69, 185 (1980).
5. D. M. Doddrell, D. T. Pegg, and M. R. Bendall, *J. Magn. Reson.* 48, 323 (1982).
6. A. Bax, R. H. Griffey, and B. L. Hawkins, *J. Magn. Reson.* 55, 301 (1983).
7. A. L. Davis, J. Keeler, E. D. Laue, and D. Moskau, *J. Magn. Reson.* 98, 207 (1992).
8. J. Cavanagh, A. G. Palmer, P. E. Wright, and M. Rance, *J. Magn. Reson.* 91, 429 (1991).
9. L. E. Kay, P. Keifer, and T. Saarinen, *J. Am. Chem. Soc.* 114, 10663 (1992).

9

BUILDING BLOCKS FOR MULTIDIMENSIONAL NMR AND SPECIAL CONSIDERATIONS FOR BIOLOGICAL APPLICATIONS OF NMR

The development of multidimensional NMR (three or more frequency dimensions in a single experiment) is a logical extension from two-dimensional NMR. A third independent time variable introduced into a multipulse experiment results in a third independent frequency domain that can represent some other desired correlation between a set of spins. Although three-dimensional NMR experiments can be homonuclear (e.g. ^{1}H in all three dimensions), most often they are used to identify heteronuclear connectivity unambiguously, particularly in large macromolecules such as proteins or nucleic acids, and it is on these experiments that we will concentrate. Three-dimensional NMR experiments are most often applied to large molecules that are at low concentration relative to solvent (usually water); solvent signal and artifact suppression are therefore particularly important. Furthermore, selective excitation schemes are necessary in some experiments. These special topics will also be considered in the present chapter.

A three-dimensional NMR experiment can be thought of as a two-dimensional experiment in which the detection period (t_2) is replaced by a second two-dimensional experiment. This gives rise to a hybrid experiment in which the correlations expected in both two-dimensional experiments are observed, with three independent time variables (t_1, t_2 and t_3) leading to three frequency dimensions after FT. Two of the first and still among the most commonly used three-dimensional NMR experiments, NOESY-HSQC and TOCSY-HSQC, combine the results of the homonuclear NOESY and TOCSY experiments with an HSQC experiment that correlates directly bonded heteronuclei (Reference 1). The NOESY and TOCSY portions of the experiments connect ^1H spins either by through-space or through-bond interactions, respectively, the results of which are then edited according to the heteronucleus to which the receiver proton (the proton that is normally detected during t_2 in the two-dimensional experiment) is attached. We will first examine the NOESY–^1H, ^{15}N HSQC as applied to a protein. If more than one NH proton in the molecule has the same chemical shift, there will be some ambiguity about which NOE cross-peak is assigned to which NH proton in the two-dimensional spectrum. However, even if two different NH groups are degenerate in ^1H chemical shift (a likely occurrence in large proteins), the homonuclear NOESY will give ambiguous results: it will not be possible to assign cross-peaks to a particular NH proton. However, simultaneous degeneracy of ^1H and ^{15}N chemical shifts is much less likely than degeneracy in only one dimension. Assuming

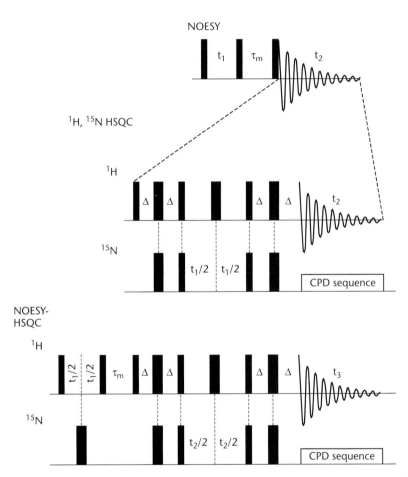

Figure 9.1 Combination of two two-dimensional sequences (NOESY and ^1H, ^{15}N HSQC) to yield a single three-dimensional sequence. Figure 9.2 shows the results of the three-dimensional experiment. Note that a ^{15}N π pulse has been added during the t_1 evolution period of the three-dimensional experiment in order to remove ^{15}N coupling to ^1H. The ^{15}N evolution time is called t_2, and the detection period is called t_3.

that the protein has been prepared in a ^{15}N-enriched form, one can imagine adding a ^{1}H, ^{15}N HSQC step that will label the observed ^{1}H coherences not only in terms of the NOEs involving that coherence, but also in terms of the ^{15}N chemical shift of the attached nitrogens. The net result is an experiment that sorts NOEs to NH protons according to the chemical shifts of their attached ^{15}N spins.

Figure 9.1 gives a conceptual picture of how the three-dimensional NOESY–^{1}H, ^{15}N HSQC experiment is put together from two separate two-dimensional experiments. Figure 9.2 shows how the three-dimensional experiment resolves overlapping correlations in three dimensions. The resulting three-dimensional spectrum can be viewed in three different ways: (1) as a stack of two-dimensional ^{1}H, ^{15}N HSQC spectra that are edited according to the chemical shifts of protons from which the NH proton detects NOEs; (2) as a stack of NOESY spectra that are edited according to the chemical shift of the ^{15}N shifts of the nitrogens to which the detected amide protons are attached; or (3) as a stack of ^{15}N, ^{1}H correlation spectra

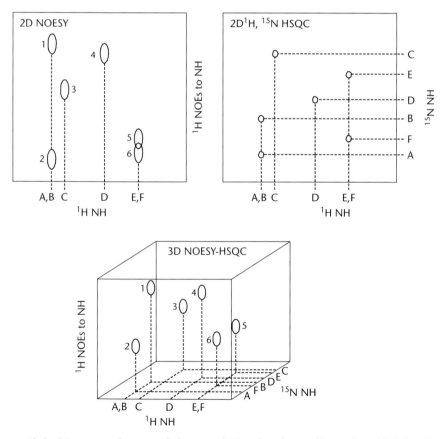

Figure 9.2 Diagram of spectral deconvolution by three-dimensional NMR. (Top left) NOESY spectrum showing ambiguous NOE connectivities; NOEs to NH protons A and B are overlapped, as are those to protons E and F. All NH correlations are unambiguously resolved in the ^{1}H, ^{15}N HSQC (upper right). The NOESY-HSQC (bottom) shows that all NOES are unambiguously resolved in the third dimension.

that show the chemical shifts of nitrogens versus the chemical shifts of protons that give NOEs to the NH attached to those nitrogens, and are edited by the chemical shift of the NH proton. Typically, the three-dimensional spectrum is analyzed in this fashion, rather than as a cube, which looks nice but is difficult to make sense of. Figures 9.3–9.7 show how such editing takes place and how it simplifies the analysis of signals that would otherwise be overlapped in two-dimensional experiments. Four-dimensional NMR, which we will not discuss in detail, takes this process one step further. In the four-dimensional case, we can imagine that the cube of the three-dimensional experiment has a vernier dial attached to it, marked with the chemical shift of the fourth correlation. As the dial is turned, cross-peaks within the three-dimensional cube appear and vanish according to whether or not they correlate with the chemical shift of the fourth dimension.

Although many three-dimensional experiments look (and are) quite complicated, they are generally put together from building blocks with which we are already familiar. Furthermore, the coherence transfer from protons to heteronuclei and

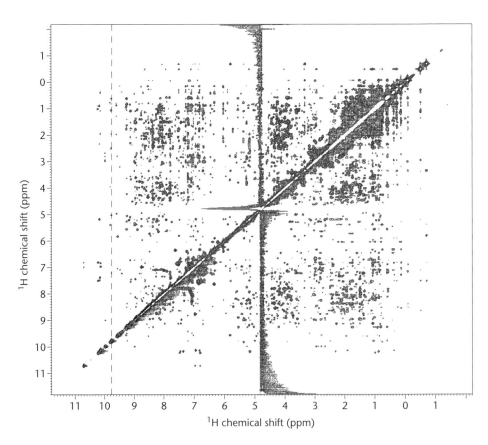

Figure 9.3 NOESY spectrum of 1 mM ^{15}N-labeled acireductone dioxygenase (ARD), a 179-residue metalloenzyme, in 90/10 H_2O/D_2O acquired at 750 MHz with ^{15}N decoupling applied during acquisition. Dashed line indicates an amide chemical shift where two amino acid residues (Q123 and A152) are overlapped (see Figure 9.4).

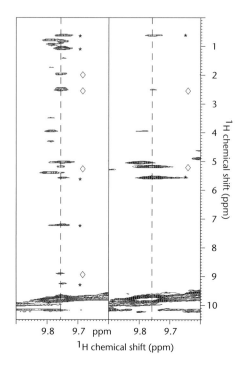

Figure 9.4 Two strips from the two-dimensional NOESY (left) and two-dimensional TOCSY (right) spectra of ^{15}N-labeled ARD in 90/10 H$_2$O/D$_2$O. The strips were extracted at the chemical shift indicated in Figure 9.3. *: Resonance correlated to A152; ◇: resonance correlated to Q123.

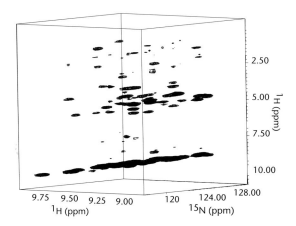

Figure 9.5 Three-dimensional (cubic) representation of the three-dimensional ^{15}N-edited NOESY spectrum of 1 mM ^{15}N-labeled ARD in 90/10 H$_2$O/D$_2$O acquired at 750 MHz shown in Figures 9.3 and 9.4.

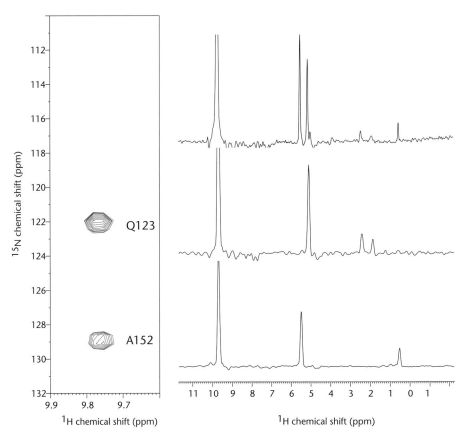

Figure 9.6 Three-dimensional ^{15}N-edited TOCSY spectrum of 1 mM ^{15}N-labeled ARD in 90/10 H$_2$O/D$_2$O acquired at 750 MHz. The two-dimensional strip (left) is extracted at 9.75 ppm in the F2 (indirect ^{1}H) and F3 (observe ^{1}H) dimensions. The two amide ^{15}N chemical shifts can clearly be seen. On the right are one-dimensional slices through the three-dimensional spectrum at 129 ppm in ^{15}N (bottom), and at 122 ppm in ^{15}N (middle). The top slice is taken from the two-dimensional TOCSY shown in Figure 9.4.

back tends to simplify the phase cycling required in order to select exclusively for particular coherence transfer pathways. In these cases, if the transfer is unsuccessful, no signal will result, so the pulse sequences themselves act as efficient filters for spurious signals. Most three-dimensional heteronuclear experiments are designed to require a maximum of four steps in a phase cycle, no more than is required for difference spectroscopy in HMQC or HSQC (i.e. subtraction of signals is required for the appropriate isotope editing). If gradient coherence selection is used, even that minimum phase cycle is not required, although quadrature detection in indirectly detected dimensions still requires some phase cycling. However, as concentrations of the proteins and polynucleic acids that are typically the subject of three-dimensional experiments are often quite low, multiple scans at each (t_1, t_2) combination are often necessary anyway in order to obtain sufficient signal-to-noise ratio in the experiment.

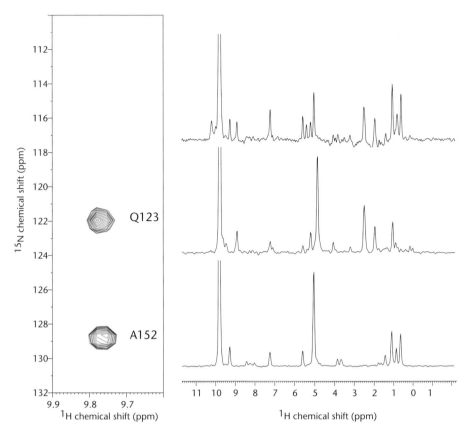

Figure 9.7 Three-dimensional ^{15}N-edited NOESY spectrum of 1 mM ^{15}N-labeled ARD in 90/10 H_2O/D_2O acquired at 750 MHz. The two-dimensional strip (left) is extracted at 9.75 ppm in the F2 (indirect ^{1}H) and F3 (observe ^{1}H) dimensions. The two amide ^{15}N chemical shifts can clearly be seen. On the right are one-dimensional slices through the three-dimensional spectrum at 129 ppm in ^{15}N (bottom), and at 122 ppm in ^{15}N (middle). The top slice is taken from the two-dimensional NOESY shown in Figure 9.3.

There are some general design guidelines that are followed in most three-dimensional NMR experiments currently in use. First, coherence is transferred from ^{1}H to an X spin at the beginning of the experiment (polarization transfer). After evolution on X (and perhaps further transfer to Y), the coherence is returned to ^{1}H for detection (**inverse detection**). Since the coherence transfer rate between two spins is proportional to their mutual coupling constant, the most efficient three-dimensional NMR experiments take advantage of coherence transfer between spins coupled with the largest J values. Often, these are one bond couplings. Also, as coherences in large molecules such as proteins or polynucleic acids tend to have short T_2 relaxation times, the time between the first pulse and acquisition is at a premium, and the shortest experiments are usually the most efficient.

As of the time of writing (2006), advances in probe technology and method modifications such as sample deuteration have made these design guidelines somewhat less general than they were in the 1990s. The use of cryogenically-operated probes (in which the detector coil and preamplifier electronics are helium-cooled to ~35 K in order to reduce thermal noise) provides the potential for direct observation of ^{13}C in multidimensional experiments, so that the back-transfer to ^{1}H is not always necessary for detection in a three-dimensional experiment. Secondly, many large proteins (>30 kDa in size) are now often prepared with uniform deuteration of non-exchanging ^{1}H spins, so that transfer from ^{13}C to ^{1}H is not always possible. Such considerations will be dealt with separately. First, we will summarize some of the most common components of multidimensional multinuclear NMR experiments. Generally, these components can be classified as polarization transfers, evolution (frequency labeling) steps, isotropic mixing, refocusing steps or filters.

Polarization transfer

Probably the single most common component of a multidimensional NMR experiment is a polarization transfer from one type of spin to another. This can be accomplished using either an HSQC- or HMQC-type transfer. Generally, HSQC transfers are preferred over HMQC transfers, because of the advantages discussed in Chapter 8. Polarization transfers occur in precisely the fashion described above for the one- and two-dimensional versions of these experiments, with delay times set to optimize polarization transfer for the particular coupling constant of the nuclei involved in the transfer.

The polarization transfer may either be forward (away from proton) or reverse (back to proton). Unless there is more than one proton dimension in the experiment (i.e. an indirectly detected ^{1}H dimension, usually t_1, as well as the directly detected ^{1}H dimension t_3), the first step in a heteronuclear three-dimensional experiment is usually a polarization transfer from ^{1}H to a heteronucleus. This can involve transfer either to ^{15}N or ^{13}C and provides the sensitivity enhancement inherent in INEPT. In many of the standard experiments for sequential assignment of proteins, the first transfer is from the amide ^{1}H to the directly bonded ^{15}N. Polarization transfer steps between nuclei other than ^{1}H are used to correlate two different types of heteronuclei (e.g. ^{15}N and ^{13}C).

Solvent suppression

A typical protein sample for NMR studies contains ~0.5 mM protein and nearly 100 M solvent protons in the form of H_2O! This presents a serious dynamic range problem, as the digitizer will be filled almost exclusively by the water signal unless something is done to minimize that signal. As such, efficient methods for the suppression of the water signal are essential. Multidimensional NMR experiments can be divided into two groups, homonuclear and heteronuclear experiments, and these generally employ different water suppression techniques.

We will first consider the simpler case of heteronuclear experiments (at least simpler from the point of view of solvent suppression). Such experiments contain polarization transfer steps involving magnetization transfer from ^{1}H to directly attached heteronuclei (usually ^{15}N or ^{13}C). Since water signals cannot pass through this transfer filter, the heteronuclear filter provides excellent water suppression, at least when gradient coherence selection is employed (Figure 9.8). It is important to remember that coherence selection by phase cycling requires the acquisition of FIDs in which all of the signals are present, including solvent, and it is only after the FIDs are subtracted by phase cycling that the unwanted signals are removed. This means that dynamic range problems still exist in each FID and must be dealt with, as described below.

Homonuclear experiments must employ some active solvent suppression technique. Originally, **presaturation** (applying a long low power pulse at the water resonance) was the method of choice. The result of this is to equalize the populations of spin states in the solvent, resulting in zero net magnetization for the solvent signal. Although presaturation provides excellent water suppression, the signals of exchangeable protons are also attenuated by presaturation by exchange with the saturated water protons, and this will, through spin diffusion, attenuate even the signals of non-exchangeable protons. While amide NH exchange is suppressed under acidic conditions, many proteins are unstable to acid, so this is not an ideal solution to the problem. Furthermore, protons that resonate near the solvent frequency are

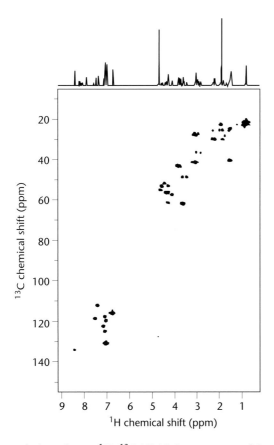

Figure 9.8 Natural abundance ^1H, ^{13}C HMQC spectrum of 5 mM LHRH (leuteinizing hormone-releasing hormone, a 10-residue peptide) in 90/10 H_2O/D_2O. Gradient coherence selection of protons bound to ^{13}C effectively suppresses the water signal (4.8 ppm).

also saturated, including those of non-exchangeable protons of interest in the sample. For proteins, this often includes many of the $C_\alpha H$ backbone protons that are critical for sequential assignments. For this reason, presaturation is rarely used in modern pulse sequences. The most widely used alternative to presaturation is WATERGATE (a lovely descriptive name for people of a certain age, since one is trying to "plug" the leak of a water signal!) (Reference 2, Figure 9.9). In WATER-GATE, a $\pi/2$ pulse is applied, bringing magnetization into the transverse plane. This pulse is followed by a gradient that dephases all transverse magnetization, including that due to water. An *antiselective* composite π pulse that is designed to have a null at the water frequency is then applied, followed by a second gradient pulse that has the effect of refocusing coherences that have been affected by the π pulse, while the water signal will remain dephased, and will not appear in the detected signal. The WATERGATE element is fairly simple to implement in multidimensional experiments and provides an excellent level of suppression (see Figure 9.10). The details of the WATERGATE sequence will be the focus of a Problem at the end of this chapter.

Another commonly employed improvement in water suppression is the water **flip-back** pulse (Reference 3). Historically, water suppression methods were designed to prevent transverse coherence from developing from the water spins, and there

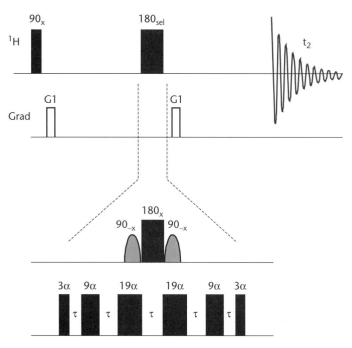

Figure 9.9 Basic WATERGATE element for pulse sequences. Two possible ways of producing the antiselective 180° pulse (exciting all frequencies over a range except that of water) are indicated by the dashed lines. The first employs two soft 90° pulses at the water frequency surrounding a hard 180° pulse. The second employs a binomial-like pulse train (3–9–19) to achieve the null at the water frequency.

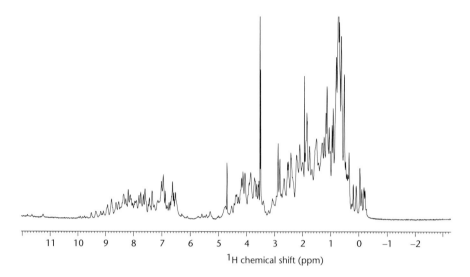

Figure 9.10 600 MHz ^{1}H spectrum of 1 mM adrenodoxin, a 108-residue ferredoxin, in 90/10 H_2O/D_2O using WATERGATE for solvent suppression.

was not much interest in where the water magnetization ended up in the course of a pulse sequence as long as it did not produce a signal during acquisition. One problem that can result from improper placement of the water magnetization is **radiation damping**, a phenomenon that is observed when an inherently intense coherence is stored along the –**z** axis (i.e. as $-\hat{\mathbf{I}}_z$). This operator represents a population inversion, and the energy that is stored in this inversion can be released coherently as a result of a self-induced precession in the laboratory frame. This rotates the coherence at the Larmor frequency of the resonance through the **x**, **y** plane, until the equilibrium population is restored along +**z**. This coherent emission results in a strong low-frequency ringing in the receiver coil and generates a distorted solvent peak in the spectrum. The process of radiation damping is precisely analogous to what occurs in a laser, in which energy stored as an inverted population over an electronic transition is released as a coherent pulse. A way to avoid radiation damping and other effects that might occur as the result of the state of the water coherence is to apply water-specific soft flip-back pulses to the water signal in order to maintain the water magnetization along the +**z** axis throughout the pulse sequence. Not only does the flip-back prevent radiation damping, but it also prevents unwanted attenuation of the signals of interest from spin diffusion or chemical exchange with water (as described above for presaturation). An example of an HSQC experiment incorporating water flip-back pulses is shown in Figure 9.11, with a comparison of the results obtained using presaturation and WATERGATE suppression with flip-back pulses shown in Figures 9.12 and 9.13.

Frequency-labeling periods and constant time NMR experiments

The second and third dimensions of three-dimensional experiments require that two indirectly detected time domains, t_1 and t_2, both independent of the acquisition time (t_3), be built up over the course of the experiment. As each data point in the t_1

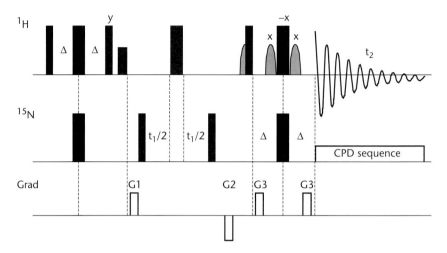

Figure 9.11 Pulse sequence for ^{1}H, ^{15}N HSQC experiment using WATERGATE for solvent suppression and employing a water flip-back pulse. The water-flip back pulse is indicated by the shaded pulse immediately before the WATERGATE element.

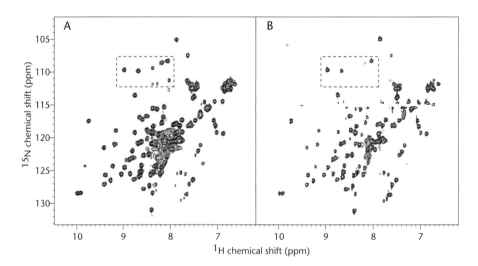

Figure 9.12 ^{1}H, ^{15}N HSQC spectra of 2 mM putidaredoxin (106 residues, 12.4 kDa molecular weight) in 90/10 H_2O/D_2O. (A) Spectrum was acquired with the pulse sequence shown in Figure 9.11. (B) Spectrum was acquired with presaturation for water suppression. Boxed regions are expanded in Figure 9.13.

and t_2 dimensions requires a separate acquisition, the number of points collected for the indirectly detected dimensions determines the duration of the experiment. In some cases, this can be as long as a week, depending on the sensitivity of the sequence, sample concentration and other variables that control signal-to-noise ratio. Fortunately, large numbers of data points in the indirectly detected dimensions are rarely necessary. While two-dimensional experiments are often acquired with as many as 512 complex points in t_1, three-dimensional experiments rarely

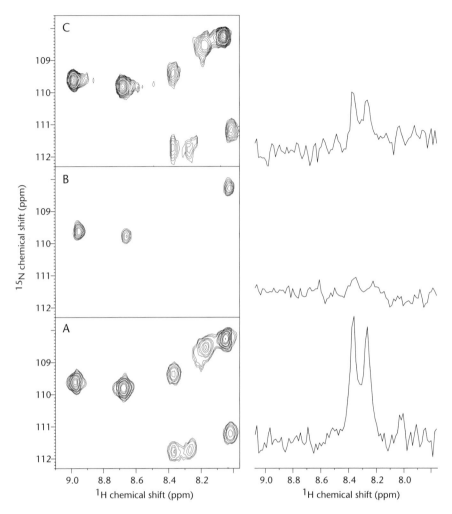

Figure 9.13 Portions of the ^{1}H, ^{15}N HSQC spectra of 2 mM putidaredoxin in 90/10 H$_2$O/D$_2$O are shown. (A and B) The two-dimensional spectra are the corresponding regions shown with dashed lines in Figure 9.12. (C) The spectrum was acquired with the pulse sequence described in Figure 8.22, employing gradient coherence selection as the only means of water suppression. The one-dimensional slices (^{15}N chemical shift = 111.8 ppm) from each of the two-dimensional spectra clearly demonstrate the improved sensitivity that can be obtained when water is kept along the **z** axis.

exceed 128 complex points in t_1 or t_2 in order to obtain sufficient resolution for most purposes, and 32 complex points in an indirectly detected dimension are often sufficient.

As with two-dimensional NMR experiments, pure phase and quadrature detection in the indirectly detected dimensions requires either States–Haberkorn or TPPI cycling in these dimensions. Of course, instead of a 2×2 set of data matrices, $2 \times 2 \times 2$ sets are required using the States method (rrr, rri, rir, irr, rii, iri, iir, iii, or the functional equivalent in TPPI). A variation on the States method, commonly called

States–TPPI, modulates the receiver phase by 180° at each increment of the evolution time for a given dimension. This results in axial peaks being shifted from the center of the spectrum to one edge, an advantage of the traditional TPPI method.

The problem with frequency-labeling periods, or evolution periods, is that they get longer as the three-dimensional experiment progresses. This exaggerates relaxation in the latter parts of the experiment, when both t_1 and t_2 (indirectly detected time domain) delays are long. This means that the signal-to-noise ratio of FIDs acquired later in the experiment will suffer from signal loss due to relaxation relative to the earlier FIDs. However, a scheme for minimizing this problem called **constant time evolution** has come into general use (Reference 4). This scheme combines the necessary evil of a relatively long J-coupling evolution for polarization transfer with the chemical shift evolution required for frequency labeling of the spin from which polarization transfer is to occur, thereby saving a step in the experiment. Furthermore, there is no increase in relaxation through the evolution time, so the interferogram that results from the constant time evolution is not affected by relaxation during the constant time evolution, making linear prediction of various types (including "negative time" linear prediction) more useful.

The heart of the constant time evolution scheme is a π pulse on the transverse coherence that "floats" through the J-evolution period during the course of the experiment. In many cases, the constant time evolution period t_1 immediately follows a polarization transfer step from ^{1}H. Consider the evolution time t_1 that occurs during a J-evolution (polarization transfer) period of duration $2T$. The time before the π pulse has a duration $T + t_1/2$, while the time after the pulse has a duration $T - t_1/2$, so that the first increment occurs with time periods of equal duration before and after the π pulse. The first period increases in duration as t_1 increments, while the second period decreases. It may not be immediately obvious why this results in frequency-labeling of the transverse coherence, so it is helpful to calculate the effect on a single transverse operator, $\hat{\mathbf{I}}_x$. During the first period $T + t_1/2$, $\hat{\mathbf{I}}_x$ evolves as expected with chemical shift (ignoring any coupling evolution):

$$\hat{\mathbf{I}}_x \to \hat{\mathbf{I}}_x \cos\Omega(T + t_1/2) + \hat{\mathbf{I}}_y \sin\Omega(T + t_1/2) \tag{9.1}$$

The application of a π_x pulse at this point gives:

$$\hat{\mathbf{I}}_x \to \hat{\mathbf{I}}_x \cos\Omega(T + t_1/2) - \hat{\mathbf{I}}_y \sin\Omega(T + t_1/2) \tag{9.2}$$

Using the definitions of the Cartesian operators in terms of raising and lowering operators, we can describe the evolution of each term during the second time period (leaving out the modulation during the first period for the moment) with:

$$\hat{\mathbf{I}}_x = \tfrac{1}{2}(\hat{\mathbf{I}}_+ + \hat{\mathbf{I}}_-) \xrightarrow{T - t_1/2} \left(\hat{\mathbf{I}}_+ \exp(-i\Omega(T - t_1/2)) + \hat{\mathbf{I}}_- \exp(i\Omega(T - t_1/2))\right)$$

$$\hat{\mathbf{I}}_y = \tfrac{1}{2i}(\hat{\mathbf{I}}_+ - \hat{\mathbf{I}}_-) \xrightarrow{T - t_1/2} \left(\hat{\mathbf{I}}_+ \exp(-i\Omega(T - t_1/2)) - \hat{\mathbf{I}}_- \exp(i\Omega(T - t_1/2))\right) \tag{9.3}$$

Returning to the Cartesian basis:

$$\frac{1}{2}\left(\left[\hat{\mathbf{I}}_x + i\hat{\mathbf{I}}_y\right]\exp(-i\Omega(T-t_1/2)) + \left[\hat{\mathbf{I}}_x - i\hat{\mathbf{I}}_y\right]\exp(i\Omega(T-t_1/2))\right)$$
$$-\frac{1}{2i}\left(\left[\hat{\mathbf{I}}_x + i\hat{\mathbf{I}}_y\right]\exp(-i\Omega(T-t_1/2)) - \left[\hat{\mathbf{I}}_x - i\hat{\mathbf{I}}_y\right]\exp(i\Omega(T-t_1/2))\right) \quad (9.4)$$

Collecting terms, and using the relationships between trigonometric and complex exponentials to convert to trigonometric functions:

$$\hat{\mathbf{I}}_x \cos(\Omega(T-t_1/2)) + \hat{\mathbf{I}}_y \sin(\Omega(T-t_1/2))$$
$$+\hat{\mathbf{I}}_x \cos(\Omega(T-t_1/2)) - \hat{\mathbf{I}}_y \sin(\Omega(T-t_1/2)) \quad (9.5)$$

These two equations represent the evolution expected after the π pulse. Multiplying by the modulation terms that evolved prior to the π pulse, we obtain:

$$\left\{\hat{\mathbf{I}}_x \cos(\Omega(T-t_1/2)) + \hat{\mathbf{I}}_y \sin(\Omega(T-t_1/2))\right\}(\cos(\Omega(T+t_1/2)))$$
$$+\left\{\hat{\mathbf{I}}_x \sin(\Omega(T-t_1/2)) - \hat{\mathbf{I}}_y \cos(\Omega(T-t_1/2))\right\}(\sin(\Omega(T+t_1/2))) \quad (9.6)$$

Using the trigonometric relationships that relate multiplication of sines and cosines, the following terms are obtained:

$$\hat{\mathbf{I}}_x\left[\tfrac{1}{2}\cos(\Omega(T+t_1/2-T+t_1/2)) + \tfrac{1}{2}\cos(\Omega(T+t_1/2+T-t_1/2))\right]$$
$$+\hat{\mathbf{I}}_x\left[\tfrac{1}{2}\cos(\Omega(T+t_1/2-T+t_1/2)) - \tfrac{1}{2}\cos(\Omega(T+t_1/2+T-t_1/2))\right]$$
$$\hat{\mathbf{I}}_y\left[\tfrac{1}{2}\sin(\Omega(T+t_1/2-T+t_1/2)) + \tfrac{1}{2}\sin(\Omega(T+t_1/2+T-t_1/2))\right]$$
$$-\hat{\mathbf{I}}_y\left[\tfrac{1}{2}\sin(\Omega(T-t_1/2-T-t_1/2)) - \tfrac{1}{2}\sin(\Omega(T-t_1/2+T+t_1/2))\right] \quad (9.7)$$

The last terms in each of the four lines in the preceding equation cancel, leaving only the sums of the first terms, that yield:

$$\hat{\mathbf{I}}_x \cos(\Omega t_1) + \hat{\mathbf{I}}_y \sin(\Omega t_1) \quad (9.8)$$

i.e. the evolution expected of a normally incremented t_1 period.

The choice of the length of the constant time evolution delay is critical. Usually, the constant time evolution involves product operators of the type $\hat{\mathbf{I}}_x\hat{\mathbf{S}}_z$ that are generated by polarization transfer to spin $\mathbf{I}$ from spin $\mathbf{S}$ at the beginning of the evolution period. The length of the constant time evolution ($2T$) is chosen so as to preserve a particular characteristic of the coherence as it was in the beginning of the constant time evolution. For example, in the CT-HNCA experiment shown in Figures 9.14 and 9.15, $2T$ is set to $1/J_{NH}$, so that the antiphase character of the $\hat{\mathbf{N}}_x\hat{\mathbf{H}}_z$ coherence

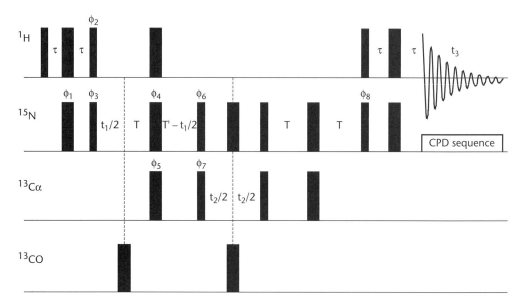

Figure 9.14 Diagram of a constant time phase-cycled HNCA experiment. Narrow lines represent $\pi/2$ pulses, broad lines represent π pulses. Delays are set as follows: $\tau = 1/4J_{NH}$, $2T = 1/J_{NH}$, $T' = T$ + length of the π pulse on on C = O. Phase cycling is: $\phi1$: x, –x; ϕ 2: y, –y; ϕ 3: x; ϕ 4: r(x), 4(y), 4(–x), 4(–y); ϕ 5: 16(x), 16(–x); ϕ 6: 16(y), 16(–y); ϕ 7: x, –x, –x, x; ϕ 8: y; Acq: =2(x, –x, –x, x, –x, x, x, –x), 2(–x, x, x, –x, x, –x, –x, x).

that is generated after the first polarization transfer (from ^{1}H to ^{15}N) is maintained during the subsequent periods and is refocused only at the end of the sequence, in time for detection during t_3.

A limitation of the constant time method is that, as the frequency-labeling period is limited by the polarization transfer step, digital resolution in the corresponding dimension is limited by the length of the coupling evolution time $2T$. The length of the t_1 increment (the dwell time) is determined by the spectral width, so only a fixed number of points can be acquired during $2T$, determined by $2T/(\text{dwell})$.

Shaped and selective pulses

Since the chemical-shift range covered by various types of ^{13}C nuclei (e.g. aliphatic, carbonyl, etc.) is wide and the shift differences between these various types can be large, most heteronuclear experiments treat carbonyl, aromatic and sp^3-hybridized ^{13}C separately. ^{13}C pulses are usually designed to be sufficiently selective so that only one type of carbon spin is excited per pulse. For this reason, one often sees polarization transfer steps between different types of carbon atoms in multidimensional NMR experiments. The simplest way to achieve this selectivity is to apply a rectangular pulse centered on the frequency of interest, with the pulse length and power adjusted to produce an excitation null at the other carbon frequencies of interest. For example, in an experiment that correlates carbonyl carbons with C$_\alpha$ carbons, an ideal pulse would affect a $\pi/2$ rotation for the ^{13}CO spins while resulting in a 2π

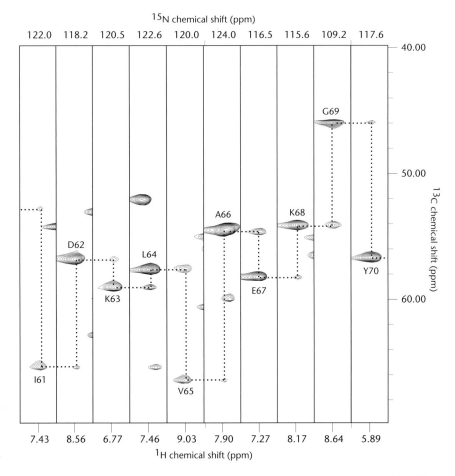

Figure 9.15 Strips from the constant time HNCA spectrum of 1 mM ^{15}N, ^{13}C-labeled ARD (acireductone dioxygenase) in 90% H_2O/D_2O acquired at 500 MHz. A sequential walk from isoleucine 61 through tyrosine 70 is shown.

(null) rotation for the $^{13}C_\alpha$ spins. Using the frame of reference shown in the figure in Problem 3.11 and performing the calculations as in Chapter 3 (Problem 10.3) we obtain the following results:

$$PW_{90} = \frac{\sqrt{15}}{4\Delta B} \quad \text{and} \quad PW_{180} = \frac{\sqrt{3}}{2\Delta B} \tag{9.9}$$

where ΔB is the chemical-shift separation in Hz between ^{13}CO and $^{13}C_\alpha$. The expression for PW_{180} describes the requirement for a π pulse on ^{13}CO with a null at $^{13}C_\alpha$.

Implicit in the above discussion is the assumption that one can switch the center frequency of a pulse rapidly in the course of a pulse sequence, so that if we wish to excite carbonyl carbons with the carrier frequency set at $^{13}C_\alpha$, we could employ a frequency jump immediately prior to center the excitation on the ^{13}CO region.

Theoretically, this is possible; however, on the majority of spectrometers, this would result in a loss of phase coherence, thereby leading to spectral artifacts (see Figure 9.16). A cleaner approach is to apply a phase ramp to the selective pulse to achieve an effective frequency change. Practically, the selective pulse is created digitally in the electronics by dividing it into small increments (i.e. points) that can be phase- and amplitude-modulated individually. If we desire to excite carbonyls while our transmitter remains at the frequency of $^{13}C_\alpha$, we use the following equation to calculate the phase advance of each of the pulse units:

$$\left(\frac{\theta}{360}\right)\left(\frac{1}{\tau}\right) \equiv \Delta\omega \qquad (9.10)$$

where $\Delta\omega$ is the difference between the transmitter and the desired frequency, τ is the length of the selective pulse divided by the number of points used to create the pulse digitally, and θ is the phase increment necessary to achieve the frequency shift (Figure 9.17). Construction of such a pulse is the subject of Problems 9.3 and 9.4.

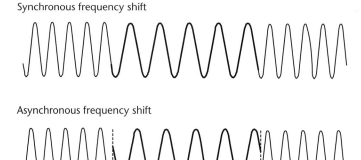

Synchronous frequency shift

Asynchronous frequency shift

Figure 9.16 Synchronous versus asynchronous frequency-shifting. Pulses generated at different frequencies (or different power levels, as in decoupling) can either be synchronous with each other (that is, continuous phase advance) or asynchronous (discontinuous with respect to phase).

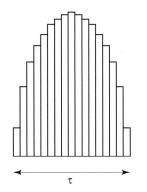

Figure 9.17 A selective pulse can be divided into small elements (points), with the minimum length of each element determined by the spectrometer hardware. The sum of the elements (τ) is the total length of the pulse. To achieve off-resonance excitation, each element will have a phase increment (θ) determined as in Equation 9.10.

From the basic elements described above, virtually all of the published triple-resonance experiments have been built. An analysis of these elements found in a standard triple resonance experiment HN(CO)CA is shown in Figure 9.18.

Composite pulse decoupling and spin-locking

One final but very important consideration in multidimensional NMR experiment design is the use of *composite pulse decoupling* (CPD). This topic was introduced in Chapter 7, but it is worth talking about it here too, since the efficiency of decoupling can determine the success of a given experiment and indeed the safety of the NMR sample and probe. It is often necessary to decouple one type of spin from another for reasons of reducing spectral complexity and improving the signal-to-

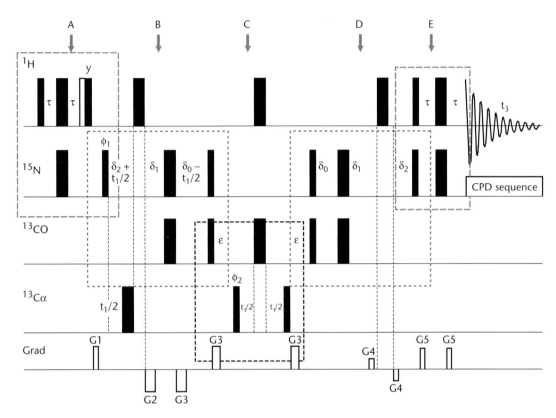

Figure 9.18 HN(CO)CA pulse sequence illustrating the incorporation of basic elements such as polarization transfer into multidimensional experiments. (A) INEPT transfer from amide protons to amide nitrogen, with crusher gradient. (B) Constant time evolution for ^{15}N coherence, with coherence transfer to carbonyl carbon, while refocusing coupling due to ^{1}H and C_α. (C) HMQC transfer from carbonyl carbon to C_α, followed by frequency labeling at C_α shift, and return to carbonyl carbon. (D) Reverse transfer from carbonyl carbon to ^{15}N, followed by (E) reverse INEPT to ^{1}H for detection. Pulse sequence details are as follows: $\tau = 2.25$ ms $[2\tau = 1/(2J_{NH})]$, $\delta_0 = 12$ ms $[2\delta_0 = 1/(3J_{NCO})]$, $\delta_1 = 9.25$ ms, $\delta_2 = 2.75$ ms $[2\delta_2 = 1/(2J_{NH})]$, $\varepsilon = 6.6$ ms $[\varepsilon = 1/(3\ J_{C_\alpha CO})]$, $G1 = G5 = 150$ μs at 9 G/cm, $G2 = 1.5$ ms at 9 G/cm, $G3 = 450$ μs at 9 G/cm, $G4 = 100$ μs at 3.5 G/cm. $G1$ is a "crusher" gradient that dephases any coherence in the transverse plane while the desired coherences are stored along the **z** axis. Quadrature detection is achieved by incrementing ϕ_1 and ϕ_2 (in t_1 and t_2, respectively).

noise ratio. In almost every heteronuclear two- and three-dimensional experiment, heteronuclei are decoupled from ^{1}H during acquisition using broadband CPD sequences, so that only ^{1}H singlets are observed. However, even prior to acquisition, CPD is often applied for a variety of reasons, including spin-locking. In the HCCH-TOCSY experiment, for example, coherence is generated first on ^{1}H and then transferred via a polarization transfer step to a directly bonded ^{13}C. At this point, a CPD sequence is applied that spin-locks the appropriate ^{13}C spins with the frequency-labeling information from the original ^{1}H passed to other ^{13}C atoms in the spin system via Hartmann–Hahn transfer. (Because direct ^{13}C–^{13}C couplings are used for magnetization transfer rather than two- and three-bond couplings, spin-lock times are considerably shorter in ^{13}C Hartmann–Hahn transfers than for ^{1}H transfers over the same spin systems.) After the spin-locking CPD is turned off, the ^{13}C spins are released to evolve with free precession during t_2, and then coherence is transferred back to the attached ^{1}H spins via a back polarization transfer for detection during t_3. The resulting three-dimensional spectrum correlates ^{1}H spins with multiple carbons in the same spin system (e.g. all part of the same amino acid side chain in a protein), providing a powerful tool for making identification of side chain spin systems (Figure 9.19).

A wide variety of CPD sequences are available, and some are more suited to particular tasks than others. The schemes can be sorted by whether they are **synchronous** (i.e. phase-coherent with pulses that precede and follow the CPD sequence) or **asynchronous**. For spin-locking, as in TOCSY and HCCH-TOCSY, it is necessary to insure that CPD is synchronous, because the states of the spins before and

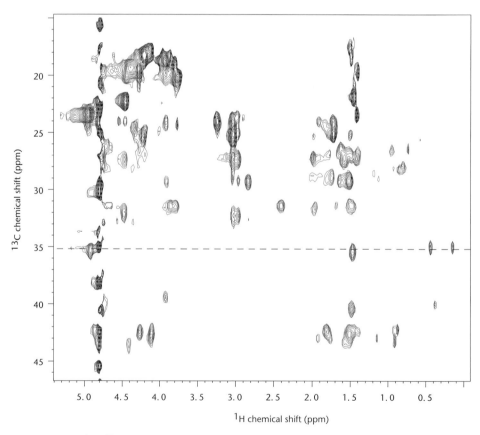

Figure 9.19 ^{1}H, ^{13}C-Plane taken at 1.44 ppm (indirect proton dimension) from the three-dimensional HCCH-TOCSY spectrum of 1 mM ^{15}N, ^{13}C-labeled ARD (acireductone dioxygenase) in 90% H_2O/D_2O acquired at 600 MHz. The dashed line indicates the V122 spin system. The ^{13}C shift of 35.3 ppm corresponds to the C_β resonance ($^1H_\beta$ is at 1.44 ppm). Correlations can be seen from the C_β to the H_γ and H_α.

after the CPD sequence are critical for the success of the experiment. That is, the spins that are detected are those that have been subjected to the spin lock. However, when all one wants to do is remove the effects of coupling, synchronous CPD can result in artifacts, sometimes called "decoupling sidebands", that result from the generation of unwanted coherences. In this case, the use of asynchronous decoupling (random phase shifts relative to what occurs before or after the CPD) insures that any artifacts resulting from the decoupling are not reinforced during signal averaging.

All CPD sequences incorporate "inversion elements" (pulse trains that result in population inversions of the decoupled spins) that are repeated or combined so as to return spins to their starting configurations, so that the time-averaged magnetization of the decoupled spins is zero. The sequences are typically designed to provide uniform excitation profiles over a desired chemical-shift range. The particular sequence employed depends on the spin to be decoupled and the frequency range over which the decoupling is to be deployed. For ^{1}H decoupling and spin locking, some long-standing sequences that are still widely used are MLEV and WALTZ decoupling schemes (References 5 and 6). Because the ^{1}H chemical shift range is fairly narrow, the correct bandwidth profile is relatively easy to achieve, and the major concern for ^{1}H decoupling is to insure that only as much power is applied as is required for a particular task. Too much power can result in sample heating, particularly for samples that are at high ionic strength.

For other nuclei, particularly ^{13}C, decoupling is a much more demanding business. Because ^{13}C chemical-shift ranges are large and increase with field, simple rectangular pulse decoupling (i.e. the pulses composing the CPD are of a length so as to excite a particular bandwidth) can be problematic at higher fields, requiring shorter pulses at high powers in order to obtain the needed excitation profile. Off-resonance effects (tip angles near the edge of the excitation envelope different from that at the center) also become more serious. With the introduction of ^{13}C spin-locking experiments such as HCCH-TOCSY, it was found that if composite pulses were used to compensate for off-resonance effects, excitation profiles were more even over the effective range of the decoupling. One commonly employed CPD sequence that incorporates composite pulses is DIPSI (Reference 7), that has been employed with great success for both ^{1}H and ^{13}C spin-locking.

As magnetic fields get larger, it becomes more difficult to use hard pulse trains for CPD. The increased bandwidth requires shorter pulses at higher power to obtain the same spectral excitation envelope, so arcing of probes (i.e. sparking across the coils during a pulse) becomes a significant problem. In answer to this, a new generation of CPD sequences based on **adiabatic** pulses has entered into use. Unlike hard pulses that are formed from a single frequency and whose excitation envelope depends upon pulse length, adiabatic pulses incorporate a frequency sweep along with amplitude modulation to insure that spins of different chemical shifts are excited to the same degree. Adiabatic pulses can be used to perform relatively uniform excitation for a given group of spins despite variations in chemical shift. Such pulses are named for the adiabatic condition, in which the rate of tip of the spin due to the pulse is slow compared to the spin's precessional frequency. This "slow passage" allows the precession of spins to track the B_1 field accurately due to the applied RF, with less sensitivity to chemical-shift effects.

The most popular CPD sequences incorporating adiabatic pulses are the plumply named WURST series (Reference 8). The WURST sequences provide fairly wide and uniform excitation bandwidths with lower power requirements than for hard pulse sequences with similar excitation envelopes. These have become the CPD sequences of choice for applications requiring broadband decoupling such as HCCH-TOCSY at high magnetic fields.

Dealing with very large biomolecules in solution: deuteration and direct ^{13}C detection

In addition to all of the other problems that are encountered in biological NMR (i.e. large solvent signals, chemical exchange, low sample concentrations and signal overlap), the very nature of the observed NMR signals from biological macromolecules hampers the experiment. In Chapter 4, the importance of the correlation time in determining the relaxation behavior of nuclear spins was discussed. As molecules increase in size, they tumble more slowly and the spin–spin interactions that result in T_2 relaxation become more efficient. In turn, this leads to line-broadening and decrease in the effective lifetimes of coherences that are generated at the beginning of a multiple-pulse NMR sequence. A more complete discussion of the theory and interpretation of relaxation phenomena as applied to biological molecules will be found in Chapter 11. However, for now it is enough to point out that for protein molecules more than ~30 kDa effective molecular weight, ^{1}H T_2 relaxation rates are short enough to seriously degrade the efficiency of coherence transfer for most multidimensional NMR experiments.

A number of approaches have been used to circumvent the apparent 30 kDa limit on NMR characterization of proteins. The most straightforward, at least in principle, is the use of random deuteration to dilute the number of ^{1}H spins present in the molecule. ^{1}H spins are the primary source of dipole–dipole relaxation for other protons as well as ^{13}C and ^{15}N spins. If a sample of protein is prepared using perdeuterated growth medium (usually by overexpression in bacterial or other cultured cells), most of the nonexchangeable protons (i.e. those bonded to carbon) will be replaced by ^{2}H. ^{2}H has a much lower gyromagnetic ratio than ^{1}H, and is not as efficient at promoting dipole–dipole relaxation in nearby nuclei. This results in a significant line-narrowing for the remaining ^{1}H spins (typically the amide NH groups that exchange with buffer during purification/refolding in protonated medium). While ^{2}H does exhibit observable couplings with other spins (and those couplings show up as triplets, since the $\mathbf{I} = 1$ ^{2}H nucleus has three accessible spin states) one-bond coupling constants of ^{2}H are much smaller than the corresponding one-bond couplings to ^{1}H. Furthermore, the chemical-shift range of ^{2}H is considerably smaller than that even of ^{1}H, so decoupling of ^{2}H during evolution and detection periods is usually a simple matter.

The most obvious drawback to perdeuteration is that the same spins that are usually used for polarization transfer in standard triple-resonance NMR experiments (i.e. those correlating ^{1}H, ^{15}N and ^{13}C) as well as those used for NOEs in structure determination are no longer there. As such, some important and useful NMR experiments that require a high density of ^{13}C-attached ^{1}H spins (e.g. HCCH-TOCSY and ^{1}H,^{13}C HSQC-NOESY) are not useful for such samples. Furthermore, for proteins that are not amenable to refolding, even nominally exchangeable protons attached to ^{15}N that are not solvent accessible will not be detected. In recent years, a large number of sequences that are specifically designed for use with deuterated proteins have been developed.

Another important development has been the introduction of probes in which transceiver coils and preamplifiers are cryogenically cooled by helium gas ("cold probes" or "cryoprobes", depending on the manufacturer). This cooling (to ~35 K) results in the reduction of thermal noise in critical detector/preamplifier circuitry, and in the case of ^{1}H detection, this can give up to 4-fold improvement in S/N, so lower concentrations of sample can be used. This is important for experiments with biological macromolecules of low intrinsic solubility (e.g. membrane proteins) or of high molecular weight. The use of cryogenically cooled probes equipped with a cooled ^{13}C preamplifier has permitted the introduction of a new class of two-dimensional and three-dimensional NMR experiments that employ direct detection of ^{13}C. Typically used with uniformly ^{13}C and ^{15}N labeled and deuterated or partially deuterated macromolecules, such experiments bypass the time-consuming polarization steps from and to ^{1}H required by ^{1}H-detected experiments, yielding a series of relatively simple two-dimensional and three-dimensional pulse sequences that correlate ^{13}C and ^{15}N directly (References 9 and 10). These experiments take advantage of the relatively long relaxation times of deuterated ^{13}C spins, permitting multiple-spin correlations even in large (>40 kDa) proteins.

Interference patterns in heteronuclear relaxation: TROSY

In Chapter 4, we discussed the primary mechanisms for relaxation of $\mathbf{I} = 1/2$ nuclei in solution and the dependence of these mechanisms upon the correlation time, τ_c. The correlation time provides a measure of the rate at which the molecule reorients in the magnetic field, and can be thought of as the mean time required for a molecule to reorient 1 rad in any direction. Recall that nuclear spin relaxation is induced by the time-dependent fluctuations in the local electromagnetic environment, and that slow tumbling encourages low frequency transitions, while faster tumbling can also encourage higher frequency transitions. [One can make the analogy with pulses: short pulses, corresponding to rapid motions of a molecule, have broad excitation bandwidths, and long pulses (slower motions) have fewer frequency components.] Spin–spin interactions that result in line-broadening (T_2) in large molecules such as proteins are typically dominated by the low-frequency terms [zero-quantum transitions $k_0 \propto r_{IS}^{-6} J(\omega_I - \omega_S)$ as shown in Equation 4.11]. These transitions result in the incoherent transfer of magnetization between spins and consequently a loss of macroscopic signal at the time of acquisition.

Another important source of relaxation, particularly at higher magnetic fields such as those commonly used for NMR of macromolecules, is *chemical-shift anisotropy* (CSA). Recall that CSA is the result of incomplete averaging of the chemical-shift tensor to a scalar quantity (the observed chemical shift) as a function of molecular tumbling. As the molecule reorients and the chemical shift of a spin is perturbed, time-dependent fluctuations of the spacing of the energy levels of that spin take place, resulting in relaxation. Again, the rate at which the molecule tumbles, as well as the spacing of the energy levels (a function of the strength of the magnetic field employed) modulate the efficiency of CSA relaxation. CSA relaxation is particularly effective for nuclei that are not in a symmetric electronic environment (more p-orbital than s-orbital character of the bonding electrons) such as ^{13}C and ^{15}N. These nuclear spins typically show large differences in chemical shift for different types of environments, and as field strengths increase, CSA may become a dominant relaxation mechanism for such spins.

The fluctuating fields due to CSA and dipole–dipole (dd) interactions at a given spin have directionality as well as frequency. As a molecule tumbles, the CSA and dd fields superimpose at the position of the spin under consideration and can interfere with each other in a constructive or destructive fashion, a phenomenon known as **cross-correlation**. For example, if two spin 1/2 nuclei, I and S, are dipole-coupled to each other, spin I can detect spin S in two possible states, $|\alpha\rangle$ and $|\beta\rangle$. The sign of the dipolar field at I is determined by the spin state of S, and as such, the cross-correlation terms between the CSA and dd terms will have opposite signs depending upon which transition of spin I is being observed, that in which the coupled S spin is $|\alpha\rangle$ and that in which spin S is $|\beta\rangle$. For one I transition (corresponding to one of the two lines in the I doublet resulting from scalar coupling to S), the cross-correlation term will enhance relaxation, but for the other I transition, it will make relaxation less efficient. The net result of this is different line widths for the two lines of the IS doublet, reflecting the differential relaxation of the two transitions of I. An example of such differential broadening is shown in Figure 9.20, and a readable

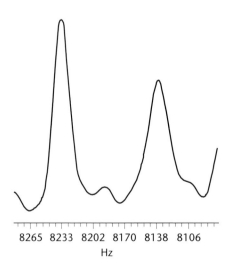

8265 8233 8202 8170 8138 8106

Hz

Figure 9.20 A slice taken through the ^1H dimension of the ^1H, ^{15}N HSQC spectrum of 1 mM ^{15}N-labeled ARD in 90% H_2O/D_2O acquired at 600 MHz showing a ^1HN signal split by J_{NH} (~ 95 Hz). The HSQC was acquired without decoupling during t_1, so the ^{15}N coupling is clearly visible. The line widths are considerably different.

description of this phenomenon as applied to a ^{15}N–^{1}H pair can be found in Reference 11.

The differential broadening becomes particularly interesting when one considers the implications for a two-dimensional heteronuclear experiment such as HSQC. Under normal circumstances, refocusing of the coupling during t_1 evolution on I by placing a π pulse on S at $t_1/2$ as well as decoupling of I during acquisition (t_2) results in singlet structure in both dimensions for the observed correlation between I and S in the transformed two-dimensional spectrum. As a result, any differential line broadening is lost. However, if one performs the HSQC experiment without refocusing or decoupling in either dimension, one would obtain, for a simple coupled IS pair, a set of four peaks, split by $^1J_{IS}$ in both dimensions. As in the one-dimensional experiment shown in Figure 9.21, the peaks would show the effects of differential line-broadening in both dimensions. For one of the four peaks, the cross-terms in both dimensions would reinforce the CSA and dd relaxation, resulting in very efficient relaxation (with a correspondingly broad line). For two of the peaks, the cross-terms would be opposite in sign during t_1 and t_2, resulting in an intermediate line width after transformation. For the fourth peak, the cross-terms would both interfere destructively with the other relaxation terms, resulting in a narrower line width in both dimensions. If one could select for just this line in the quartet, the size of molecules amenable to multidimensional NMR methods using HSQC-type correlations for detection could be considerably increased. Such selection is the basis of the TROSY (*transverse relaxation optimized spectroscopy*) experiment first described by Pervushin *et al.* (Reference 12). The TROSY pulse sequence (Figure 9.21) is designed to select only the transition in which the cross-correlation terms suppress T_2 relaxation in both time domains. Figures 9.22 and 9.23 show the implementation of TROSY.

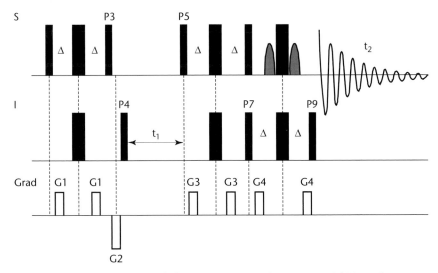

Figure 9.21 Pulse sequence of the TROSY experiment. $\Delta = 1/4J_{NH}$, phases on all pulses are **x** except as follows (eight-step phase cycle). P3: 8{*y*}. P4: {*y*, –*y*, –*x*, *x*, *y*, –*y*, –*x*, *x*}. P5: 4{*y*}, 4{–*y*}. P7: 8{*y*}. P9: 4{*x*}, 4{–*x*}. Receiver: {*x*, –*x*, –*y*, *y*, *x*, –*y*, *y*}. Gradients with the same label have the same intensity and duration. The gradient between P3 and P4 is a "crusher gradient".

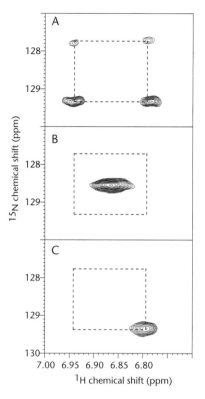

Figure 9.22 (A) ^{1}H, ^{15}N HSQC correlation from oxidized cytochrome P450$_{cam}$, a 414-residue monomeric heme-containing enzyme, acquired without refocusing of the $^1J_{NH}$ coupling in ω_1 and without decoupling of ^{15}N in ω_2, showing the different line widths of individual components of the multiplet due to interference of dd and CSA relaxation (see text). (B) Same correlation, but with refocusing of the $^1J_{NH}$-coupling in ω_1 and decoupling of ^{15}N in ω_2. (C) Same correlation but with TROSY phase cycling to remove all but the narrowest line of the multiplet.

In order to analyze the TROSY sequence, we need to expand the density matrix in terms of a new basis set that is convenient for describing single transitions (corresponding to single lines in the NMR experiment). Recall that the Cartesian basis is convenient for describing the evolution of coherences under the influences of pulses and **z**-directed operators, while the single-element basis allows one to determine the evolution of particular orders of coherence under a variety of influences. However, in TROSY, we are concerned with differential effects upon single transitions that would otherwise behave identically under the influences of pulses and **z**-directed operators. For this purpose, we need to observe the evolution of the density matrix using the **single-transition basis** set.

Consider the two $I = 1/2$ spins of the IS spin system with transitions as diagrammed in Figure 9.24. We wish to consider individual transitions within this diagram, for

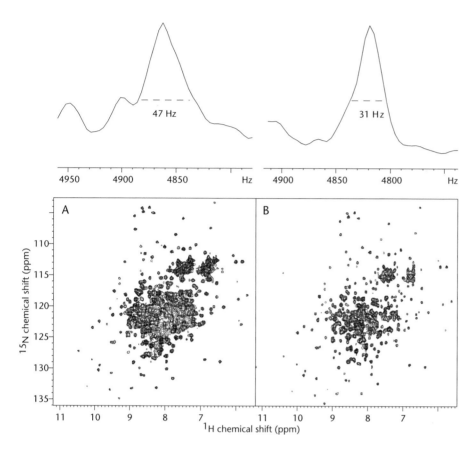

Figure 9.23 (A) ^{1}H, ^{15}N HSQC spectrum of oxidized cytochrome P450$_{cam}$ obtained at 600 MHz ^{1}H with the line width of a single correlation in ω_2 shown above the spectrum. (B) TROSY version of the same spectrum, showing the reduced line width for the same correlation as in (A).

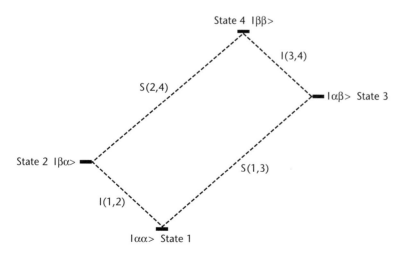

Figure 9.24 Diagram of transitions in the single-transition basis for the *IS* doublet.

example $\mathbf{I}(1,3)$ or $\mathbf{S}(1,2)$. The single-transition operators are defined relative to the Cartesian and single-element operator basis sets as:

$$\hat{\mathbf{S}}_x^{(1,2)} = \hat{\mathbf{I}}_\alpha \hat{\mathbf{S}}_x \quad \hat{\mathbf{S}}_y^{(1,2)} = \hat{\mathbf{I}}_\alpha \hat{\mathbf{S}}_y \quad \hat{\mathbf{S}}_+^{(1,2)} = \hat{\mathbf{I}}_\alpha \hat{\mathbf{S}}_+ \quad \hat{\mathbf{S}}_+^{(1,2)} = \hat{\mathbf{I}}_\alpha \hat{\mathbf{S}}_+$$

$$\hat{\mathbf{S}}_x^{(3,4)} = \hat{\mathbf{I}}_\beta \hat{\mathbf{S}}_x \quad \hat{\mathbf{S}}_y^{(3,4)} = \hat{\mathbf{I}}_\beta \hat{\mathbf{S}}_y \quad \hat{\mathbf{S}}_+^{(3,4)} = \hat{\mathbf{I}}_\beta \hat{\mathbf{S}}_+ \quad \hat{\mathbf{S}}_+^{(3,4)} = \hat{\mathbf{I}}_\beta \hat{\mathbf{S}}_+$$

$$\hat{\mathbf{I}}_x^{(1,3)} = \hat{\mathbf{I}}_x \hat{\mathbf{S}}_\alpha \quad \hat{\mathbf{I}}_y^{(1,3)} = \hat{\mathbf{I}}_y \hat{\mathbf{S}}_\alpha \quad \hat{\mathbf{S}}_-^{(1,2)} = \hat{\mathbf{I}}_\alpha \hat{\mathbf{S}}_- \quad \hat{\mathbf{I}}_-^{(1,3)} = \hat{\mathbf{I}}_- \hat{\mathbf{S}}_\alpha$$

$$\hat{\mathbf{I}}_x^{(2,4)} = \hat{\mathbf{I}}_x \hat{\mathbf{S}}_\beta \quad \hat{\mathbf{I}}_y^{(2,4)} = \hat{\mathbf{I}}_y \hat{\mathbf{S}}_\beta \quad \hat{\mathbf{S}}_-^{(3,4)} = \hat{\mathbf{I}}_\beta \hat{\mathbf{S}}_- \quad \hat{\mathbf{I}}_-^{(2,4)} = \hat{\mathbf{I}}_- \hat{\mathbf{S}}_\beta \qquad (9.11)$$

The superscripted (i, j) indicate the particular transitions that each operator represents, as shown in Figure 9.24. Note that all of these product operators are the result of the combination of a population operator on one spin and a raising or lowering operator on the other. We will also need the relationships between polarization operators and the identity operator given in Equation 6.15:

$$\hat{\mathbf{I}}_z = \tfrac{1}{2}(\hat{\mathbf{I}}_\alpha - \hat{\mathbf{I}}_\beta)$$

$$\tfrac{1}{2}\hat{\mathbf{I}} = \tfrac{1}{2}(\hat{\mathbf{I}}_\alpha + \hat{\mathbf{I}}_\beta)$$

$$\hat{\mathbf{I}}_\alpha = \tfrac{1}{2}\hat{\mathbf{I}} + \hat{\mathbf{I}}_z$$

$$\hat{\mathbf{I}}_\beta = \tfrac{1}{2}\hat{\mathbf{I}} - \hat{\mathbf{I}}_z \qquad (9.12)$$

With these definitions in hand, we can consider how the TROSY experiment results in cancellation of the broader components of the *IS* multiplet, while reinforcing the narrow component. At first, the experiment proceeds as the normal HSQC experiment. The initial $\pi/2$ pulse on ^{1}H (S) gives $-\hat{\mathbf{S}}_y$ that evolves through the first polarization transfer (up through P4) to $2\hat{\mathbf{I}}_x\hat{\mathbf{S}}_z$. Expanding this in terms of raising and lowering operators, one obtains $\hat{\mathbf{I}}_+\hat{\mathbf{S}}_z + \hat{\mathbf{I}}_-\hat{\mathbf{S}}_z$. Using the relationships shown in Equations 9.11, this can in turn be expanded into the single-transition basis as:

$$\hat{\mathbf{I}}_\pm^{(1,2)} = \tfrac{1}{2}\hat{\mathbf{I}}_\pm + \hat{\mathbf{S}}_z\hat{\mathbf{I}}_\pm$$

$$\hat{\mathbf{I}}_\pm^{(3,4)} = \tfrac{1}{2}\hat{\mathbf{I}}_\pm - \hat{\mathbf{S}}_z\hat{\mathbf{I}}_\pm \qquad (9.13)$$

The time evolution of the single-element operators on the right-hand sides of Equation 9.13 during t_1 results in the following set of terms (ignoring relaxation):

$$\tfrac{1}{2}\hat{\mathbf{I}}_-\left(\exp\left[i\omega_{12}t_1\right] - \exp\left[i\omega_{34}t_1\right]\right)$$

$$+\hat{\mathbf{S}}_z\hat{\mathbf{I}}_-\left(\exp\left[i\omega_{12}t_1\right] + \exp\left[i\omega_{34}t_1\right]\right)$$

$$\tfrac{1}{2}\hat{\mathbf{I}}_+\left(\exp\left[-i\omega_{12}t_1\right] - \exp\left[-i\omega_{34}t_1\right]\right)$$

$$+\hat{\mathbf{S}}_z\hat{\mathbf{I}}_+\left(\exp\left[-i\omega_{12}t_1\right] + \exp\left[-i\omega_{34}t_1\right]\right) \qquad (9.14)$$

The subsequent polarization transfer back to spin S, prior to evolution during acquisition (t_2) results in the following observable terms:

$$\frac{1}{2}\hat{\mathbf{S}}_- \left(i\cos\left[\omega_{12}t_1\right] - \sin\left[\omega_{12}t_1\right] \right)$$
$$-\hat{\mathbf{S}}_-\hat{\mathbf{I}}_z \left(i\cos\left[\omega_{12}t_1\right] - \sin\left[\omega_{12}t_1\right] \right)$$
$$+\frac{1}{2}\hat{\mathbf{S}}_- \left(i\cos\left[\omega_{34}t_1\right] + \sin\left[\omega_{34}t_1\right] \right)$$
$$+\hat{\mathbf{S}}_-\hat{\mathbf{I}}_z \left(i\cos\left[\omega_{34}t_1\right] + \sin\left[\omega_{34}t_1\right] \right) \qquad (9.15)$$

Not explicitly shown in Expressions 9.14 and 9.15 are the associated relaxation exponentials, that differ between the terms associated with transition ω_{12} and ω_{34}. Notice that the two terms associated with ω_{12} have opposite signs, while those associated with ω_{34} have the same sign. This difference can be exploited via phase cycling to select only for ω_{12}-associated terms, so that after the eight steps of the phase cycle shown in Figure 9.21, only the following terms remain (now with the relaxation exponentials shown explicitly):

$$-4\hat{\mathbf{S}}_- \sin\left[\omega_{12}t_1\right]\exp\left[-R_{12}t_1\right]$$
$$+8\hat{\mathbf{S}}_-\hat{\mathbf{I}}_z \sin\left[\omega_{12}t_1\right]\exp\left[-R_{12}t_1\right] \qquad (9.16)$$

This will correspond to the hyper-real portion of the interferogram that will eventually yield the ω_1 dimension of the TROSY experiment. A $\pi/2$ phase shift of P4 at each value of t_1 gives the hyper-imaginary portion, as usual with States quadrature detection experiments. The two operators in Equation 9.16 will evolve during t_2 to give the following expressions:

$$\hat{\mathbf{S}}_- \rightarrow \left(\exp\left[i\omega_{13}t_2 - R_{13}t_2\right] + \exp\left[i\omega_{24}t_2 - R_{24}t_2\right] \right)$$
$$2\hat{\mathbf{S}}_-\hat{\mathbf{I}}_z \rightarrow \left(\exp\left[i\omega_{13}t_2 - R_{13}t_2\right] - \exp\left[i\omega_{24}t_2 - R_{24}t_2\right] \right) \qquad (9.17)$$

Combining expressions 9.16 and 9.17 gives the density operator element for the single desired component of the IS multiplet:

$$\sigma_{12,24} \rightarrow 8\left(\exp\left[i\omega_{12}t_1 + i\omega_{24}t_2\right] + \exp\left[-\left(R_{12}t_1 + R_{24}t_2\right)\right] \right) \qquad (9.18)$$

The TROSY sequence can be incorporated into virtually any three-dimensional experiment in which an HSQC-type transfer is used for polarization transfer prior to detection, yielding a suite of experiments with TROSY modifications that can be used for high-resolution NMR work with very large proteins. At present, high-resolution structure determination using these methods is still problematic; nevertheless the availability of sequential assignments provides a way of examining details of large protein structure and dynamics in solution.

Problems

9.1   Show that $2\hat{\mathbf{I}}_x\hat{\mathbf{S}}_z$ can be expanded into Equation 9.15 using the definitions of the single element operators. Show the evolution of the single element terms during t_1 of TROSY, and then recombine the terms back into the Cartesian basis.

*9.2   Triple resonance experiments based on the ^{1}H, ^{15}N HSQC pulse sequence often incorporate "crusher gradients" to remove unwanted transverse magnetization while maintaining the desired coherence transfer pathways. Such gradients are usually very short, and placed between the two 90 degree pulses used to convert $2\hat{\mathbf{S}}_x\hat{\mathbf{I}}_z$ to $2\hat{\mathbf{S}}_z\hat{\mathbf{I}}_x$ prior to t_1 evolution on ^{15}N. Why is this? Use operators to explain your answer.

*9.3   Using Equations 9.9 and 9.10, design an off-resonance pulse square pulse consisting of 32 steps, that will invert ^{13}C carbonyl resonances (~180 ppm) without affecting C_α carbons (~60 ppm) at 600 MHz ^{1}H frequencies. What is the phase increment at each step?

9.4   If one wanted to make the pulse in Problem 9.3 a 90° pulse (tip angle of $\pi/2$) rather than an inversion pulse at the carbonyl carbon, but still a null at the C_α resonance, how long should the total pulse be? What is the phase increment for each step?

9.5   A problem that one often runs into while setting up two-dimensional and three-dimensional experiments is setting the proper phase corrections in the indirectly detected dimensions for processing. Not only can this mean a tedious processing and reprocessing until each dimension is correctly phased, but first-order phase corrections can result in a significant baseline distortion (called "baseline roll"). The origins of the phenomenon lie in the fact that RF pulses have a finite width, and signals of different chemical shifts will change their phase under the influence of the pulse at different rates, resulting in a frequency-dependent phase advance by the time the end of the first sampling period $[t_1(0)]$ is reached. In the old days, this problem was dealt with by scaling the first FID [corresponding to $t_1(0)$] prior to transformation in t_1, often by multiplication by 0.5, or by back-predicting that point using linear prediction. However, some simple considerations in setting up the experiment can minimize this problem without post-acquisition processing [Bax *et al.*, *J. Magn. Reson.* 91, 174–178 (1991)]. If the acquisition delay τ is known, the appropriate first-order phase correction ϕ_1 in the ω_1 dimension can be calculated by $\phi_1 = \tau/\Delta t_1 \times 360°$. So for an acquisition delay τ that is 1/2 the dwell time Δt_1, a linear phase correction of 180° will give pure-phase spectra in ω_1 without any need to scale the first point. The value of τ can be calculated for a simple 90° pulse of duration t_{90} as $\tau = 4t_{90}/\pi + \Delta t_1(0)$, where $\Delta t_1(0)$ is a programmable delay time for the first t_1 increment. Rearrangement gives this expression $\Delta t_1(0) = 1/(2SW) - 4t_{90}/\pi$ for the desired phase correction. For a NOESY experiment with a t_{90} of 7.3 μs and a 10 kHz spectral width (*SW*), what should the $\Delta t_1(0)$ be set to in order to have a preset first-order phase correction of 180°? How about for an HMQC experiment with a ^{15}N sweep width of 2000 Hz and a ^{15}N t_{90} of 35 μs? Don't forget to include the ^{1}H 180° pulse width (you can use 14.6 μs) in the calculation, since the t_1 period is divided by the ^{1}H 180° pulse.

9.6   Setting correct RF pulse lengths and powers is critical for maximum sensitivity in three-dimensional experiments. Since direct observation of ^{15}N and ^{13}C (X) is often inconvenient for dilute biological samples, 90° pulses for such nuclei are often set by using the simple sequence: $\pi/2_y$ (^{1}H)–$\tau(1/2J_{XH})$–pulse (X)–observe (^{1}H). The $\pi/2$ X pulse is determined by systematically changing the length (at fixed power) or power (at a fixed length) of the X pulse and watching for a null in the signal of ^{1}H coupled to X. Using product operators, describe what one is observing in this experiment.

*9.7 Most modern NMR spectrometers are equipped with linear RF amplifiers, for which simple linear calibrations allow one to predict the $\pi/2$ pulse length as a function of amplifier output. Amplifier output is usually described in decibels (dB), which is a logarithmic measure of the ratio of the total available power to the output power (measured in watts, or joule/s), i.e. dB = 10 (log P_2/P_1). Based on this scale, halving output power results in a +3 dB change in attenuation. It is a commonly used rule of thumb that a +6 dB change in attenuation doubles the length of the $\pi/2$ pulse. Based on this, what is the change in $\pi/2$ pulse length one might expect as one goes from 50 W to 35 W pulse power if the pulse length at 50 W is 7 µs?

9.8   Analyze the HN(CO)CA sequence shown in Figure 9.18 in terms of product operator evolution at the end of each component of the sequence as indicated by the dotted lines.

*9.9 Analyze the WATERGATE pulse train involving two soft water pulses flanking the central hard π pulse shown in Figure 9.9. In particular, consider what is happening to water magnetization in the presence of the field gradients applied during the sequence.

*9.10 The HN(CO)CA experiment is relatively more sensitive at 600 MHz (^{1}H) than it is at 800 MHz (^{1}H), all parameters being otherwise equivalent. The HNCA experiment does not undergo as large a drop in relative sensitivity upon changing fields. Speculate on a reason for this.

References

1. D. Marion, L. E. Kay, S. W. Sparks, D. A. Torchia, and A. Bax, *J. Am. Chem. Soc.* 111, 1515 (1989).
2. M. Piotto, V. Saudek, and V. Sklenár, *J. Biomol. NMR* 2, 661 (1992).
3. S. Grzesiek and A. Bax, *J. Am. Chem. Soc.* 115, 12593 (1993).
4. G. W. Vuister and A. Bax, *J. Magn. Reson.* 98, 428 (1992).
5. R. Freeman, T. Frenkiel, and M. H. Levitt, *J. Magn. Reson.* 50, 345 (1982).
6. A. J. Shaka, J. Keeler, T. Frenkiel, and R. Freeman, *J. Magn. Reson.* 52, 335 (1983).
7. A. J. Shaka, C. J. Lee, and A. Pines, *J. Magn. Reson.* 77, 274 (1988).
8. E. Kupce and R. Freeman, *J. Magn. Reson. Ser. A* 115, 273 (1995).
9. A. Fletsky, O. Moreira, H. Kovacs, and K. Pervushin, *J. Biomol. NMR.* 26, 167 (2003).
10. W. Bermel, I. Bertini, I. C. Felli, M. Piccioli, and R. Pierattelli, *Prog. NMR Spectr.* 48, 25 (2006).
11. M. Guerón, J. L. Leroy, and R. H. Griffey, *J. Am. Chem. Soc.* 105, 7262 (1983).
12. K. Pervusin, R. Riek, G. Wider, and K. Wuthrich, *Proc. Natl. Acad. Sci. USA* 94, 12366 (1997).

10

NMR UNDER ANISOTROPIC CONDITIONS: NMR IN THE SOLID STATE AND ORDERED FLUIDS

The detail and ease of interpretability of a solution NMR spectrum compared with its solid-state counterpart (Figure 10.1) illustrates why solid-state NMR has long been less widely applied than solution-state methods. Although the first experimental observation of NMR was of the ^{1}H signals from solid paraffin wax (Reference 1), the technical problems associated with solid-state NMR are formidable, and non-specialists have usually opted for the more accessible solution state experiments. However, high-resolution solid-state NMR has undergone a renaissance in the past decade. Spectrometer manufacturers routinely provide accessories necessary for solid-state experiments with spectrometer purchases, the repertoire of solid-state experiments is growing, and the wealth of information inherent in solid-state NMR is becoming accessible to more scientists. In this chapter, concepts relevant to the solid-state NMR of $\mathbf{I} = 1/2$ nuclear spins will be discussed, as well as phenomena associated with partial ordering in fluids.

Anisotropy in NMR: chemical shielding and dipolar coupling

The bane and boon of solid-state NMR arise from the same source, the incomplete time averaging of spatially anisotropic (i.e. orientationally dependent) phenomena. The most obvious of these is chemical shielding. We noted in Chapter 2 (and then subsequently ignored) the fact that the chemical shielding (which produces the observed chemical shift of a spin) is a second-rank tensor, i.e. its value for a particular nuclear spin depends upon the orientation of the surrounding molecule with

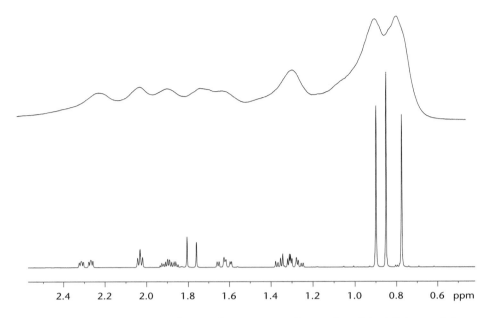

Figure 10.1 (Top) 400 MHz ^{1}H NMR spectrum of camphor "slush" (essentially a solid-state spectrum) in deuterochloroform. (Bottom) Solution state 400 MHz ^{1}H spectrum of camphor in deuterochloroform.

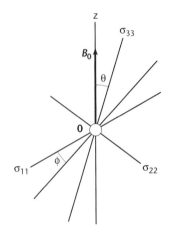

Figure 10.2 Representation of the principal axes σ_{11}, σ_{22} and σ_{33} of the chemical shift tensor $\bar{\bar{\sigma}}$ for an atom "O" in one orientation relative to the applied field $\vec{B}_0$ as defined by the angles θ and ϕ to a polar coordinate system fixed in the laboratory frame. The tensor axes are fixed in the molecular frame of reference (i.e. any bond to "O" will always have the same orientation with respect to the tensor axes).

respect to the applied magnetic field. Rapid isotropic motion in solution results in the averaging of the shielding tensor to yield the scalar and orientation-independent chemical shift (see Equation 2.12). By replacing the scalar chemical shift constant σ in Equation 2.12, $\omega_i = \gamma B_0(1 - \sigma_i)$ with the chemical-shielding tensor $\bar{\bar{\sigma}}$, we obtain an expression for the variation in chemical shielding of the nuclear spin as a function of the orientation of the principal axes of the shielding tensor (that is fixed in the molecular frame of reference) with respect to the applied field $\vec{B}_0$ (see Figure 10.2). The quantum mechanical expression of the Hamiltonian representing the shielding (Equation 5.51) must include this spatial dependence:

$$\hat{\mathbf{E}}_{CS} = -\hbar\gamma\hat{\mathbf{I}} \cdot \vec{B}_{induced}$$
$$\vec{B}_{induced} = \bar{\bar{\sigma}} \cdot \vec{B}_0 \tag{10.1}$$

The shielding of each spin is independent, i.e. there is a unique shielding tensor for each nuclear spin in the molecule.

The appropriate representation of $\bar{\bar{\sigma}}$ (or any second-rank tensor) is a 3×3 matrix. Of particular interest is the molecular frame of reference in which $\bar{\bar{\sigma}}$ is diagonal, i.e. all off-diagonal elements are zero. In this reference frame, the chemical shielding component of the Hamiltonian is determined as a function of the angles θ and ϕ made by the principal axes of the shielding tensor with respect to the applied field as shown in Figure 10.2:

$$\hat{\mathbf{E}}_{CS} = -\hbar\gamma\hat{\mathbf{I}}(1 - \sigma_{zz})\vec{B}_0 \tag{10.2}$$

where:

$$\sigma_{zz} = \sigma_{11} \sin^2 \theta \cos^2 \phi + \sigma_{22} \sin^2 \theta \sin^2 \phi + \sigma_{33} \cos^2 \theta \qquad (10.3)$$

and σ_{11}, σ_{22} and σ_{33} are the values of $\overline{\overline{\sigma}}$ associated with each of the orthogonal elements (see Figure 10.2). By convention, σ_{33} is taken as the principal axis of the shielding tensor and represents the most upfield component. The nature of the sample will determine how the chemical shielding tensor is represented in the observed spectrum. If a single crystal is placed in the field in a fixed orientation relative to the field, the resonances will appear as sharp lines at the shifts determined by Equation 10.3 appropriate to the particular orientation of the molecules in the crystal with respect to the applied field. The orientation dependence of these shifts becomes clear if the sample is mounted on a goniometer and rotated as shown in Figure 10.3. The observed shift of each line is a function of the value of σ_{zz} determined from Equation 10.3 for a particular orientation of the molecule in the

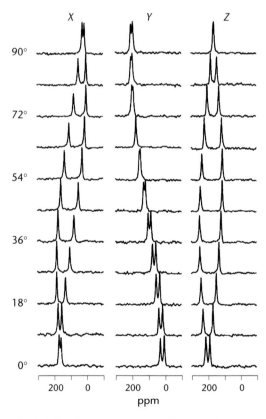

Figure 10.3 Chemical shift of ^{31}P resonances from a single crystal of Ru$_2$(CO)$_6$(μ_2-η_2-C≡C–Ph) (μ_2-PPh$_2$) as a function of rotation of the crystal around three orthogonal axes *X*, *Y* and *Z*. The crystal contains four P atoms per unit cell in two symmetry-related pairs that are magnetically equivalent, so two P resonances are observed. Reprinted with permission from K. Eichele *et al.*, *J. Am. Chem. Soc.* 124, p. 1547 (2002).

applied field. On the other hand, if the sample is a powder or frozen solution, all possible orientations of the molecules will be represented, resulting in a **powder pattern** showing the upfield and downfield extremes of the chemical shielding at either end of the pattern: σ_{11} (downfield), σ_{33} (upfield) and the most common shielding σ_{22} at the maximum of the signal (Figure 10.4). If for symmetry reasons (or accidental overlap) either σ_{11} or σ_{33} has the same value as σ_{22}, the powder pattern will have the appearance shown in parts B and C of Figure 10.4. The isotropic shift (σ_{iso}) observed in solution is the average of the three principal components of the shift tensor $\sigma_{iso} = 1/3(\sigma_{11} + \sigma_{22} + \sigma_{33})$ and the **span** of the chemical shielding is defined as $\Omega = \sigma_{33} - \sigma_{11}$. Ω is a measure of the **shielding anisotropy**, that results in the **chemical shift anisotropy** (**CSA**) seen in solution NMR and is determined largely by the distribution of electrons around the nucleus being observed. For ^1H, Ω tends to be small, reflecting the relatively spherical distribution of electrons in the s orbital of hydrogen. On the other hand, the anisotropy can be quite large for ^{13}C and ^{15}N (into the hundreds of ppm), particularly in the presence of π orbitals, which have very different shielding effects depending upon their orientation with respect to the applied field.

The other important anisotropic phenomenon that is observed directly in solid-state NMR experiments is dipolar coupling. This was discussed in Chapter 4, and reflects the direct through-space interactions between two spins $\hat{\mathbf{I}}_i$ and $\hat{\mathbf{I}}_j$. In the classical picture, two magnetic dipoles $\vec{\mu}_i$ and $\vec{\mu}_j$ separated by a distance r have an interaction energy given by:

$$E = \mu_0 / 4\pi \left[\left(\vec{\mu}_i \cdot \vec{\mu}_j \right) r^{-3} - 3(\vec{\mu}_i \cdot \vec{r})(\vec{\mu}_j \cdot \vec{r}) r^{-5} \right] \qquad (10.4)$$

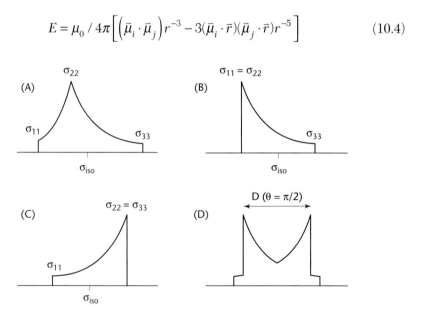

Figure 10.4 (A) Diagram of the expected powder pattern for a solid-state NMR spectrum of a single spin for which $\sigma_{11} \neq \sigma_{22} \neq \sigma_{33}$. (B) Same as (A), but $\sigma_{11} = \sigma_{22} \neq \sigma_{33}$ for reasons of symmetry or accidental overlap. (C) Same as (B), but $\sigma_{11} \neq \sigma_{22} = \sigma_{33}$ for reasons of symmetry or accidental overlap. (D) Doublet arising from a single dipolar coupling to the observed spin, with the coupling at a maximum when the angle θ between the internuclear vector and the applied field is $\pi/2$.

where μ_0 is the permeability of free space. This can be converted to the equivalent quantum mechanical Hamiltonian representing dipolar coupling between two spins by replacing the magnetic dipoles $\vec{\mu}$ with the appropriate spin operators $\vec{\mu} = \gamma\hbar\hat{\mathbf{I}}$:

$$\hat{\mathbf{E}}_{dip} = \left(\frac{\mu_0\gamma_1\gamma_2\hbar^2}{4\pi}\right)\left[\left(\hat{\mathbf{I}}_i\cdot\hat{\mathbf{I}}_j\right)r^{-3} - 3\left(\hat{\mathbf{I}}_i\cdot\vec{r}\right)\left(\hat{\mathbf{I}}_j\cdot\vec{r}\right)r^{-5}\right] \qquad (10.5)$$

Expansion of Equation 10.5 in terms of the individual scalar products (see Problem 10.1) and rearranging gives:

$$\hat{\mathbf{E}}_{dip} = D_{ij}\left(\hat{\mathbf{I}}_i\cdot\bar{\bar{D}}\cdot\hat{\mathbf{I}}_j\right) \qquad (10.6)$$

where D_{ij} is the dipolar coupling constant $D_{ij} = \mu_0\hbar^2\gamma_i\gamma_j/4\pi r_{ij}^3$ and $\bar{\bar{D}}$ is a second-rank (3×3) tensor with elements given by $[\delta_{kl} - 3\varepsilon_k\varepsilon_l]$. The k and l are the $\mathbf{x}$, $\mathbf{y}$ and $\mathbf{z}$ axes of the laboratory Cartesian coordinate system, δ_{kl} is the Kronecker delta function (1 when $k = l$, otherwise 0) and the ε_k, ε_l are the k and l components of the unit vector along $\vec{r}$, the internuclear axis between spins i and j. The $\bar{\bar{D}}$ tensor is "traceless", meaning that the sum of the diagonal elements is zero. As a result, isotropic averaging over time results in a net zero dipolar coupling. (Note that the same is not true of the chemical shift tensor!)

In practice, it is more convenient to convert Equation 10.6 to the spherical polar coordinate system (see Figure 10.5), and to replace the transverse operators $\hat{\mathbf{I}}_x$ and $\hat{\mathbf{I}}_y$ with their equivalent raising and lowering operators. Once this is done (see Problem 10.1), the expansion of Equation 10.6 yields a series of expressions sometimes called the "dipolar alphabet":

$$\hat{\mathbf{E}}_{dip} = D_{ij}(A + B + C + D + E + F)$$

$$A = \left(1 - 3\cos^2\theta\right)\hat{\mathbf{I}}_{iZ}\hat{\mathbf{I}}_{jZ}$$

$$B = -\frac{1}{4}\left(1 - 3\cos^2\theta\right)\left(\hat{\mathbf{I}}_{i+}\hat{\mathbf{I}}_{j-} + \hat{\mathbf{I}}_{i-}\hat{\mathbf{I}}_{j+}\right)$$

$$C = -\frac{3}{2}\left(\sin\theta\cos\theta\exp^{-i\phi}\right)\left(\hat{\mathbf{I}}_{i+}\hat{\mathbf{I}}_{jZ} + \hat{\mathbf{I}}_{iZ}\hat{\mathbf{I}}_{j+}\right)$$

$$D = -\frac{3}{2}\left(\sin\theta\cos\theta\exp^{i\phi}\right)\left(\hat{\mathbf{I}}_{i-}\hat{\mathbf{I}}_{jZ} + \hat{\mathbf{I}}_{iZ}\hat{\mathbf{I}}_{j-}\right)$$

$$E = -\frac{3}{4}\left(\sin^2\theta\exp^{-i2\phi}\right)\left(\hat{\mathbf{I}}_{i+}\hat{\mathbf{I}}_{j+}\right)$$

$$F = -\frac{3}{4}\left(\sin^2\theta\exp^{i2\phi}\right)\left(\hat{\mathbf{I}}_{i-}\hat{\mathbf{I}}_{j-}\right) \qquad (10.7)$$

We will focus only on the first two terms of this expansion. This is allowed by the **secular approximation**, which states that higher-order interactions between two objects (spins, planets, electrons) tend to average with time and can be ignored in determining the motional behavior of the system. (Note that the Hamiltonian expression in Equation 10.2 also invokes this approximation.)

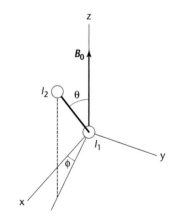

Figure 10.5 Spherical coordinate system applicable to the dipolar coupling "alphabet" for two dipolar coupled nuclei $\mathbf{I}_1$ and $\mathbf{I}_2$ at a distance **r** from each other (see Equations 10.7).

Term A of Equation 10.7 describes the splitting observed in the solid-state spectrum as a function of the angle θ between the applied field and the ij internuclear vector (see Figure 10.4D). Term B contains the "flip-flop" operators analogous to those in Equation 5.57 (Chapter 5). These terms are significant when the coupling energies are comparable to the difference in chemical shielding between the two spins (homonuclear coupling) or when the Hartmann–Hahn condition is fulfilled (see below).

Resolving the solid-state NMR spectrum: magic angle spinning (MAS) and high-power ^{1}H decoupling

In a powdered solid sample all possible molecular orientations are randomly represented. As a result, all possible values of the chemical shielding (as well as all possible values of any active dipolar couplings) will be observed for a given spin type, as shown in Figure 10.4. The spins in molecules that are oriented with the downfield principal axis σ_{11} parallel to the applied field $\vec{B}_{applied}$ will give rise to the downfield extreme of the signal, while those oriented with σ_{33} parallel to $\vec{B}_{applied}$ provide the upfield extreme. However, if the sample is spun rapidly around an axis that lies at an angle θ with respect to the applied field $\vec{B}_{applied}$ molecular orientations orthogonal to the rotation axis would be averaged (i.e. if the **z** axis is chosen as the axis of rotation, the **x** and **y** components are averaged) (Figure 10.6, Reference 2). The details of this averaging are the subject of Problem 10.2, but the result is that the chemical shielding term of the Hamiltonian becomes:

$$\hat{\mathbf{E}}_{CS} = -\hbar\gamma\hat{\mathbf{I}}_z \left[\sigma_{iso} + \frac{1}{2}(3\cos^2\theta - 1)(\sigma_{33} - \sigma_{iso}) \right] \qquad (10.8)$$

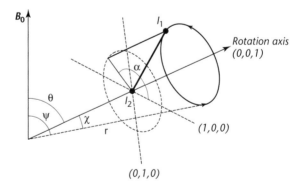

Figure 10.6 Averaging of dipolar coupling and chemical shift by magic angle spinning (MAS). The internuclear vector between two coupled spins $\mathbf{I}_1$ and $\mathbf{I}_2$ has a component perpendicular to the rotation axis (0,0,1) in the plane defined by axes (0,1,0) and (1,0,0) that is averaged by spinning around the rotation axis (making the angle ψ a function of time), leaving only the component parallel to the rotation axis. If the angle θ between the applied field $\vec{B}_0$ and the rotation axis is 54.73°, the term $(1-3\cos^2\theta) = 0$ and the dipolar coupling term drops out of the Hamiltonian. The observed transition frequencies are modulated around the isotropic shift by ω_r as described in Problem 10.3.

at rotor frequencies that exceed the span of the shielding anisotropy. An interesting feature of this expression is that when the axis of rotation is at an angle $\theta = 54.73°$ or 0.95 rad, the second term equals zero, and the observed signal is centered on the isotropic chemical shift frequency σ_{iso}. This is the **magic angle**, and is the angle at which the axis of rotation is placed in most solid-state NMR probes relative to the applied field. (Note that the magic angle is the angle made by a vector to the three axes of the Cartesian coordinate system if it is equidistant from all three axes, as demonstrated in Problem 10.4.) It turns out that spinning frequencies ω_r that are less than the span of the shielding anisotropy for a given spin result in the presence of side bands around the isotropic shift line spaced at integral multiples of ω_r (see Figure 10.7). One can think of this as analogous to the Doppler effect, in which the

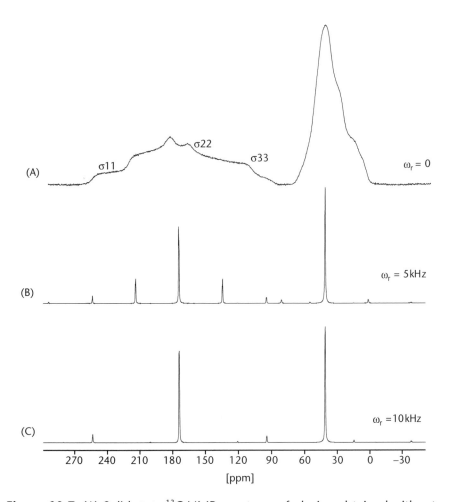

Figure 10.7 (A) Solid-state ^{13}C NMR spectrum of glycine obtained without MAS. Contributions from chemical shift anisotropy and dipolar coupling are observed (positions of the principal components of the chemical shift tensor of the carbonyl carbon are noted as σ_{11}, σ_{22} and σ_{33}. (B) Same sample with MAS sample rotation at a frequency of $\omega_r = 5$ MHz. Side bands are observed offset from the center frequency (isotropic shift) at spacings of ω_r. (C) Same sample, with $\omega_r = 10$ MHz. Figure courtesy of Prof. Beat Meier and Dr. Mattias Ernst, ETH, Zurich.

frequency of rotation is either added to or subtracted from the incident RF, making more than one frequency component of the RF able to excite a particular transition. Whether or not side bands are present, MAS breaks up the powder pattern, yielding resolved signals with enhanced signal-to-noise ratio.

Rapid spinning at the magic angle also zeros the secular terms of dipolar coupling (see terms A and B of Equation 10.7) so that the only terms remaining in the Hamiltonian describing the MAS experiment are the same ones seen for solution-state NMR: the Zeeman, chemical shift and J-coupling terms. In practice, even the highest MAS speeds are not sufficient to remove the large one-bond dipolar couplings involving ^1H. Also, spin–spin (T_2) relaxation between abundant spins such as ^1H is very efficient in the solid state, leading to broad lines. This broadening extends to the lines of dilute spins, such as ^{13}C, to which the ^1H spins are coupled via dipolar interactions. (A "dilute" spin in solid-state NMR terminology is any other spin than ^1H, even if that spin is isotopically enriched.) For this reason, it is usually necessary to apply high-power broadband ^1H decoupling during acquisition of the dilute spin signal in order to remove dipolar broadening even in MAS experiments. Because dipolar coupling between directly bonded ^1H and ^{13}C is in the range of 20 kHz (three orders of magnitude greater than J-coupling!) the RF field required for effective decoupling in solids is considerably larger than what is needed for decoupling in solution: the RF field needs to have an excitation bandwidth greater than the splitting in order to accomplish the required decoupling. One of the earliest examples of a pulse train for removing dipolar couplings is the WAHUHA sequence (from the inventors, *Wa*ugh, *Hu*ber and *Ha*eberlen; see Reference 3).

On the other hand, J-coupling is usually such a small contribution that it can be ignored in considering the appearance of a solid-state NMR spectrum, although very high-speed spinning can allow J-splitting to be observed in favorable cases. MAS has been applied to the analysis of solid-phase synthesis samples (often generated combinatorially) to obtain high resolution spectra. An example of this is shown in Figure 10.8. It is important to note that MAS does not average out conformational heterogeneity on the molecular scale: if a line is broadened due to the presence of multiple conformations of a molecule in the same sample, MAS will not narrow it.

Cross-polarization for signal enhancement of dilute spins and spin–spin correlations

Although T_2 values are usually short in solid samples due to efficient spin–spin couplings, the longitudinal (T_1) or spin–lattice relaxation times of dilute spins are often quite long due to the absence of the necessary spectral density elements that promote such transitions. As ^1H T_1 relaxation is generally much more efficient that that of the dilute spins in a solid sample (and ^1H polarization larger), it is usually advantageous to use ^1H spins as a polarization and relaxation reservoir to enhance the signals of dilute spins by the application of **cross-polarization** (**CP**). The concept of polarization transfer is familiar to solution NMR spectroscopists because of HSQC and HMQC experiments, but solid-state CP is actually related to the energy

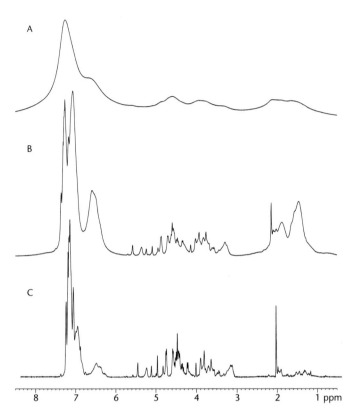

Figure 10.8 High-resolution MAS (HRMAS) spectroscopy applied to a resin-bound trisaccharide. (A) Static ^{1}H spectrum of solvent-swollen sample. (B) ^{1}H Spectrum with MAS spinning at 3.5 kHz. (C) ^{1}H Spectrum with 3.5 kHz MAS spinning and spin-echo detection to remove short T_2 components.

transfer used in the TOCSY experiment, and depends upon reaching the Hartmann–Hahn condition (**HH matching**):

$$\gamma_I B_{1I} = \gamma_S B_{1S} \tag{10.9}$$

described in Chapter 7. The B_1 fields (called spin-locking fields) are the result of RF applied at appropriate frequencies to both the abundant (usually ^{1}H) and the dilute spins to generate HH matching. Cross-polarization has two benefits: it shortens the recycle delay between experiments (by using ^{1}H cross-relaxation to relax the dilute spins) and improves signal-to-noise ratio in a single acquisition by polarization transfer. The maximum improvement in sensitivity in the CP experiment is determined by the ratio of γ values of the two nuclei involved, so for a ^{1}H/^{13}C polarization transfer, the potential improvement in signal is a factor of 4. The duration of the spin-lock period of CP is determined by the time required for the proton spin population to equilibrate in the spin-lock field, a characteristic relaxation time known as $T_{1\rho}$ that will be discussed more thoroughly in Chapter 11. Typically, CP is combined with MAS to provide simplified high-resolution NMR spectra of solid-state samples (CP/MAS). A diagram of the CP experiment is shown in Figure 10.9.

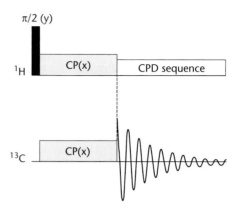

Figure 10.9 Cross-polarization (CP) experiment. After a $\pi/2$ **(y)** pulse on ^{1}H (abundant spin), spin-locking fields are applied along **x** on both abundant and dilute (^{13}C in this case) spins that force the Hartmann–Hahn condition. After the CP period (the length of which is determined by the ^{1}H relaxation time in the rotating frame, $T_{1\rho}$), the dilute spin FID is acquired with ^{1}H composite pulse decoupling (CPD) to remove ^{1}H dipolar coupling to the dilute spins.

Technically, the simplest CP methods are those involving ^{1}H spins. Because of the large dipolar couplings between ^{1}H spins in most solid-state samples and the resulting broad lines, the frequency-matching requirements that satisfy Equation 10.9 are relatively loose. This means that the spin-locking fields do not need to be very precisely controlled in order to obtain polarization transfer from ^{1}H to other spins. However, if one desires to obtain polarization transfer between lower-gamma nuclei, e.g. to obtain correlations and distance information between dilute spins, the problem is trickier. First, the dipolar coupling between spin i and spin j is weaker for low-gamma nuclei (due to the $\gamma_i\gamma_j$ dependence of the dipolar coupling), particularly if the coupled spins are further than one bond length apart. Secondly, as MAS frequencies increase, the broad powder pattern envelopes are resolved into sets of fairly narrow lines spaced by the frequency of MAS spinning ω_r (see Figure 10.7). Obtaining HH matching under these conditions means that the amplitude of the RF field used to obtain the locking must be precisely coordinated with the frequency of sample spinning (ω_r). The requirement of the HH condition between two spins A and X becomes:

$$\gamma_A B_{1A} = \omega_A$$
$$\gamma_X B_{1X} = \omega_X$$
$$\omega_A = \omega_X \pm n\omega_r \, (n = 1, 2) \tag{10.10}$$

Finally, the large chemical shift ranges of ^{13}C and ^{15}N means that HH matching will be chemical shift dependent (the precise amplitude of the B_1 field needed to give efficient HH transfer will depend on the chemical shift of the ^{13}C spin of interest). There are several different ways in which this matching may be accomplished, as discussed below.

Selective reintroduction of dipolar couplings between dilute spins: rotational resonance, RFDR, and REDOR

As discussed in Chapter 2, the great value of NMR rests in large part upon the ability to observe coupling between spins. One of the most important uses of polarization transfer in solution state NMR, as we have seen, is to generate and follow coherences between spins. In this respect, MAS throws out the baby with the bathwater, since the dipolar couplings between low γ spins are lost via magic angle nulling. Much of the effort in solid-state NMR over the last 20 years has been directed at selectively reintroducing dipolar coupling and detecting polarization transfer between dipolar-coupled nuclei. Besides aiding in resonance assignment, the dipolar coupling between two nuclei is exquisitely sensitive to the internuclear distance (see Equation 10.6). If two signals in a solid-state NMR spectrum can be reliably assigned and their dipolar coupling measured relative to a standard over a known distance, very precise internuclear distances can be obtained.

One way of accomplishing homonuclear recoupling of two dilute spins (most commonly ^{13}C) is called rotational resonance, or R^2 (Reference 4). This takes advantage of precise control of MAS rotor speeds so that the chemical shift difference between the two dipolar-coupled spins is an integer multiple of the rotor frequency (in hertz): $\Delta v_{12} = n v_r$. This results in the superposition of spinning sidebands of the two coupled resonances, or superposition of the sideband of one resonance upon the central line (isotropic shift) of the other. The pulse sequence for R^2 is shown in Figure 10.10. After the initial polarization transfer from ^1H to ^{13}C, generating polarized $\hat{\mathbf{I}}_x$ coherence on all ^{13}C spins, a $-\mathbf{y}$ $\pi/2$ pulse is applied to return the polarization-enhanced ^{13}C coherence to $+\mathbf{z}$. Immediately, a selective π pulse is used to invert one of the R^2-superimposed dipolar-coupled ^{13}C resonances, creating a situation similar to that observed in a transient NOE experiment. Magnetization transfer between the two spins then occurs during the mixing time, τ_m. The efficiency of this transfer is greatest when the transition frequencies are the same (hence the requirement for superposition of lines), in analogy to the HH effect. A final $\pi/2$ read pulse is applied, and the change in signal intensity at the coupling

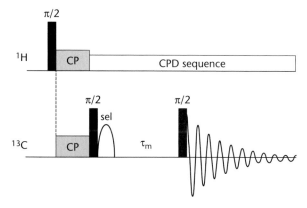

Figure 10.10 Pulse sequence for rotational resonance, or R^2 (see Reference 4). "Sel" indicates selective inversion pulse.

partner of the inverted spin is followed as a function of mixing time. This signal intensity change can be related quite precisely (± 0.5 Å) to the distance between the two spins using the relationship for the dipolar coupling constant. R^2 is generally useful only for homonuclear interactions, and there is a requirement that the chemical shift difference between the two spins of interest be great enough so that one spin can be inverted selectively, but not so far apart that the difference in shift exceeds the maximum rotor frequency by more than a factor of two.

A more general solution to detecting homonuclear dipolar couplings between dilute spins is to use **rotor-synchronized pulses** to recouple spins regardless of chemical shift. RFDR (**radio-frequency-driven recoupling**) is an example of such an experiment. In the two-dimensional version of RFDR for correlating ^{13}C spins, the first step is cross-polarization between ^{1}H to ^{13}C, generating transverse $\hat{\mathbf{I}}_x$ (^{13}C) coherence that is frequency-labeled during t_1. A $\pi/2$ pulse on ^{13}C places the carbon coherences along the $-\mathbf{z}$ axis at the start of the mixing time, τ_m, which is set to be a doubled integer multiple of the time required for a single rotor cycle, τ_r. That is, $\tau_m = 2n\tau_r$, where n is an integer. At the middle of each subsequent rotor cycle, a π pulse is applied. This pulse prevents the full averaging of the dipolar coupling between ^{13}C spins that would normally occur during the rotor cycle. The scaled (but non-zero) dipolar coupling generates zero-quantum coherence in the form of flip-flop terms between the coupled spins (term B in Equation 10.7). (Remember that the mechanical rotor frame of reference is not the same as the RF-determined rotating frame!) Because the components of the dipolar coupling tensor that are averaged by MAS are always at the same point in the rotor cycle when the π pulses are applied (since the pulses are synchronized with sample rotation), the π pulse train gives rise to a nonaveraged exchange of coherence via the dipolar coupling.

Two-dimensional spectra generated by homonuclear RFDR are reminiscent of NOESY spectra, and indeed contain information concerning the dipolar (through-space) connectivity of ^{13}C spins. Magnetization exchange curves obtained by varying mixing times and measuring cross-peak intensities can be fitted to obtain inter-atomic distances of considerable accuracy. In favorable cases, homonuclear RFDR spectra obtained with high rotor spinning rates (13 kHz) are comparable in terms of resolution with solution spectra of proteins (see Figure 10.11).

The problem of extending RFDR methods to heteronuclear correlation experiments is technically more challenging, since the dependence of the recoupling via zero-quantum coherence transfer upon the rotor frequency is given by the following expression:

$$\sqrt{\omega^2_{(eff)A} + \Omega^2_A} - \sqrt{\omega^2_{(eff)X} + \Omega^2_X} = n\omega_r \, (n = 1, 2) \tag{10.11}$$

Clearly, the effective fields $\omega_{(eff)i}$ required for obtaining the recoupling condition will depend upon the chemical shift offset of a particular spin Ω_i. The easiest way to accomplish this for heteronuclear polarization transfer is to apply an amplitude-ramped pulse to one of the two nuclei of interest. This will change the effective field for the spin with the ramped pulse, and will allow the generation of ZQ coherence

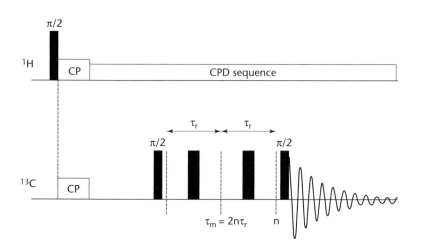

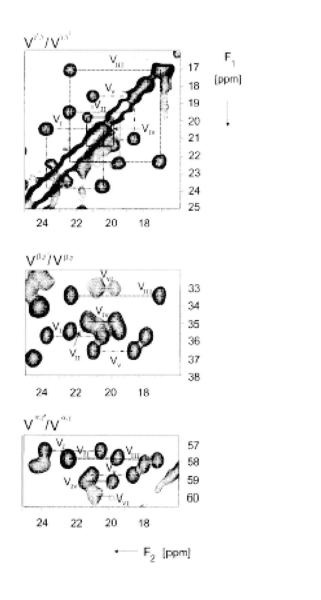

Figure 10.11 (Top)
Sequence of the two-
dimensional RFDR (*radio-
frequency-driven recoupling*)
experiment. (Bottom) Spin
systems of valine *gem*-
dimethyl groups obtained
from two-dimensional
^{13}C–^{13}C RFDR experiment on
uniformly ^{13}C-labeled
precipitated SH3 domain of
Fyn tyrosine kinase.
[Reprinted by permission
from J. Pauli, B. van Rossum,
H. Forster, H. J. M. de Groot
and H. Oschkinat, *J. Magn.
Reson.* 143, 411–416 (2000),
Figure 4, by permission of
Elsevier.]

for different spins during the course of the mixing pulses. Examples of this type of experiment as applied to biological samples will be discussed in the next section.

Another experiment designed to detect heteronuclear dipolar correlations (between ^{13}C and ^{15}N spins, for example) is REDOR, that stands for **rotational echo double resonance**. As originally formulated, REDOR is a difference experiment that monitors the change in signal intensity of a given spin (often ^{13}C) in the presence and absence of π pulses on a heteronuclear dipolar coupling partner such as ^{15}N (Reference 5). The original pulse scheme for REDOR is shown in Figure 10.12. In REDOR, again, the first step is cross-polarization between ^{1}H and ^{13}C, giving rise to transverse coherence on ^{13}C. During the mixing time, two ^{15}N π pulses are applied per rotor period, one at no more than half of a period, and the other at the end of each rotor period. At the middle of the sequence, the ^{15}N π pulse is replaced by a ^{13}C π pulse in order to refocus ^{13}C shift evolution without affecting the dipolar coupling. The effect of the ^{15}N π pulses is to prevent complete refocusing of the rotational echo due to dipolar coupling, thereby changing the observed signal intensity. By comparing ^{13}C signal intensities in the presence and absence of the ^{15}N pulse train, the magnitude of the coupling constant between the two spins can be determined, and from that, an interatomic distance obtained.

A simple (if somewhat inaccurate) way to think about REDOR is by analogy to J-coupling. If J-coupling is allowed to evolve during a mixing time τ, the result will be (from $\hat{\mathbf{I}}_x$ coherence) a modulated term $\hat{\mathbf{I}}_x \cos\pi J_{12}\tau$. In the absence of such coupling, the signal due to $\hat{\mathbf{I}}_x$ is unmodulated. The difference between $\hat{\mathbf{I}}_x$ and $\hat{\mathbf{I}}_x \cos\pi J_{12}\tau$ (with decoupling applied during acquisition) would be observed as an amplitude change

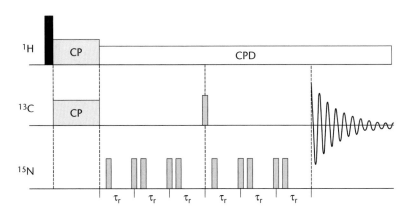

Figure 10.12 Pulse sequence for REDOR (*r*otational *e*cho *d*ouble *r*esonance, Reference 5). After cross-polarization from ^{1}H to ^{13}C, two rotor-synchronized ^{15}N π pulses are applied per rotor cycle, one at the end of each cycle, and one at some fractional period of the cycle. In the middle of the mixing period defined by the ^{15}N pulses, the ^{15}N pulse is replaced by a refocusing π pulse on ^{13}C. A difference spectrum is obtained by subtracting the result from a spectrum obtained in the absence of the ^{15}N pulses. The amplitude modulation observed can be related to the magnitude dipolar ^{15}N–^{13}C coupling and hence the internuclear distance.

upon spectral subtraction. In the case of REDOR, the *J*-coupling is replaced by the dipolar coupling, introducing a reliable distance dependence into the modulation.

Heteronuclear two-dimensional techniques in solid-state NMR

In recent years, efforts to bring the benefits of multinuclear NMR to solid-state NMR have intensified. As of this writing, a number of complete or nearly complete sets of ^{13}C and ^{15}N resonance assignments have been determined for small proteins in the solid state, and solid-state NMR-derived (or at least NMR-assisted) protein structures are beginning to appear. While the technical difficulties and theoretical complexity of solid-state NMR remain a considerable barrier to the casual user, improvements in hardware and pulse sequences have made it almost inevitable that the next important developments in NMR of complex systems will come in the solid state.

As discussed in the previous section, one of the greatest challenges in applying heteronuclear NMR methods to solid samples is the difficulty of generating the conditions required for polarization transfer across the wide range of chemical shifts available to spins such as ^{15}N and ^{13}C. In many experiments, this is dealt with by the use of **ramped pulses** on one of the two types of spins to be correlated. **Amplitude-ramping** (changing the magnitude of B_1 across the duration of the pulse) results in a changing $\vec{B}_{eff}$ so that the HH condition will apply to all of the spins of interest at some point during the pulse regardless of chemical shift. Direct **frequency-ramping** can also be accomplished by introduction of a continuously changing phase shift across the duration of the pulse. As we have seen, this results practically in a frequency shift, in the same manner that phase shifts are used to generate an off-resonance frequency from a carrier frequency. In either case, the ramps are designed to excite only particular regions of the spectrum, and in this way, correlations between ^{15}N amides and carbonyl ^{13}C or ^{13}C$_\alpha$ in proteins exclusively can be obtained. Some of the experiments that have been used to make resonance assignments in proteins are shown in Figures 10.13 and 10.14.

Another important tool for characterizing macromolecules in the solid state is **proton-driven spin diffusion** (PDSD). This technique is used to obtain both connectivity and through-space information (usually for ^{13}C). The original pulse sequence is shown in Figure 10.15 (Reference 6). The basis of the experiment is the generation of transverse ^{13}C magnetization by cross-polarization from ^{1}H, followed by a t_1 evolution period. At the end of the evolution, the ^{13}C coherence is placed along the $-\mathbf{z}$ axis by a $\pi/2$ pulse, and magnetization transfer occurs in a fashion analogous to the NOESY experiment for protons during the mixing time, τ_m. Mechanisms for transfer include spin diffusion (hence the name) and chemical exchange. After mixing, the magnetization is returned to the transverse plane by another $\pi/2$ pulse, and the ^{13}C signal is detected during t_2. By varying the mixing time, buildup curves can be generated and calibrated using known carbon–carbon distances in the sample to generate distance restraints that are precisely analogous to NOESY restraints used in determination of solution structures of macromolecules. Because ^{1}H decoupling is not applied during τ_m, this time is not a limitation on the experiment, and mixing times of up to 500 ms have been used for protein

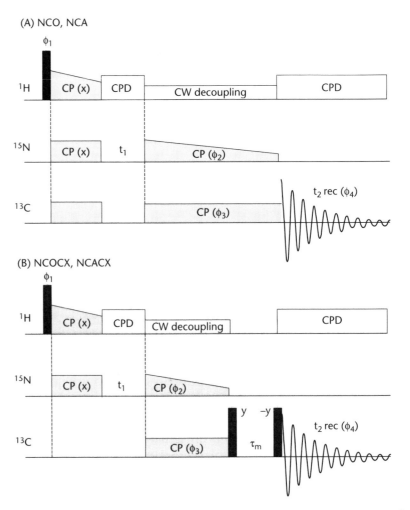

Figure 10.13 Two- and three-dimensional heteronuclear pulse sequences for correlation of protein ^{13}C and ^{15}N resonances. Sloped CP sequences indicate RF amplitude-ramping for cross-polarization of dilute spins. (A) NCO and NCA sequences using RF fields of 30 kHz (^{15}N) and 15 kHz (^{13}C) respectively for polarization transfer. Choice of ^{13}C carrier frequency determines the type of transfer. (B) NCOCX and NCACX sequences are similar, but include a ^{13}C, ^{13}C longitudinal mixing time τ_m prior to detection (bracketed by the two 90° y and $-y$ pulses) in order to extend transfer to the adjacent carbons. Phase cycles are as follows: ϕ_1, y, y, $-y$, $-y$; ϕ_2, x, $-x$, x, $-x$, x, $-x$, x, $-x$; ϕ_3, x, x, x, x, $-x$, $-x$, $-x$, $-x$; ϕ_4, x, $-x$, $-x$, x. CPD stands for composite-pulse decoupling. Sequences are described in Reference 9.

PDSD spectra, allowing long (up to 7 Å) distance correlations to be identified between dipolar-coupled ^{13}C spins. The biggest problem with PDSD spectra is their complexity and overlap. Correlations between directly bonded ^{13}C atoms are strong and tend to swamp the more important through-space restraints required for structure determinations. This problem can be circumvented by using selectively ^{13}C-labeled samples, so that directly bonded atoms are not both labeled (see Figure 10.16).

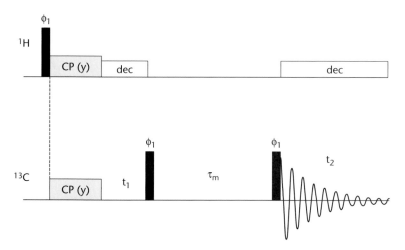

Figure 10.14 Two-dimensional ^{13}C, ^{13}C PDSD sequence. The sequence begins with CP from ^{1}H to ^{13}C, and, after frequency-labeling of t_1 of ^{13}C, undergoes mixing via dipolar interactions during τ_m with decoupling of attached protons during transverse ^{13}C evolution (t_2). A weak ^{1}H decoupling field is often applied during τ_m in order to improve the ^{13}C mixing.

Solid-state NMR using oriented samples: PISEMA

The dipolar coupling is intrinsically rich in information, both in terms of distance between coupled nuclei and the orientation of the internuclear vector with respect to the applied field. However, because MAS is required for resolution of signals from microcrystalline powders (the most common type of sample for solid-state NMR), the orientational component of the dipolar coupling is generally less accessible than the distance component. It is possible to use oriented samples without MAS to obtain information regarding the relative orientations of particular structural features via dipolar couplings. The best-studied examples of this are helical proteins in association with phospholipid bilayers. Samples are prepared in thin multiple layers (e.g. stacked between microscope slide covers) and, due to the alignment of the membrane-associated helices with respect to the bilayers as well as the orientation of the bilayer with respect to the magnetic field, the relative magnitudes of dipolar couplings from individual spin pairs (typically from the N–H groups in peptide bonds) in these complex samples provide information regarding their orientation within the membrane. Complementary information is available from the chemical shielding, as the observed shielding also reflects orientations. An experiment for extracting this information from oriented samples is called PISEMA (***polarization inversion spin exchange at the magic angle***) (Reference 7). PISEMA is somewhat analogous to *J*-resolved two-dimensional NMR experiments in liquids, with dipolar splittings represented in the indirectly detected dimension of a two-dimensional NMR spectrum. The pulse sequence of PISEMA is shown in Figure 10.17, as well as a two-dimensional PISEMA spectrum. High-power heteronuclear decoupling applied throughout the sequence narrows lines in the absence of MAS, and the orientation of the samples reduces the broadening due to chemical shielding anisotropy.

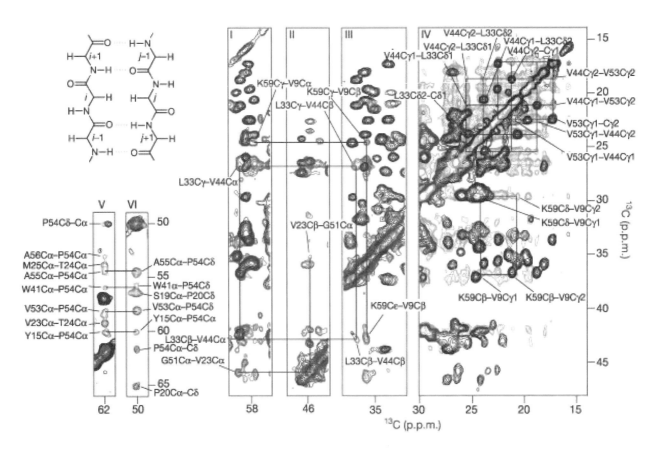

Figure 10.15 High-resolution solid-state two-dimensional PDSD spectra of a protein, SH3 domain from α-spectrin. Shaded areas correspond to contour plots of two-dimensional ^{13}C, ^{13}C PDSD spectra recorded with a mixing time of 500 ms at 17.6 T and a spinning frequency of 8.0 kHz on a PDSD spectrum of uniformly ^{13}C-labeled SH3 domain (black) recorded with a short mixing time (15 ms). A combination of selective and uniform labeling permits a reduction in the number of correlations and improves resolution for assignment purposes. Connectivities for two residues, Leu 33 and Val 44, are shown. The correlations between the C and C signals of Val 44 and C and C signals of Leu 33 are observed in panels I, II and III for SH3 prepared with 2-^{13}C glycerol, whereas correlations between the methyl groups appear in the spectrum of SH3-labeled using 1,3-^{13}C glycerol (panel IV). Figure reprinted by permission from Castellani *et al.*, *Nature* 420 pp. 98–102 (2002), Macmillan Publishers Ltd.

The difficulty that PISEMA addresses is that in order to observe splitting due to one-bond heteronuclear (^{1}H–^{15}N) dipolar couplings, one needs to remove the effects of longer range homonuclear ^{1}H dipolar couplings as much as possible so that line widths are reasonably narrow in the indirectly detected dimension. That is, one must decouple long-range (smaller) dipolar interactions while leaving the one-bond (short range) dipolar couplings reasonably intact during the t_1 evolution of the dipolar coupling. In PISEMA, this is accomplished using Lee–Goldburg off-resonance decoupling (Reference 8) for spin-locking during t_1. The Lee–Goldburg decoupling scheme places the decoupler frequency $\vec{B}_1$ off-resonance at a frequency and power so that $\vec{B}_{eff}$ lies at the magic angle relative to the chemical shift offset (parallel to $\vec{B}_0$). There are two points in the spectrum that fulfill this requirement, one upfield and one downfield from the ^{1}H carrier frequency. The sequence rapidly alternates the

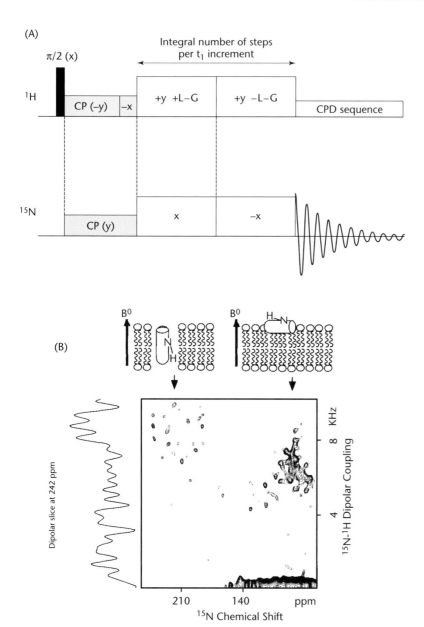

Figure 10.16 (A) PISEMA pulse sequence, as described by Wu *et al.* (Reference 7). Each value of t_1 contains an integral number of Lee–Goldburg cycles, each cycle of which contains a positive phase/frequency jump followed by a negative phase/frequency jump. (B) Solid-state two-dimensional PISEMA spectrum (700 MHz ^{1}H) of an oriented sample of uniformly ^{15}N-labeled colicin El in a hydrated phospholipid bilayer. The sample was oriented on a stack of 25 glass plates in a flat coil probe. The PISEMA spectrum correlates ^{15}N chemical shielding with the magnitude of the ^{1}H–^{15}N dipolar coupling constant for NH pairs. A dipolar slice taken at 242 ppm is indicated at the left of the two-dimensional spectrum. The region of the spectrum around 230 ppm represents intensity from amides in the transmembrane helix, while the intensity around 80 ppm are due to amides in a helix in the plane of the bilayers as indicated in the diagram at top. Figure courtesy of Prof. Stanley Opella.

Figure 10.17 PISA wheels for analysis of membrane protein helix orientation from two-dimensional PISEMA data. The ^{1}H–^{15}N dipolar coupling for a particular N–H bond in the helix is calculated as a function of the angle τ the long axis of the helix $\vec{h}_z$ makes with respect to the applied field $\vec{B}_0$ and the angle ρ around $\vec{h}_z$. The ^{15}N chemical shift is the σ_{zz} value obtained from Equation 10.3 using values obtained from the predicted chemical shift tensor for ^{15}N for the same amide. (A) PISA wheel calculated for an ideal α-helix using an averaged chemical shift tensor value. (B) Experimental resonances from the transmembrane peptide of M2 protein of the influenza A virus. Ideal values (A29, S31, G31, H37) were used for the missing experimental data. Figure reprinted with permission from J. Wang *et al.*, *J. Magn. Reson.* 144, 162–167, Figs 2A, 4B and 4C (2000).

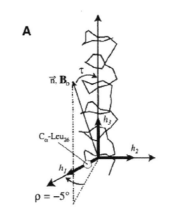

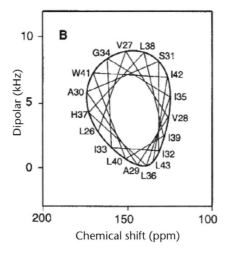

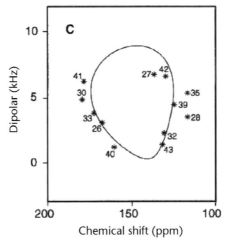

frequency between these two positions, accompanied by a π phase shift at each step, at a rate that permits one complete revolution of the magnetization around $\vec{B}_{eff}$ at each position. The result of this decoupling is a dramatic narrowing of lines in the indirectly detected (dipolar coupling) dimension, with only a relatively small reduction (by a factor of 0.82) in the desired heteronuclear dipolar splitting.

The resulting PISEMA spectrum (Figure 10.16) can be analyzed in order to obtain information regarding the angles made by α-helices with respect to the applied field and the lipid bilayer. For a given angle τ relative to the applied field $\vec{B}_0$, plotting the averaged ^{15}N chemical shift tensor value and expected ^1H–^{15}N dipolar coupling results in an off-set circular pattern of resonances in the two-dimensional PISEMA spectrum known as a "PISA wheel" (Figure 10.17), which is reminiscent of the helical wheels used to determine the relative positions of side chains on an α-helix. A given angle τ will plot to a PISA wheel in a well-defined region of the PISEMA spectrum. Not only does the PISA wheel give information regarding the orientations of helices in the bilayer, but permits resonance assignments to be made based on positions of signals within the wheel.

Bringing a little order to solution NMR: residual dipolar couplings and CSA in ordered fluids

The assumption underlying our discussion of solution NMR in previous chapters is that the motion of molecules in solution completely averages the anisotropy of chemical shift and dipolar couplings, resulting in an isotropic shift and an averaged dipolar coupling of zero for all spins. The primary manifestations of CSA and dipolar coupling in solution are relaxational. As a molecule tumbles, the corresponding modulations of chemical shift (CSA) encourage relaxation, as do the modulations of dipolar couplings due to the angular dependence of D_{ij} with respect to the applied field. However, the assumption of completely isotropic tumbling is appropriate only for a spherically symmetric species. If a molecule (or complex) is magnetically anisotropic, i.e. it has a preferred orientation in an applied field, it will tumble in a slightly anisotropic manner, with some orientations with respect to the applied field that are lower in energy than others. This anisotropy is reflected by the **magnetic susceptibility tensor**, $\overline{\overline{\chi}}$. $\overline{\overline{\chi}}$ is a second-rank tensor like the CSA and dipolar coupling, but magnetic susceptibility is a property of the molecule or complex treated as a rigid body rather than of individual spins or spin pairs. In the appropriate frame of reference fixed relative to the molecular frame, $\overline{\overline{\chi}}$ is diagonal, with elements defined such that $\|\chi_{zz}\|>\|\chi_{yy}\|>\|\chi_{xx}\|$. In practice, the absolute magnitudes of the components of the tensor are less important than their relative magnitudes, so experimentally we are concerned with the **axial anisotropy**, $\Delta\chi_{ax} = \chi_{zz} - 1/2(\chi_{xx} + \chi_{yy})$ and the **rhombic anisotropy**, $\Delta\chi_{rh} = (\chi_{xx} - \chi_{yy})$.

Given the presence of magnetic isotropy in a molecule, the probability of a particular orientation being populated in solution is determined by the relative energy of that orientation. The energy of a particular orientation is determined by the magnitude of the applied field B_0, the magnitudes of the axial and rhombic anisotropies and the angles θ and ϕ that determine the orientation of the anisotropy tensor relative to the applied field (see Figure 10.18). A Boltzmann distribution provides the relative probabilities of available orientations:

$$P(\theta,\phi) = N^{-1}\left[1 + B^2(4kT)^{-1}\left[\Delta\chi_{ax}(2/3)(3\cos^2\theta - 1) + \Delta\chi_{rh}(\sin^2\theta\cos 2\phi)\right]\right] \quad (10.12)$$

where N^{-1} is a constant of normalization. In the absence of anisotropy (or an applied field), the second term goes to zero, leaving the probability of any orientation

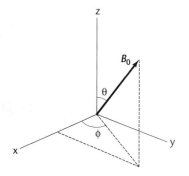

Figure 10.18 Relationship between the molecular frame of reference in which the magnetic anisotropy tensor is diagonal and the applied field $\vec{B}_0$ as used to determine relative energies of molecular orientations as described in Equation 10.12.

equally likely. For an anisotropic molecule, some orientations of the angles θ and ϕ with respect to the applied field are higher in energy, and some are lower, leading to a small but often measurable weighting of particular orientations. This preferential orientation has a number of interesting consequences. The one that is arguably of the most practical importance is the reappearance of the dipolar coupling, albeit in highly attenuated form. The apparent dipolar coupling constant between two spins 1 and 2 as a function of preferential orientations in solution is given by:

$$D_{12} = D_{12}^{\max} \left\langle \tfrac{1}{2}(3\cos^2\theta - 1) \right\rangle \qquad (10.13)$$

The angled brackets indicate that the angle is a time or ensemble average of all the orientations. D_{12} is the **residual *dipolar* coupling** (RDC) between spins 1 and 2, whereas $D_{12}^{\max}$ is the static dipolar coupling constant given in Equation 10.6, and θ is the angle between the internuclear vector connecting the coupled spins and the applied field. The RDC appears in solution state NMR as a variation in the magnitude of the normally field-independent *J*-coupling as a function of magnetic field strength (Figure 10.19).

Even for the most anisotropic molecules (that tend either to contain a paramagnetic center or to be rod- or disk-shaped), the excess probability of a particular orientation relative to the average rarely exceeds a factor of 10^{-4}. As a result, the magnitude of the RDC is usually no more than a few hertz even at high fields for molecules with significant anisotropy. However, the information contained in the RDC is useful: the RDC detected between two spins can be directly related to the orientation of the internuclear vector in the single molecular frame of reference provided by the anisotropy tensor. This makes RDC determination a valuable complement to the NOE and *J*-coupling restraints typically used for macromolecule structure determination in solution NMR. NOE and *J*-coupling restraints are local in nature, and although they can define small regions in great detail, NMR-determined structures often lack accuracy in representing the relative positions of remote structure features such as domain interactions and relative orientations of helices. This is particularly true for elongated structures (proteins bound to DNA, for example), which can have bends over long distances and local constraints are sparse. RDCs help to overcome this problem, and permit structures to be refined in a manner that has more in common with X-ray crystallography (i.e. refinement with respect to a single frame of reference) than standard NMR structure determinations.

Even the small RDCs resulting from intrinsic anisotropy have been used successfully for structure refinement. However, to make the technique more general, it has become common practice to introduce small amounts of substances with large magnetic anisotropies that interact with the molecule of interest in order to induce an orientational preference with respect to the applied field. Many different substances have been used for this purpose, each with their own advantages and drawbacks. Among the most commonly used are nematic liquid crystal phases that form flat micelles (**bicelles**) in solution. Solutions of macromolecules in the presence of a few percent of such bicelles often exhibit RDCs on the order of tens of hertz, easily measurable in standard two-dimensional and three-dimensional spectra

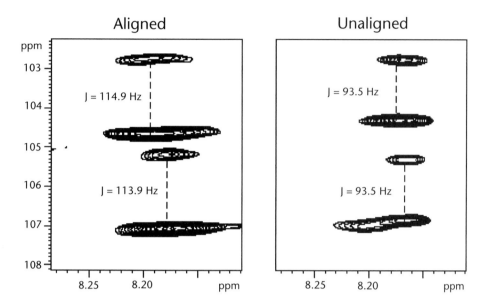

Figure 10.19 Modulation of ^{15}N–^{1}H doublet splitting due to residual dipolar couplings induced by a nematic liquid crystal additive. Two-dimensional spectrum is a ^{1}H, ^{15}N HSQC experiment acquired on a sample of 1 mM acireductone dioxygenase (20 kDa) without ^{1}H decoupling during the ^{15}N evolution period. The difference in splitting between the sample aligned in 5% $HO(CH_2)_{12}$–$(O$–$CH_2CH_2)_5OH$–octanol mixture (left) and unaligned (right) spectra is due to the presence of residual dipolar coupling.

obtained without decoupling in the dimension of interest (Figure 10.19). Also common is the use of virus particles, particularly filamentous phage, to induce order. Phage has the benefit of being largely concentration-independent in being magnetically alignable, so it can be titrated into solutions, whereas liquid-crystal phases tend to be sensitive to concentration (in that the desired phase may require a minimum concentration to form) and doping (additives are often required to obtain the desired nematic phase).

The primary difference between RDCs obtained from intrinsic anisotropy and those obtained using orienting additives is that the intrinsic RDCs reflect properties of the molecule under observation, whereas those obtained in the presence of orienting agents are determined by the interactions between all of the species present. This means that intrinsic information (such as magnetic susceptibility tensors) are unobtainable from additive-induced alignment. As this is rarely of practical importance except for paramagnetic molecules, the use of liquid crystalline additives is by far the most common method of obtaining RDC data for structural and dynamic studies.

There are a number of different experiments that can be used to measure residual dipolar couplings. The simplest method is to leave the coupling of interest intact during the evolution time of a two-dimensional experiment, so that splittings are observed in the indirectly detected dimension. An example of this is shown in

Figure 10.19. In this case, the 180° ^{1}H pulse used to refocus ^{1}H–^{15}N couplings during t_1 in a standard HSQC experiment was removed. Splittings are then measured with and without an orienting medium present and the difference between those splittings represents the RDC for the one-bond ^{1}H–^{15}N pairs. This method works best for smaller proteins and large RDCs (i.e. a highly orienting medium) but suffers in that the resolution is limited by relaxation in the indirectly detected dimension. For more complicated systems and larger proteins (or smaller RDCs), J-modulated HSQC (Reference 10) and *in-phase-anti-phase* (IPAP) (Reference 11) experiments are often used. J-modulated experiments incorporate a delay time during which the signal intensity is modulated by the coupling; if the experiment is performed at a number of different delays, the observed signal intensity can be fitted to obtain the coupling constant to a high degree of accuracy. The IPAP experiment (Figure 10.20) generates two HSQC spectra with the coupling of interest resolved in the indirectly detected dimension. In one spectrum, the two peaks of the coupling doublet are in phase and in the other they are antiphase. The additive combination of the two spectra reinforces the peak that is of the same sign in both spectra, while canceling the peak that is opposite in sign. Spectral subtraction returns the opposite result. The two combination spectra are compared; the difference between the shifts in the two spectra yields the coupling constant.

Briefly, the evolution for the in-phase spectrum during t_1 is:

$$2\hat{\mathbf{S}}_z\hat{\mathbf{I}}_y \xrightarrow{t_1} 2\cos\Omega_I t_1 \cos(\pi J_{IS}t_1)\hat{\mathbf{S}}_z\hat{\mathbf{I}}_y - 2\sin\Omega_I t_1 \cos(\pi J_{IS}t_1)\hat{\mathbf{S}}_z\hat{\mathbf{I}}_x + \ldots$$

$$\xrightarrow{\pi/2_x(I,S)} 2\cos\Omega_I t_1 \cos(\pi J_{IS}t_1)\hat{\mathbf{S}}_y\hat{\mathbf{I}}_z + \ldots \xrightarrow{1/2J_{IS}} \cos\Omega_I t_1 \cos(\pi J_{IS}t_1)\hat{\mathbf{S}}_x \quad (10.14)$$

Recall that the presence of two cosine terms results in pure in-phase peaks split with respect to the coupling between spins I and S.

The incorporation of the extra pulses (shown as empty rectangles in Figure 10.20) and decrementing the phase of the first $\pi/2$ pulse on I for the antiphase spectrum results in the following term prior to t_1:

$$-R\sin(\pi J_{IS}\Delta)\hat{\mathbf{I}}_y \tag{10.15}$$

where R is an exponential decay factor determined by the length of the delay Δ and the T_2 relaxation time of the I spin (^{15}N in this example). After evolution with respect to t_1, one obtains:

$$2R\cos\Omega_I t_1 \sin(\pi J_{IS}\Delta)\sin(\pi J_{IS}t_1)\hat{\mathbf{S}}_z\hat{\mathbf{I}}_x + 2R\sin\Omega_I t_1 \sin(\pi J_{IS}\Delta)\sin(\pi J_{IS}t_1)\hat{\mathbf{S}}_z\hat{\mathbf{I}}_y + \ldots$$

$$\xrightarrow{\pi/2_x(I,S)} -2R\sin\Omega_I t_1 \sin(\pi J_{IS}\Delta)\sin(\pi J_{IS}t_1)\hat{\mathbf{S}}_y\hat{\mathbf{I}}_z + \ldots$$

$$\xrightarrow{1/2J_{IS}} R\sin\Omega_I t_1 \sin(\pi J_{IS}\Delta)\sin(\pi J_{IS}t_1)\hat{\mathbf{S}}_x$$

$$\tag{10.16}$$

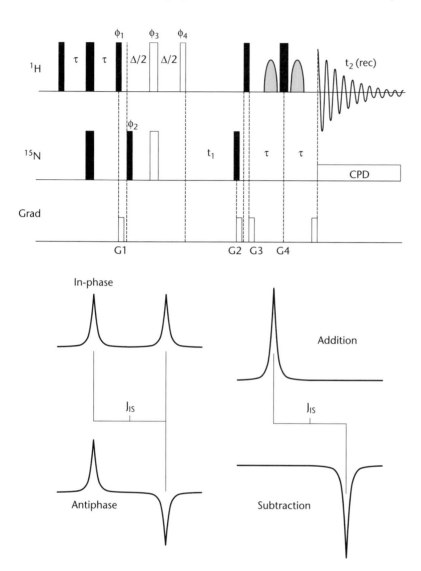

Figure 10.20 IPAP (in-phase/antiphase) pulse sequence for measurement of 1-bond ^1H, ^{15}N coupling constants. Narrower lines represent $\pi/2$ pulses, and broader lines π pulses. Open pulses are used only in the experiment that generates antiphase signals. All pulses are **x** unless otherwise noted. Delays are τ = 2.5 ms, Δ = 5.3 ms. Pulse phases are $\phi_1 = -y, y, \phi_2 = 2(x), 2 (-x)$ for in-phase spectra; $2(-y), 2(y)$ for antiphase spectra; $\phi_3 = 4(x), 4(y), 4(-x), 4 (-y); \phi_4 = 8(x), 8(-x)$; receiver $= x, 2(-x), x$ for in-phase spectra; receiver $= x, 2(-x), -x, 2(x), -x$ for antiphase spectra. Gradients are for artifact suppression and water suppression only. Coherence selection is accomplished by phase-cycling. Quadrature detection in this scheme obtained by States–TPPI. The lower left part of the scheme shows the resulting in-phase and antiphase doublets observed in t_1 for the in-phase and antiphase spectra (see text). After appropriate scaling to account for relaxation that occurs during the added delay Δ in the antiphase spectrum, the in-phase and antiphase spectra are added to obtain a spectra consisting solely of the in-phase peaks of all of the doublets, while the spectrum obtained by subtracting the in-phase from the antiphase spectrum contains only the antiphase peaks.

Again referring to Chapter 5, this will yield antiphase peaks split with respect to J_{IS}. Note that the antiphase sequence requires extra pulses and delays, and so is longer in duration. This means that prior to combination, one of the two spectra must be scaled so that the relaxation R that occurs during the extra delay times is appropriately taken into account. Also, although the result of Equation 10.14 is cosine modulated with respect to the chemical shift of I in t_1, the result of Equation 10.16 is sine modulated, and quadrature components for both datasets are required in order to obtain pure phase spectra that can be properly combined.

Besides residual dipolar couplings, orientational anisotropy is reflected in the average chemical shift observed for a given resonance. The isotropic shift represents a weighted average of the shifts of all possible orientations, and if this preference is weighted by a preferred orientation, the observed chemical shifts are also perturbed. While this would be a minor consideration in the case of ^1H, which has a low CSA, for nuclei with higher intrinsic anisotropies the effect can be detectable, and may be used to provide structural constraints with respect to a single frame of reference. An example of this type of measurement for ^{15}N amide resonances in a polypeptide can be found in Reference 12.

Analysis of residual dipolar couplings

Although a detailed description of how RDC data are used to analyze molecular structure and dynamics is beyond the scope of this text, the relationship between this type of information and anisotropic phenomena in solid-state NMR makes it worthwhile to look in some detail at how raw RDC data are interpreted. The extent and directionality of residual orientation experienced by a molecule can be represented by a 3×3 **order matrix** $\bar{\bar{S}}$, also referred to as a Saupe order matrix. The order matrix is traceless (i.e. like the dipolar coupling tensor, the sum of the diagonal elements is zero) and symmetric ($S_{ij} = S_{ji}$), so that only five elements of the nine-element matrix are independent. If the appropriate rotations are performed to diagonalize the order matrix, the diagonal elements can be considered as orthogonal Cartesian directors fixed in the molecular frame of reference, with a **principal order parameter** corresponding to S_{zz}, and an **asymmetry parameter** $\eta = (S_{yy} - S_{xx})/S_{zz}$ where $|S_{zz}| > |S_{yy}| > |S_{xx}|$. The frame of reference in which $\bar{\bar{S}}$ is diagonal is called the **principal order frame**.

In general, the rotations required to transform the original molecular frame of reference (e.g. the PDB coordinate frame) into the principal order frame are not initially known, but must be determined from the experimental data. The elements of the order matrix S_{ij} in an arbitrary frame of reference is given by:

$$S_{ij} = \left\langle \frac{3\cos\theta_i \cos\theta_j - \delta_{ij}}{2} \right\rangle \quad (10.17)$$

where i and j are **x**, **y** and **z** permutations, and δ_{ij} is the Kroenecker delta, equal to one for diagonal elements and zero for off-diagonals. The angles θ_i are those between the applied field $\vec{B}_0$ and the axes of the coordinate system (see Figure 10.21). In turn, the relationship between the elements of $\bar{\bar{S}}$ and the observed values of residual dipolar couplings (as related to Equation 10.13) are given by:

$$D_{12} = D_{12}^{max} \sum_{i,j=\{x,y,z\}} S_{ij} \cos\phi_i^{12} \cos\phi_j^{12}$$

(10.18)

Now, the angles ϕ_i are those between the vector joining spins 1 and 2 and the axes of the coordinate system. Each RDC that is measured can be used to generate a linear equation by the appropriate substitution into Equation 10.15. The individual expressions will be of the form:

$$D_{nm} = D_{nm}^{max} \left[\begin{array}{l} S_{yy}\left(\cos^2\phi_y^{nm} - \cos^2\phi_x^{nm}\right) + S_{zz}\left(\cos^2\phi_z^{nm} - \cos^2\phi_x^{nm}\right) + S_{yy}\left(\cos^2\phi_y^{nm} - \cos^2\phi_x^{nm}\right) \\ +S_{xy}\left(2\cos\phi_x^{nm}\cos\phi_y^{nm}\right) + S_{xz}\left(2\cos\phi_x^{nm}\cos\phi_z^{nm}\right) + S_{yz}\left(2\cos\phi_y^{nm}\cos\phi_z^{nm}\right) \end{array} \right]$$

(10.19)

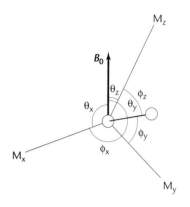

Figure 10.21 Angles used to define the orientation of the molecular frame of reference and internuclear vectors relative to the applied field $\vec{B}_0$ in Equations 10.17–10.19.

where the S_{ij} are the independent elements of the order matrix. If a series of n such linear equations are constructed, one for each measured value of D_{nm}, a $5 \times n$ matrix can be constructed that can be used to solve for the independent elements of S_{ij} (Reference 13). In principle, only five independent values of D_{nm} are required to solve this set for these values. In practice, many more RDCs are measured for a given molecule, so the order matrix is almost always overdetermined.

Despite the overdetermination of the order matrix from a single set of RDCs, there are four independent sets of Euler angles that can satisfy Equation 10.19. Thus, from one set of RDCs, there is ambiguity about the directionality of individual bond vectors. In practice, it is desirable to measure the RDCs in several alignment media with different alignments in order to remove this ambiguity. A helpful review of how RDC data are acquired and analyzed can be found in Reference 14.

Problems

10.1 Expand Equation 10.5 in terms of the scalar products of the angular momentum operators. Use the following conversions to get to the polar coordinate system (Figure 10.5):

$$x = r\sin\theta\cos\phi$$
$$y = r\sin\theta\sin\phi$$
$$z = r\cos\theta$$

Using these terms, as well as the raising and lowering operator equivalents of the cartesian operators, derive the secular terms (A and B) of the "dipolar alphabet" (Equation 10.7).

10.2 For a sample rotating at a frequency ω_r around an axis at an angle θ to the applied field $\vec{B}_0$ (see accompanying figure below), the average value of the angle α of vector $\vec{r}$ with respect to $\vec{B}_0$ is given by:

$$\langle \cos^2 \alpha \rangle = \cos^2 \theta \cos^2 \beta + 2 \sin \theta \cos \theta \sin \beta \cos \beta \langle \cos \omega_r t \rangle + \sin^2 \theta \sin^2 \beta \langle \cos^2 \omega_r t \rangle$$

where brackets represent time average values.

Show that

$$\langle 3 \cos^2 \alpha - 1 \rangle = \left[1 / 2 (3 \cos^2 \theta - 1)(3 \cos^2 \beta - 1) \right].$$

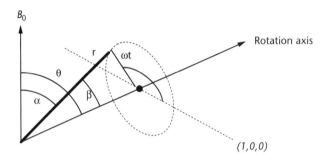

10.3 Show that a vector $\vec{r}$ that is at the same angle to all three axes in the Cartesian frame is at the magic angle.

10.4 The WAHUHA (from *Waugh–Hu*ber–*Hae*berlen) pulse train (τ–$\pi/2_x$–2τ–$\pi/2_{-x}$–τ–$\pi/2_y$ –2τ–$\pi/2_{-y}$) is used to remove the effects of dipolar coupling from solid-state MAS spectra. From A and B in the expression for the dipolar Hamiltonian, Equation 10.7, show that the WAHUHA pulse train results in a zero net change in the Hamiltonian after four pulses. (Hint: expand the expression in terms of the Cartesian components.)

10.5 (a) Show that for an isotropic molecule, Expression 10.12 results in a value of $P(\theta,\phi) = N^{-1}$. (b) The degree of alignment along an axis i for a molecule relative to the applied field can be expressed in terms of an order parameter $S_i = \Delta\chi_{ii}B_0^2/15kT\mu_0$. $S = 0$ implies random isotropic tumbling, while $S = 1$ implies complete alignment along that axis. Calculate the order parameter S_z for a benzene molecule given a 14 T field, and a value of $\Delta\chi_{zz} = -1.27 \times 10^{-33}$ m^3 for benzene at 300 K. $\mu_0 = 4\pi \times 10^{-7}$ T^2 m^3 J^{-1}.

10.6 Given a static field of 11.74 T, calculate the frequency offset required for a B_1 field of 20 kHz (on-resonance) required to generate a B_{eff} for ^{1}H that is at the magic angle ($\cos^{-1}\sqrt{3}$) (the Lee–Goldburg condition). If the placement were required to be 20 kHz off-resonance, what would be the required amplitude of B_1?

10.7 What spinning speeds are required in order to obtain rotational resonance for two spins separated in isotopic shift by 25 ppm in ^{13}C at 11.74 T?

10.8 Show the relevant product operators leading up to the terms in Equation 10.16 for the antiphase component of the IPAP sequence. Show why certain terms in both the antiphase and in-phase spectra can be ignored in Equations 10.14 and 10.16.

References

1. E. M. Purcell, H. C. Torrey, and R. V. Pound, *Phys. Rev.* 69, 37–38 (1946).
2. M. M. Maricq and J. S. Waugh, *J. Chem. Phys.* 70, 3300–3316 (1979).
3. J. S. Waugh, L. M. Huber, and U. Haeberlen, *Phys. Rev. Lett.* 20, 180–182 (1968).
4. D. P. Raleigh, F. Creuzet, S. K. Das Gupta, M. H. Levitt, and R. G. Griffin, *J. Am. Chem. Soc.* 111, 4502–4503 (1989).
5. T. Gullion and J. Schaefer, *J. Magn. Reson.* 81, 196–200 (1989).
6. N. M. Szeverenyi, M. J. Sullivan, and G. E. Maciel, *J. Magn. Reson.* 47, 462–475 (1982).
7. C. H. Wu, A. Ramamoorthy, and S. J. Opella, *J. Magn. Reson. A* 109, 270–272 (1994).
8. M. Lee and W. I. Goldburg, *Phys. Rev.* 140, A1261–A1271 (1965).
9. J. Pauli, M. Baldus, B. van Rossum, H. de Groot, and H. Oschkinat, *ChemBioChem* 2, 272–281 (2001).
10. G. W. Vuister and A. Bax, *J. Am. Chem. Soc.* 115, 7772–7777 (1993).
11. M. Ottiger, F. Delaglio, and A. Bax, *J. Magn. Reson.* 131, 373–378 (1998).
12. J. Boyd and C. Redfield, *J. Am. Chem. Soc.* 121, 7441–7442 (1999).
13. J. A. Losonczi, M. Andrec, M. W. F. Fischer, and J. H. Prestegard, *J. Magn. Reson.* 138, 334–342 (1999).
14. J. H. Prestegard, H. M. Al-Hashimi, and J. R. Tolman, *Quart. Rev. Biophys.* 33, 371–424 (2000).

11

RELAXATION REVISITED: DYNAMIC PROCESSES AND PARAMAGNETISM

Until now, we have dealt with nuclear spin relaxation only as a parameter affecting the NMR experiment. However, spin relaxation rates depend on time-dependent fluctuations in the local magnetic environment (see Chapter 4 and the discussion of the spectral density), and careful measurement and analysis of those rates can provide valuable information about molecular dynamics. Recall that the overall relaxation rates for a nuclear spin $[T_{1(obs)}^{-1}$ and $T_{2(obs)}^{-1}]$ are the sums of the individual contributions from different relaxation mechanisms (Equation 4.21). We will now examine how those individual contributions are determined and interpreted for a variety of situations. We will also discuss the special (but common) case of relaxation induced by unpaired electron spins.

Time scales of molecular motion, dynamic processes and relaxation

In Chapter 4, the importance of the spectral density function $J(\omega)$ in determining the efficiency of nuclear spin relaxation was discussed. To review, the spectral density is a measure of the frequency elements available from the local environment to cause the mixing of states required for relaxation to occur. For a given transition to occur, the spectral density must be nonzero at the frequency of that transition. The difference between longitudinal (T_1) and transverse (T_2) relaxation was described, with T_1 processes causing a return to equilibrium state populations (an enthalpic effect) and T_2 processes resulting in a loss of coherence (an entropy effect) without necessarily affecting state populations. T_2 processes represent a loss of **phase**

memory. Even though spin angular momentum is conserved in a spin flip (W_0 transition in Equation 4.11), the process results in discontinuity in the phase of the precession of the two spins, and as such they will no longer contribute to observable coherence (Figure 11.1). It is worth expanding on this discussion now. For small molecules in nonviscous solvents, T_2 processes are often no more efficient than T_1, and T_1 is equal to T_2 in such cases. As molecules become larger or solvents become more viscous, the rapid random motions that give rise to the extreme narrowing limit (i.e. all relevant transitions being equally well represented in the spectral density resulting from random motion) are no longer present. This means that the higher-frequency transitions that lead to T_1 relaxation (W_2 and W_1 in Equation 4.15) are no longer favored. On the other hand, the lower-frequency transitions that lead to spin flips (W_0) are still present, and T_2 processes remain efficient. An approximate relationship between molecular weight, T_1 and T_2 is shown in Figure 11.2. The solid state is the extreme of this progression, with low (or zero)-frequency processes favored over high-frequency transitions by the lack of overall molecular motion.

In general, two-state **stochastic** (random) reversible events that take place as:

$$R \underset{k_r}{\overset{k_f}{\rightleftharpoons}} P \qquad (11.1)$$

will affect relaxation processes with transitions on a time scale of $[R]k_f + [P]k_r$. Any event that changes the magnetic environment of a spin in a random time-dependent fashion is a potential contributor to relaxation. The time scales of such processes determine how effective they are at inducing spin relaxation. The fastest submolecular motions (bond vibrations) occur in the infrared region of the EMR spectrum (10^{12}–10^{14} Hz), and are generally not contributors to nuclear spin relaxation, since they average too quickly to produce discrete changes in the magnetic environment. The magnetic fields generated by the rotation of symmetric rotors such as methyl groups can couple with nuclear spins, and random interruptions in such motions

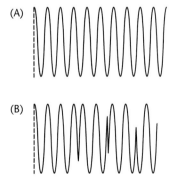

Figure 11.1 Illustration of the loss of phase coherence due to spin–spin interactions leading to T_2 relaxation. (A) Frequency and phase are maintained on a single spin throughout the observation time. (B) Spin flips between two spins with the same frequency leads to a loss of phase memory and a corresponding decrease in observable coherence.

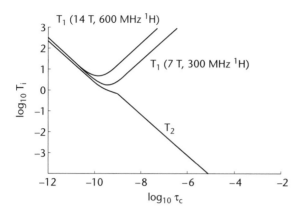

Figure 11.2 Log–log plot of relaxation time T_1 and T_2 for a pair of ^1H spins with relaxation dominated by the homonuclear (^1H–^1H) dd interaction. The lines are calculated as described in Problem 11.1. Note that T_1 goes through a minimum where $\omega_0\tau_c = 1.12$, while T_2 has an inflection point there, but continues to decrease monotonically, due to a linear dependence on $J(0)$, the spectral density at zero frequency, which is in turn linearly dependent on the correlation time.

can, under some circumstances, contribute to relaxation [**spin–rotation (sr)** relaxation]. The contribution of sr to T_1 relaxation is given by:

$$T_{1(sr)}^{-1} = 2I_r kTC^2 \tau_{sr} / \quad^2 \qquad (11.2)$$

where I_r is the moment of inertia of the rotor, T is temperature, τ_{sr} is the autocorrelation time for collisionally induced perturbations in the rotation, and C is the spin–rotation coupling constant. Two features of Equation 11.2 are worth mentioning. First, τ_{sr} is not the rotational correlation time, as is the case for intramolecular phenomena such as dipolar-coupled spins (dd) and CSA, but a measure of the mean time between collisional events involving the molecular rotor. Secondly, sr relaxation is the only common relaxation mechanism that becomes more important with increasing temperature, due to the linear dependence of Equation 11.2 on temperature.

The most important dynamic processes for nuclear spin relaxation are overall molecular tumbling (correlation times between 10^{-8} and 10^{-12} s), segmental motions of molecules, and chemical exchange (state lifetimes from 10^{-3} to 10^{-6} s). These processes result in the modulation of the orientation of internuclear vectors between dd with respect to $\vec{B}_0$, resulting in fluctuations of the dipolar coupling. They also result in time-dependent modulation of chemical shifts for spins in non-spherically symmetric environments because of CSA. At least in fluids, these fluctuations tend to be on a time scale that allows some or all of the nuclear spin transitions discussed in Chapter 4.

The fact that motions and other dynamic processes encourage relaxation of nuclear spin transitions that correspond in frequency to those motions has been exploited to examine dynamic processes since the earliest NMR studies of chemical systems. Motions of different types have different characteristic time scales and therefore give different spectral density profiles. A thorough analysis of relaxation behavior as a function of parameters such as magnetic field strength, temperature and nuclear spin type can provide details of motion and chemical dynamics in small and large molecules.

The spectral density revisited

The spectral density $J(\omega)$ provides a measure of the "white noise" available from the environment to encourage transitions required for nuclear spin relaxation. For isotropically tumbling molecules, $J(\omega)$ has the form:

$$J(\omega) = \frac{2\tau_c}{5\left(1 + \omega^2 \tau_c^2\right)} \tag{11.3}$$

Recall that the spectral density function is the frequency-domain Fourier transform of the time-domain autocorrelation function, which in turn is a measure of the "memory" of a system. When considering molecular motion (which provides spectral densities in the appropriate range for nuclear spin transitions), the autocorrelation function is a measure of how long it takes for a molecule in a defined starting orientation to reorient to a completely random orientation. For a rapidly tumbling molecule, this is not very long, and such molecules have short correlation times. The correlation time τ_c is the decay constant for the autocorrelation function, and short correlation times give rise to spectral densities that are nonzero over a wide range of frequencies. If a molecule tumbles slowly, the correlation time is long, and the spectral density spans a narrower (and lower) range of frequencies. This relationship between the correlation time and frequency range of the spectral density is completely analogous to the observation that the shorter an RF pulse is, the wider the frequency range it excites. In the case of molecular motion, the "pulse" is the reorientation of the molecule (or subunit) in the magnetic field. If the spectral density resulting from such random motions is equal (and nonzero) over all of the transitions of interest, the system is in the extreme narrowing limit.

In the simplest case, that of a spherically symmetric rigid molecule tumbling isotropically in a nonviscous medium, a single correlation time is sufficient to describe the spectral density. This is the case that is shown in Figure 11.3. However, consider a system where more than one type of motion occurs. If, for example, there is local segmental motion that is much faster than the overall tumbling of the molecule, the observed spectral density for affected spins will be the result of a biexponential decay of the autocorrelation function, as shown in Figure 11.4. This is often the case with macromolecules, and one can get a great deal of information about local motions in macromolecules by fitting the observed spectral density curve using correlation time(s) as adjustable parameters.

Of course, obtaining a complete spectral density curve is not trivial, as $J(\omega)$ is not directly observable, but must be extracted from measurable quantities such as

The spectral density revisited   307

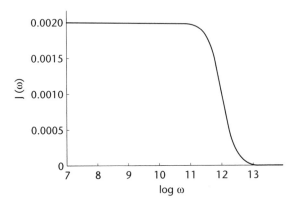

Figure 11.3 Spectral density (Equation 11.3) plotted versus the log of frequency (in radians) as a function of a single correlation time ($\tau_c = 10^{-12}$ s).

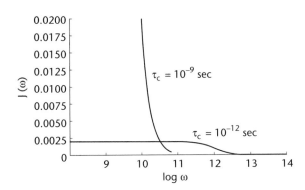

Figure 11.4 Spectral density (Equation 11.3) plotted versus a longer correlation time ($\tau_c = 10^{-9}$ s) combined with the plot Figure 11.3. Note that the spectral density at longer correlation time is of greater intensity, but covers a narrower frequency range than that due to the short correlation time.

relaxation rates and nuclear Overhauser effects. In principle, one can obtain a complete spectral density curve by measuring relaxation over a wide frequency range, varying the magnetic field continuously while holding other parameters constant. In practice, one rarely has access to a wide range of fields, and is limited to the two or three magnetic fields available in the local NMR facility. One way of getting around this problem is to use **field cycling**, i.e. rapidly modulating the magnetic field experienced by the sample in some fashion. Field cycling can be accomplished by changing the field directly using a cycling electromagnet or by rapidly moving the NMR sample in a static field. The second option is clearly preferred for modern superconducting magnets. In this case, the sample is rapidly shuttled between a high field at the center of a superconducting magnet (where the spectroscopy is done) and intermediate-to-low fields at the edge of the magnet (where relaxation processes occur). If the shuttle device is equipped with a Helmholz coil for generating controlled local fields, one can even obtain fields near zero. An example of such a map is shown in Figure 11.5. A limitation of field cycling is the time required to move the

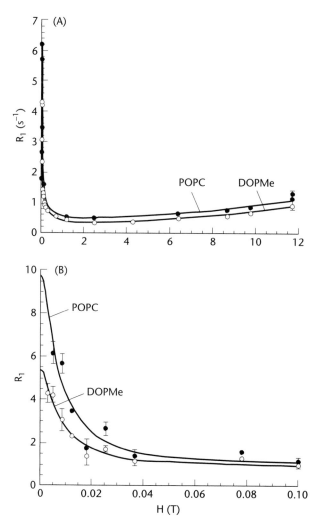

Figure 11.5 ^{31}P R_1 rate constants for the headgroup in vesicles of phosphatidylcholine (POPC) and phosphatidylmethanol (DOPMe) measured using a field cycling apparatus on a commercial 500 MHz NMR spectrometer. ^{31}P relaxation is dominated by CSA at fields >1 T, but dd contributions can be detected at lower fields accessible by field cycling. (A) Relaxation over 0–12 T range. (B) Expansion showing the range from 0 to 0.1 T. Figure courtesy of A. Redfield and M. Roberts.

sample between fields (usually in the range of fractions of a second) and the mechanical stability of the sample. Still, there is currently some effort to commercialize such devices for high-resolution NMR and make field cycling generally available.

Another way of reaching low effective fields for relaxation measurements is to apply a spin-locking field B_1, as used in TOCSY or ROESY, to create an environment where the equilibrium magnetization $M_{sl} = (B_1/B_0)M_0$ is very small and effective transition frequencies are in the kHz range. The time constant for the rate at which the initial spin-locked magnetization approaches this equilibrium is called $T_{1\rho}$. Since the effective field in a spin-locking experiment is on the order of kHz (γB_{eff}) rather than hundreds of MHz (γB_0), the transition frequencies are much smaller than for T_1 relaxation, and the relevant spectral densities are also in this range.

The experimental data available for characterizing local spectral densities in macro-molecules are the experimental relaxation rate constants, $R_1, R_2 R_{1\rho}$, (corresponding to T_1, T_2 and $T_{1\rho}$ processes, respectively) and the steady-state heteronuclear NOE. For a heteronuclear spin pair H–X (where typically X is ^{15}N or ^{13}C in proteins), in the absence of cross-correlation effects (see Chapter 9), these values are related to the spectral densities as:

$$R_{1X} = \left(d^2/4\right)\left[J(\omega_H - \omega_X) + 3J(\omega_X) + 6J(\omega_H + \omega_X)\right] + c^2 J(\omega_X) \quad (11.4)$$

$$R_{2x} = \left(d^2/8\right)\left[4J(0) + J(\omega_H - \omega_X) + 3J(\omega_X) + 6J(\omega_H) + 6J(\omega_H + \omega_X)\right] +$$
$$\left(c^2/6\right)\left[4J(0) + 3J(\omega_X)\right] \quad (11.5)$$

$$\sigma = \left(d^2/4\right)\left[6J(\omega_H + \omega_X) - J(\omega_H - \omega_X)\right], \quad (11.6)$$

$$X\{^1H\}NOE = 1 + \sigma / R_{1X} \quad (11.7)$$

$$R_{1\rho}(X) = R_{1X}\cos^2\theta + R_{2X}\sin^2\theta + R_{ex}\sin^2\theta \quad (11.8)$$

The constants in the expressions above are as follows: $d = (\mu_0 h\gamma_X\gamma_H/8\pi^2)/r_{XH}^{-3})$, where μ_0 is the permeability of free space, h is Planck's constant, and r is the distance between the two nuclei H and X. The second term in both R_1 and R_2 deals with CSA of X, where $c = \Delta\sigma\omega_X/\sqrt{3}$ and $\Delta\sigma$ is the chemical shift anisotropy, and the chemical shift tensor is assumed to be axially symmetric. In the expression for $R_{1\rho}$, θ is the angle between $\vec{B}_1$ and $\vec{B}_{eff}$ as shown in Figure 11.6, and so depends upon the chemical shift of the resonance of interest and upon the magnitude of $\vec{B}_1$. R_{ex} is the rate constant for relaxation due to chemical exchange, and will be discussed below.

Note that in Equation 11.5, the value of R_2 depends on the spectral density at zero frequency, $J(0)$, while that of R_1 does not (Equation 11.4). This makes sense intuitively, since R_2 reflects the loss of coherence via transitions (spin flips) that can be enthalpically small, while R_1 requires an interaction with the surroundings that result in a net change in the enthalpy of the spin system. This results in the observation shown in Figure 11.2, that R_2 increases with molecular weight (and increasing correlation time) but R_1 increases with molecular weight only to a certain point, and then begins to decrease again. For large molecules or viscous solutions, this gives rise to the difficult situation of very long R_1 values (requiring long delays between repetition of an experiment) while also having very short R_2 values (broad lines). In the solid state, this phenomenon reaches an extreme in the absence of molecular tumbling.

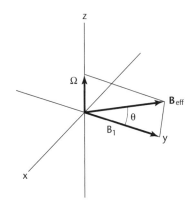

Figure 11.6 Angle θ between the applied spin-locking field **B**$_1$ and the effective field **B**$_{eff}$ defined by $\tan\theta = \Omega/B_1$ where Ω is the difference between the chemical shift frequency of the locked spin and the applied field **B**$_1$.

Experimental measurement of heteronuclear relaxation parameters in proteins

The HSQC experiment provides the basis for most heteronuclear relaxation measurements in proteins and other macromolecules (Figure 11.7). After polarization transfer from ^{1}H (e.g. $\hat{\mathbf{H}}_x\hat{\mathbf{X}}_z$) the heteronuclear magnetization is oriented appropriately for the experiment in question: for R_1 measurements, the X-nucleus term is placed along –$\mathbf{z}$ and allowed to relax during a variable time period T. Attached ^{1}H spins are saturated during T. Afterwards, frequency labeling and back-transfer occurs as usual, with the signal intensities reduced by R_1 relaxation of X spins

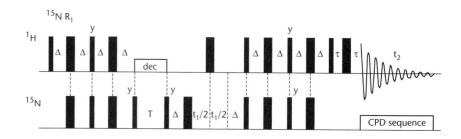

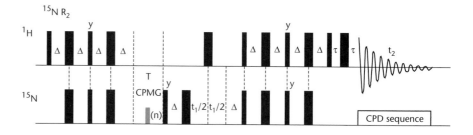

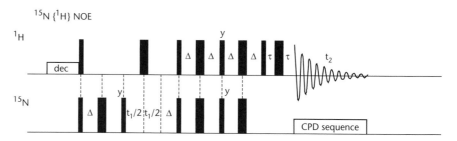

Figure 11.7 HSQC-based experiments for measurement of ^{15}N relaxation parameters. Experiments are shown simplified, excluding phase cycling and gradients used for coherence selection and artifact suppression. The value of Δ is ~$1/4J_{NH}$, and $\tau = 0.5$ ms. For the third sequence to measure heteronuclear steady-state NOE, the experiment is acquired with and without ^{1}H broadband decoupling during the pre-experiment delay to generate steady-state NOE difference experiments. For a complete description, see N. A. Farrow *et al.*, *Biochemistry* 33, 5984–6003 (1994).

during T. A series of data sets collected with appropriate variation of T can be used to construct an exponential decay curve for each peak in the HSQC spectrum.

The measurement of R_2 is technically more challenging. As we saw in Chapter 4, the standard T_2 experiment consists of a series of π pulses that produce "echoes" at time τ, where the spacing of the π pulses is 2τ (the Carr–Purcell–Meiboom–Gill or CPMG experiment). The refocusing removes broadening due to experimental effects such as field inhomogeneity, and only the irreversible loss of coherence due to intrinsic characteristics of the system contributes to the decrease in magnitude of the echo as a function of the number of repetitions of the $\tau-\pi-\tau$ series (see Figures 4.4 and 4.5). The duration of the relaxation time T is determined by the number of times the CPMG sequence is repeated. Since the net result of the CPMG sequence is to rotate the X-spin coherence around a transverse axis, the pulse train can be described in terms of the frequency of the rotation, just as a spin-locking field or RF pulse is described; typically, the CPMG pulse train corresponds to a rotation in the kHz range. As the CPMG sequence is applied to spin X, asynchronous π pulses are applied to ^{1}H (S) spins in order to eliminate cross-correlation effects on relaxation. Because the relaxation time T can last hundreds of milliseconds, with much shorter values of τ, there can be many π pulses applied to both ^{1}H and X spins during T. Such long, relatively high-power, pulse trains can result in significant sample heating, probe arcing (sparking across the probe coils) and duty-cycle problems with RF amplifiers. While many modern amplifiers are equipped with automatic duty-cycle cutoffs that shut off the amplifier if the power output (energy/unit time) exceeds a specified level, this does not always protect the probe from damage, and arcing can result in soot deposits and other damage to the coils. As such, care must be taken when setting up such experiments to prevent damage to the sample and to the spectrometer. As an aside, probe arcing most often occurs when simultaneous high-power pulses are applied using the same RF coil (e.g. simultaneous ^{15}N and ^{13}C pulses during triple resonance experiments). Erratic and nonreproducible pulse lengths can be a symptom of probe arcing. To test this possibility, one should lower the power levels and see if the problem goes away.

Model-free analysis of spin relaxation

A variety of analytical tools has been developed in order to separate relaxation effects due to local motions from those resulting from the overall tumbling of a molecule. The most commonly employed is the **model-free approach** developed by Lipari and Szabo in the early 1980s (Reference 1). The model-free approach is most useful for a combination of slow (time scale 10^{-8}–10^{-9} s) and fast (10^{-11}–10^{-12} s) motions, and is based on the assumption that macromolecular motion can be described in terms of two correlation times, a global (isotropic) τ_m that describes the overall motion of the molecule, and a local effective correlation time τ_e for the fast internal motions. The correlation between the two types of motion is described by an order parameter S^2. In the model-free approach, the spectral density is defined in terms of these parameters as:

$$J(\omega) = \tfrac{2}{5}\left[S^2 \tau_m / (1 + \omega^2 \tau_m^2) + \left(1 - S^2\right)\tau / (1 + \omega^2 \tau^2) \right] \qquad (11.9)$$

where:

$$\tau = (\tau_m^{-1} + \tau_e^{-1})^{-1} \tag{11.10}$$

Note that as S^2 approaches 1, Equation 11.9 simplifies to the spectral density expression in Equation 11.3. The experimental data values for the left-hand terms of Equations 11.4–11.7, R_{1X}, R_{2X} and $X\{^1H\}NOE$, are used in a least-squares non-linear fit using the form of the spectral density shown in Equation 11.9 to obtain values for S^2, τ_e and τ_c.

The model-free approach is so called because there are no structural or model-based assumptions implicit in the values of S^2, τ_e and τ_c other than that the overall motion of the molecule is isotropic, and that the local effective correlation time τ_e is in the extreme narrowing limit. The physical interpretation of S^2 for small-amplitude axially symmetric motions (such as rotation around a bond) is as a mean angle relative to a vector fixed in the molecular frame of reference whose motion is described by the overall correlation time τ_c (Figure 11.8). In the presence of asymmetric diffusion, as for a rod-shaped molecule, Equation 11.3, which describes the spectral density for isotropic motion, needs to be modified to include the effects of diffusional asymmetry:

$$J(\omega) = \tfrac{2}{5} \sum_{i=0}^{2} A_i \left(\frac{\tau_i}{1 + \omega^2 \tau_i^2} \right) \tag{11.11}$$

where the A_i are functions of the angle formed by the particular X–H vector in question and the unique axis of the molecular diffusion tensor, and τ_i is a modified correlation time that is also a function of the diffusion tensor. A more complete description of this modification is given in Reference 2.

The primary limitation of the model-free approach to analysis of relaxation data for macromolecules is that the time scales of τ_c and τ_e must be different, with τ_e in the extreme narrowing regime. Practically, this means that internal motions with τ_e approaching τ_c cannot be characterized using a strict model-free analysis.

An alternative way of obtaining the relevant spectral densities for a given spin pair from relaxation measurements is by **spectral density mapping**, first described by Wagner and Peng (Reference 3). In this method, besides the standard X-spin–relaxation experiments described above, experiments are performed in which the coherence order of both members of the spin pair (e.g. a ^1H–^{15}N coupled pair) is specified (for example, $\hat{H}_z\hat{X}_z$, $\hat{H}_z\hat{X}_{xy}$, etc.). The data obtained from these experiments are used to solve a set of linear equations that yield the values of spectral densities at the appropriate frequencies.

Chemical exchange and motion on slow and intermediate time scales (10^{-6}–10^{-1} s)

Dynamic processes in macromolecules that are of interest for chemical or biological processes (ligand binding, catalytically important motions) are often (usually?)

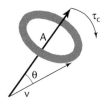

Figure 11.8 Physical interpretation of model-free parameters. The large vector **A** is fixed in the molecular frame of reference, and reorients with the overall correlation time τ_c. The local vector **v** (that might represent a bond vector) reorients with a local effective correlation time τ_e and occupies a series of angles with respect to **A** that have an axially symmetric Gaussian distribution around an average value of θ. θ is related to the order parameter S as $S^2 = 1 - \tfrac{3}{2}\langle\theta^2\rangle$.

much slower than the librations and torsions that are represented by the fast local effective correlation times represented by τ_e, or even the overall correlation time τ_c. Consider the simplest case of two-site exchange:

$$p_1 \underset{k_r}{\overset{k_f}{\rightleftharpoons}} p_2 \qquad (11.12)$$

where p_i are the fractional state populations ($p_1 + p_2 = 1$) and k_f and k_r are the forward and reverse rate constants for exchange, respectively. This could represent any number of real situations: a proline cis–trans isomerization, a ligand or substrate binding event or a protein–protein interaction. In the extreme case, slow exchange on the chemical-shift time scale results in different resonances for the same spin in different conformers ($k_{ex} < \Delta\nu_{12}$, see Figure 2.13). If exchange between the conformers in the slow-exchange regime on the chemical-shift time scale is still relatively rapid on the timescale of T_1 relaxation (i.e. the exchange will happen many times during a time equal to T_1 of the involved spins), the rate of exchange is reflected in the intensity of cross-peaks in NOESY spectra between the corresponding resonances in the two conformers relative to their diagonal peaks. For example, a cross-peak between two exchanging resonances at 1% intensity of the diagonal in a NOESY experiment with a 100 ms mixing time indicates that 1% of the conformers will exchange during 100 ms, and a rate constant for the exchange process can be determined from that information (see Figure 11.9). A more complete analysis of rates can be obtained from analysis of the buildup of the cross-peaks between the two sites as a function of mixing time.

More often, exchange is fast enough to result in only a single resonance that is at the weighted average of the chemical shifts of the individual conformers, so:

$$\omega_{obs} = \sum_i p_i \omega_i \qquad (11.13)$$

In the simple case of two-site exchange, the lifetime τ_{ex} of the exchange process is given by:

$$p_1 / k_r = p_2 / k_f = \tau_{ex} \qquad (11.14)$$

If τ_{ex} is on the order of 10^{-3}–10^{-6} s, the exchange will be reflected in a broadening of the line, indicating a contribution to R_2 from exchange, R_{ex}, so that $R_{2(obs)} = R_{2(intrinsic)} + R_{ex}$. The line-broadening will be R_{ex}/π, and the relationship between R_{ex} and τ_{ex} is given by:

$$R_{ex} = p_1 p_2 \Delta\omega_{12}^2 \tau_{ex} \qquad (11.15)$$

A simple test for the presence of an exchange contribution to R_2 in heteronuclear relaxation is to examine the ratio R_1/R_2, that in the limit of a short effective local

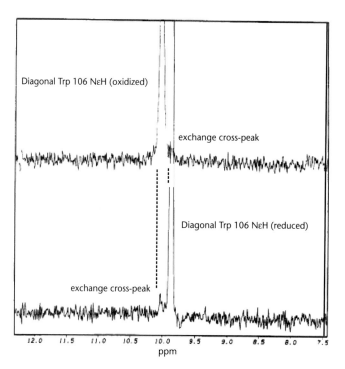

Figure 11.9 Exchange cross-peaks in NOESY. The tryptophan 106 indole NεH in putidaredoxin shows a redox-dependent chemical shift. Shown are one-dimensional slices from a 500 MHz NOESY spectrum of a 1:1 mixture of oxidized and reduced putidaredoxin (2 mM total concentration) corresponding to the chemical shifts of the oxidized and reduced forms of Trp 106 NεH. Electron exchange is slow on the chemical-shift time scale, so exchange cross-peaks (marked) are observed. The mixing time of the experiment is 100 ms, and the cross-peaks are ~1% of the diagonal peaks, indicating that ~1 of every 100 reduced putidaredoxin molecules will reoxidize every 0.1 s, yielding an electron self-exchange rate constant of 100 M^{-1} s^{-1}.

correlation time τ_e and significant order parameter (>0.8) should be a constant for a given globular protein. If R_2 is shortened by exchange, this ratio will increase.

Measurement of R_{ex}

R_{ex} can be measured by fitting the observed transverse relaxation rate as a function of effective field strength (B_{eff}) in the CPMG (T_2) or spin-locking ($T_{1\rho}$) experiment, which yields the so-called relaxation dispersion curve. The appropriate equations to fit are:

$$R_{ex} = \frac{1}{2\tau_{ex}} - \frac{1}{2\tau_{CPMG}} \sinh^{-1}\left[\frac{1}{\tau_{ex}\kappa} \sinh(\tau_{CPMG}\kappa / 2)\right] \tag{11.16}$$

$$R_{ex} = p_1 p_2 \Delta\omega_{12}^2 \tau_{ex} / \left(1 + \tau_{ex}{}^2 \omega_{eff}{}^2\right) \tag{11.17}$$

where $\omega_{eff}^2 = (\Omega^2 + \gamma^2 B_1^2)$ is the spin-lock field (γB_{eff}) in rad/s for the $T_{1\rho}$ experiment, $\kappa = (\tau_{ex}^{-2} - 4p_1p_2\Delta\omega_{12}^2)^{1/2}$, sinh is the hyperbolic sine function defined as sinh(θ) = $(\exp(\theta) - \exp(-\theta))/2$ and τ_{CPMG} is the delay between π pulses in the CPMG pulse train. Typical relaxation dispersion curves, one for an NH pair undergoing conformational exchange and one that is not, are shown in Figure 11.10. Note that for both experiments, the effect of increasing the effective field (either ω_{eff} or ω_{CPMG} = $2\pi/\tau_{CPMG}$) is to decrease the apparent value of R_{ex}, i.e. the application of an RF field suppresses the contribution of R_{ex} to R_2. For this reason, it is necessary to measure R_2 (or $R_{1\rho}$) as a function of variable effective field in order to clearly establish the lifetime of the exchange process τ_{ex}.

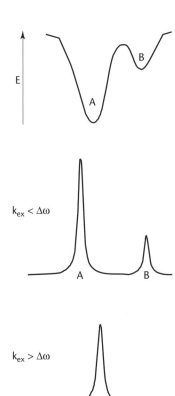

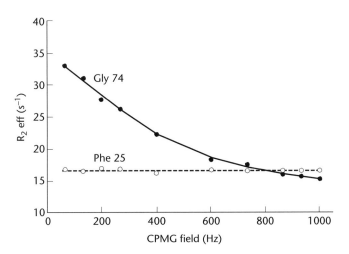

Figure 11.10 ^{15}N R_2 relaxation dispersion curves for backbone amide NH groups in cyclophilin A. Glycine (Gly) 74, within a mobile loop adjacent to the active site, exhibits chemical exchange, while phenylalanine (Phe) 25, in a more restricted region of the protein, does not. Note that as the CPMG field increases in frequency, the exchange contribution to R_2 is suppressed. Figure courtesy of E. Eisenmesser.

Figure 11.11 Exchange between conformations of unequal standard free energies. The rate of exchange is determined by the height of the barrier between the two conformations A and B, while the relative populations p_A and p_B (such that $p_A + p_B = 1$) are determined by a Boltzmann distribution. If exchange is fast on the chemical-shift time scale ($k_{ex} > \Delta\omega$), only a single peak occurs at a weighted time average, with a line width increased as $\Delta v_{1/2} = (rate)_{ex}/\pi$. If exchange is slow ($k_{ex} < \Delta\omega$), a line is seen at each chemical shift, with relative integrations determined by the ratio of p_A to p_B.

If two populations related by exchange are unequal ($p_1 \neq p_2$), it may be difficult to distinguish whether the exchange is slow or fast on the chemical-shift time scale, since the peak from the minor conformer could be just too low in intensity to distinguish (see Figure 11.11). An experimental approach to distinguishing slow from fast exchange for skewed populations is to examine the effect of changing static field B_0 on R_{ex}. Millet *et al.* have defined a parameter α that has specified values between 0 and 2 for different exchange regimes (Reference 4).

$$\alpha = \frac{d(\ln R_{ex})}{d(\ln(\Delta\omega))} = \frac{\delta R_{ex} / R_{ex}}{\delta B_0 / B_0} \qquad (11.18)$$

For two exchanging populations that are skewed such that $p_1 > 0.7$, in the fast exchange regime, $1 \leq \alpha \leq 2$. For intermediate exchange, $\alpha \approx 1$, and for slow exchange, $0 \leq \alpha \leq 1$. Experimentally, α is obtained by measuring R_{ex} at two different static fields $B_{0(1)}$ and $B_{0(2)}$ using the relaxation dispersion curve fit described above, and is given by:

$$\alpha = \left(\frac{B_{0(2)} + B_{0(1)}}{B_{0(2)} - B_{0(1)}} \right) \left(\frac{R_{ex(2)} - R_{ex(1)}}{R_{ex(2)} + R_{ex(1)}} \right) \qquad (11.19)$$

Of course, real life is often not simple, and when one observes two unequally populated states at slow exchange on the NMR time scale in a macromolecule, there is often the possibility that more such states exist. For example, a flexible loop of polypeptide in a protein is likely to have more than two possible conformations, and if exchange between these conformations is slow, the application of Equation 11.19 is no longer possible (Figure 11.12).

Quadrupolar relaxation

In keeping with the scope of this text, we will avoid an extensive discussion of quadrupolar relaxation, other than to note that it tends to be the dominant relaxation mechanism for nuclear spins with $I > 1/2$. In the presence of a nonspherical charge distribution around a quadrupolar nucleus (e.g. nonsymmetric ligation of the atom), the energy of a particular orientation of the nucleus with respect to an electric field gradient $eq_{zz} = d^2V/dz^2$ is given by:

$$E_Q = e^2 q_{zz} Q \left[3\mathbf{m}_I^2 - I(I-1) \right] / \left[4I(2I-1) \right] \qquad (11.20)$$

In this expression, e is the charge on a proton, and q_{zz} is a measure of the asymmetry of the local electronic environment, as determined by such factors as orbital hybridization and ligand identity. Q, the nuclear quadrupole moment, like the gyromagnetic ratio, is a constant for a given nuclide. The spin quantum numbers $\mathbf{m}$ and I have their usual meaning. Equation 11.20 means that a nuclear quadrupole can occupy multiple energy levels with respect to the electric field gradient at the nucleus. Transitions between these orientations can be detected spectroscopically in solids, with transition frequencies dependent upon the value of Q and q_{zz} (usually in the RF regime). This is known as pure nuclear quadrupole resonance (NQR). Because the electric field gradient is fixed in the molecular frame of reference, the tendency of the nucleus to orient in a magnetic field that results in Zeeman splitting for NMR (orienting the spins with respect to the macroscopic applied field) is modulated by the competing tendency of the nuclear quadrupole to orient with respect to the local field gradient. In solids, this results in a splitting of the NMR signal due to the quadrupolar coupling, with the quadrupolar coupling constant χ defined by:

$$\chi = e^2 q_{zz} Q / h \qquad (11.21)$$

In solution, the tumbling of the molecule results in a rapid fluctuation of the angle between the applied field B_0 and the electric field gradient q_{zz}, averaging the

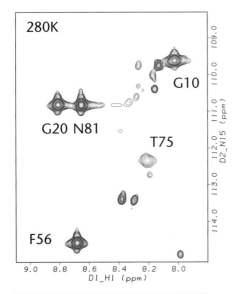

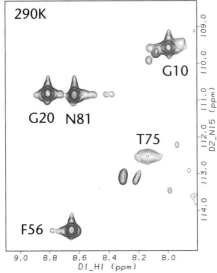

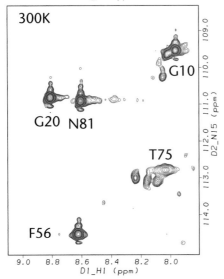

Figure 11.12 14 T (600 MHz ¹H) HSQC spectra of a mutant of putidaredoxin, a 2-Fe 2-S ferredoxin, in which exchange between conformers is slowed sufficiently that minor conformations can be detected as discrete peaks. As temperature is increased, exchange between some conformational states becomes fast on the chemical-shift time scale, and multiple peaks collapse into fewer peaks. Note this especially with the peaks marked G10 (glycine 10) and T75 (threonine 75). [Used with permission from T. C. Pochapsky *et al.*, *Biochemistry* 40, 5602–5614 (2001).]

quadrupolar coupling to zero. Unlike dipolar coupling or chemical shift anisotropy (which are often in the kHz range) the quadrupolar coupling constant can be in the MHz–GHz range, and is often incompletely averaged even by fast molecular tumbling. As such, NMR lines of quadrupolar nuclei can be very broad. In mobile isotropic liquids, quadrupolar relaxation is given by:

$$T_{2Q}^{-1} = T_{1Q}^{-1} = \frac{3}{10}\pi^2 \frac{2I+3}{I^2(2I-1)}\chi^2\tau_c \qquad (11.22)$$

If χ is small, either due to an intrinsically small Q (as is the case for ^{2}H or ^{6}Li) or high symmetry at the quadrupolar nucleus (no electric field gradient), quadrupolar relaxation can be significantly suppressed, to the extent that couplings to dipolar nuclei can be observed and other relaxation mechanisms can become important (Figure 11.13).

The primary effect of quadrupolar relaxation on the spectra of $I = 1/2$ spins to which the quadrupolar spin is J-coupled is dynamic uncoupling. Depending upon the efficiency of the quadrupolar relaxation, the dipolar spin may see a complete averaging of the spin states of the quadrupolar nucleus, in which case the dipolar spin resonance will be a singlet. If the quadrupolar relaxation occurs at an intermediate rate on the coupling time scale, multiplet structure corresponding to J-coupling will be observed to the $I = 1/2$ spin, with uncertainty broadening of the multiplet lines occurring in a manner completely analogous to that observed due to chemical

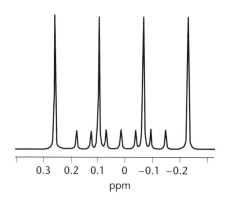

0.3   0.2   0.1   0   −0.1   −0.2
ppm

Figure 11.13 Coupling between ^{1}H and a quadrupolar nucleus (boron) at a site of high symmetry in the borohydride anion (BH$_4^-$). The relaxation of the boron spin due to quadrupolar interactions is usually very efficient, leading to very broad resonance lines and typically unresolved couplings. The borohydride anion is tetrahedral, so the average electric field gradient at the boron nucleus is zero, and quadrupolar relaxation is suppressed. The larger quartet is due to coupling between ^{11}B ($I = 3/2$, 81% natural abundance, four spin states, $^1J_{BH}$ = 90 Hz) while the smaller slightly off-center septet is due to ^{10}B ($I = 3$, 19% natural abundance, seven spin states, $^1J_{BH}$ = 30 Hz). The offset is due to a small boron isotope shift to the ^{1}H spins. Note that even in this situation, quadrupolar relaxation still contributes significantly to T_1 relaxation of the boron nucleus (~4/5 of R_{1obs}).

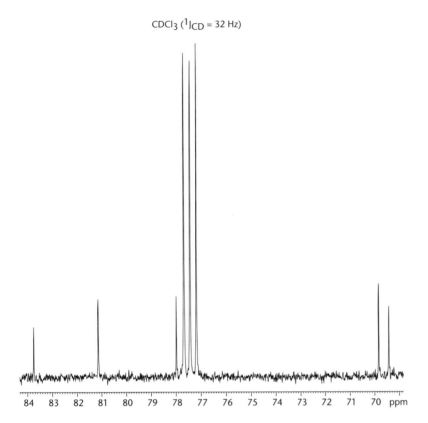

CDCl$_3$ (^{1}J$_{CD}$ = 32 Hz)

Figure 11.14 ^{2}H coupling to ^{13}C in CDCl$_3$ in a ^{1}H-decoupled ^{13}C spectrum. Deuterium has a relatively small quadrupolar moment, and resolved couplings are often observed to other spins.

exchange decoupling (Chapter 3). This situation is often observed for couplings between ^{2}H and ^{1}H or ^{13}C (see Figure 11.14), which gives rise to a distinct triplet from coupling to the I = 1 ^{2}H nucleus. In the case of ^{1}H attached to ^{14}N, the broadening is usually sufficient to lose the multiplet structure, but the ^{1}H lines often still are broadened by the intermediate exchange rate between ^{14}N spin states.

Hyperfine interactions and paramagnetic shifts of nuclear spins

Recall that electrons have a spin of 1/2, and thus are subject to the same considerations as nuclear spins. The electronic equivalent to the gyromagnetic ratio, i.e. the proportionality between magnetic moment and the spin angular momentum of the electron, is given by $-g\mu_B/\hbar$, where g is the **Lande splitting factor** (or just **g-factor**) and μ_B is the Bohr magneton. The value of g for a given electron depends upon the coupling of electronic motion and the electron spin (**spin–orbit coupling**). The corresponding relationship for a nuclear spin is $\gamma = g_N\mu_N/\hbar$. It turns out that the ratio of the nuclear magneton μ_N to the Bohr magneton is inversely proportional to the ratio of the mass of the proton to that of the electron. As such, the electronic magnetic moment is roughly 1000 times greater than that of the proton, and one expects to see proportionally larger Zeeman splittings and Boltzmann population differences across electronic spin transitions than nuclear spin transitions. However, the fact that electrons in closed-shell molecules are spin-paired within the orbitals in which the electrons reside cancels the effects of the electronic spin, and electronic magnetic spectra (EPR, for *e*lectron *p*aramagnetic *r*esonance, or ESR, for *e*lectron *s*pin *r*esonance) are not observed for such **diamagnetic** species. However, if an electron is unpaired (i.e. in a singly occupied orbital), as in a free radical or in many transition metal ions, an EPR signal can often be observed for such **paramagnetic** species.

A complete discussion of EPR is well beyond the scope of this text. Still, unpaired electronic spins can have large effects on the NMR spectrum of a paramagnetic molecule, both in terms of chemical shift and relaxation behavior, and this must be considered. As with two nuclear spins, unpaired electronic spins can interact with nuclear spins via through-bond (**Fermi contact**) and through-space (**dipolar**) mechanisms. The coupling between a nuclear spin and an electronic spin in an EPR spectrum results in splitting of the EPR signal, a phenomenon known as **hyperfine splitting**.

Such splitting is never directly observed in the corresponding NMR spectrum of the coupled nucleus, as electronic relaxation times are always fast enough so that the nuclear spin only detects an averaged electronic spin. Instead, the nuclear spin resonance occurs at the weighted average of the two components of the spin doublet (Figure 11.15). Since the Zeeman splitting of the electronic spin is larger than that of a nuclear spin for a given static field, the Boltzmann population of the electronic

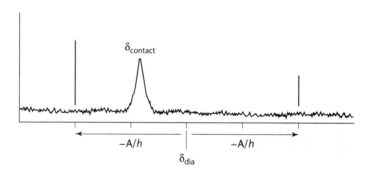

Figure 11.15 Diagram showing the origin of the hyperfine (contact) shift in the NMR spectra of paramagnetic species. The hyperfine coupling constant (A) between the observed nuclear spin and the unpaired electronic spin determines the expected positions of the lines of the hyperfine doublet ($\Delta v = A/h$). However, electronic relaxation is usually fast enough to dynamically decouple the nuclear spin, resulting in a singlet. The position of that singlet ($\delta_{contact}$) is shifted from the diamagnetic position (δ_{dia}) by the Boltzmann weighting of the lower energy electronic spin state, so that the actual position of the line is given by $\delta_{contact} - \delta_{dia} = (\Delta E_{zeeman}A)/(4kTh)$ where ΔE is the Zeeman energy splitting of the electronic spin states. In the absence of a population difference (e.g. as $T \to \infty$), $\delta_{contact} \to \delta_{dia}$ (Curie shift).

spin is more heavily weighted to the lower energy state of the electronic spin than is observed in nuclear spin systems. Also, the hyperfine coupling constant A between a nuclear and electronic spin is typically much larger than J-coupling between nuclei. As such, the weighted average position of the affected nuclear spin transition can often be well outside the normal chemical shift range of that nucleus. This phenomenon is called a **hyperfine shift**. An example of hyperfine-shifted resonances in a ^1H spectrum is shown in Figure 11.16. In the high-field approximation and for an isotropically tumbling molecule, the Fermi contact contribution to the observed chemical shift of a nuclear spin I due to a well-defined electronic state with a spin S in an isotropically tumbling molecule is given by:

$$\delta_{contact} = \frac{\Delta B}{B_0} = -\frac{A}{\hbar}\frac{g\mu_B S(S+1)}{3\gamma_I kT} \tag{11.23}$$

where g is the electronic g-factor, A is the hyperfine coupling constant, and the other constants are as defined previously. Note that as temperature increases, the Boltzmann difference between the electronic spin states will decrease, so that the contact shift will decrease with increasing temperature such that $\delta_{contact} \to 0$ as $T \to \infty$. This inverse temperature trend toward the diamagnetic shift with increasing temperature is called a **Curie shift**. Equation 11.23 is the simplest form for the contact shift, and is only applicable when Curie temperature dependence is observed. **Anti-Curie** behavior ($\delta_{contact}$ increasing with increasing temperature) is evidence that multiple electronic spin states with different inherent $\delta_{contact}$ values are occupied, and that states with larger values of $\delta_{contact}$ become more populated as temperature increases. In such cases, a more complicated calculation for $\delta_{contact}$ is required, one that defines separate hyperfine coupling constants for each occupied electronic state (Reference 5).

The primary requirement for the observation of a contact shift is that some unpaired electron spin density be delocalized directly onto the nuclear spin in

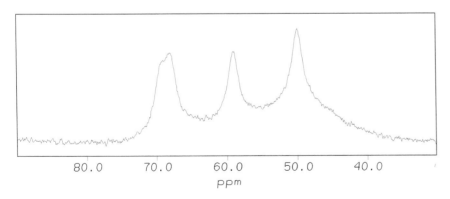

Figure 11.16 Contact-shifted ^1H resonances in acireductone dioxygenase, a Ni(II)-containing metalloenzyme. Note that the chemical shifts are well outside the normal 0–10 ppm range of ^1H spins in diamagnetic environments. The shifted signals are imidazole ring protons of the histidine residues that ligate the paramagnetic Ni(II) ion.

question. That is, there must be some through-bond or close contact interaction between the unpaired electronic spin and the nuclear spin, and s-orbital hybridization must be involved in the orbital containing the unpaired spin density (other types of orbitals have nodes at the nucleus). The hyperfine coupling constant A is a measure of the extent of that delocalization. However, even nuclear spins that are not connected directly to a paramagnetic center via bonding interactions can be affected by paramagnetism through magnetic dipole interactions. If an unpaired electron spin occupies orbitals that are not of high three-dimensional symmetry (i.e. not tetrahedral or octahedral symmetry), the interaction of the electronic spin with the applied field varies depending on the relative orientation of the paramagnetic molecule and the applied static field $\vec{B}_0$. This produces shielding effects on nearby nuclei analogous to (although often larger than) the effects observed from ring currents in aromatic rings discussed in Chapter 2. Such shifts are called **dipolar shifts** because they result from the interaction of the magnetic dipole of the unpaired electron and the nuclear spin dipole. As with any shielding effect, the dipolar shift has a tensor associated with it, and in solids, the static orientation of the molecule with respect to the applied field determines the magnitude and direction of the dipolar shift for a given nucleus. In the case of an axially symmetric magnetic susceptibility anisotropy tensor, the dipolar shift is given by:

$$\delta_{dipolar} = \frac{\Delta B_{dip}}{B_0} = \frac{1}{4\pi r^3}\left[\begin{array}{c} \chi_{\|}\cos^2\alpha(3\cos^2\theta-1)+ \\ \chi_{\perp}\sin^2\alpha(3\sin^2\theta\cos^2\varsigma-1) \\ +\frac{3}{4}(\chi_{\|}+\chi_{\perp})\sin2\alpha\sin2\theta\cos2\varsigma \end{array}\right] \quad (11.24)$$

where α is the angle between the molecular **z** axis and the applied field B_0, θ is the angle between the electron–nuclear vector **en** and the molecular **z** axis, ζ is the angle between the projection of **en** in the **x**–**y** plane of the molecular frame of reference, and $\chi_{\|}$ and $\chi_{\perp}$ are the components of the magnetic susceptibility anisotropy tensor parallel and perpendicular to the applied field $\vec{B}_0$ (see Figure 11.17). In isotropic solutions, the observed dipolar shift is an average of the shift over all orientations, and because it reduces to a scalar quantity in isotropic solution like the contact shift, this averaged shift is often called a **pseudocontact shift.** The pseudocontact shift can be related to the overall magnetic anisotropy of the paramagnetic center (which in turn tends to dominate the magnetic anisotropy of the entire molecule) via the magnetic susceptibility anisotropy tensor, $\overline{\overline{\chi}}$ (see Chapter 10). In the case of an axially symmetric magnetic susceptibility anisotropy tensor ($\chi_{zz} = \chi_{\|}$, $\chi_{xx} = \chi_{yy} = \chi_{\perp}$), the pseudocontact shift term is:

$$\delta_{pc} = \frac{1}{12\pi r^3}\left(\chi_{\|}-\chi_{\perp}\right)\left(3\cos^2\theta-1\right) = \frac{\mu_0}{4\pi r^3}\frac{\mu_B^2 S(S+1)}{9kT}\left(g_{\|}^2-g_{\perp}^2\right)\left(3\cos^2\theta-1\right)$$

$$(11.25)$$

where the angles are as shown in Figure 11.18. Using Equation 11.25, the pseudocontact shift can be related to either of two experimentally determined tensor quantities, the g-tensor or the magnetic susceptibility anisotropy tensor. The difference

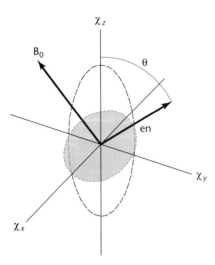

Figure 11.17 Angular relationships between the molecular χ tensor and the **en** (electron–nuclear) vector that can be related to the pseudocontact shift as shown in Equation 11.25.

term in Equation 11.25, $(\chi_\parallel - \chi_\perp)$, exhibits a T^{-2} dependence on temperature (not shown), which provides an experimental means for distinguishing the Fermi contact contribution from the pseudocontact shift.

Pseudocontact shifts are commonly observed in solution for paramagnetic heme proteins such as myoglobin and cytochrome b_5 (Figure 11.19), because of the planar delocalization of the unpaired electron over the heme. Such shifts are smaller and more difficult to detect for paramagnetic proteins with tetrahedral or octahedral

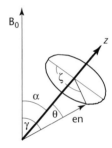

Figure 11.18 Angular relationships describing the dipolar shift in relation to the applied field B_0 and the molecular frame of reference (see Equation 11.24 and associated discussion for the definition of angles and vectors).

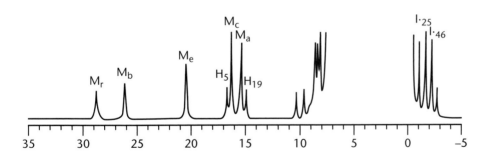

Figure 11.19 Combination of contact and pseudocontact shifts in the 11.74 T (500 MHz) ^{1}H NMR spectrum of oxidized cytochrome b_5 reconstituted with a modified hemin. Oxidized cytochrome b_5 contains a low-spin Fe(III) ($S = 1/2$). The downfield-shifted resonances between 30 and 10 ppm (marked as M, H) arise from heme substituent and axial Fe-ligand histidine protons onto which unpaired electron density is delocalized (contact shifts). The upfield signals (marked I) are pseudocontact-shifted resonances of amino acid side-chain methyls that are near to the heme. [Modified with permission from Figure 2 of K. B. Lee *et al.*, *J. Am. Chem. Soc.* 113, 3576–3583 (1991).]

coordination of paramagnetic centers, such as ferredoxins or rubredoxins. Since both pseudocontact shifts and residual dipolar couplings can be related to the magnetic susceptibility anisotropy tensor, they can both be used as long-range structural restraints in protein structure determinations (see Chapter 10 and Reference 6).

Paramagnetic relaxation of nuclear spins

Time-dependent modulation of dipolar and contact electron–nuclear couplings typically lead to very efficient nuclear spin relaxation. This results in line broadening (in the kHz range for affected ^{1}H resonances in some cases) and rapid equilibration to Boltzmann populations for affected nuclear spins. The precise effect depends upon a number of factors, the most important of which is the electronic spin relaxation time. Electronic spin relaxation times τ_e are typically very fast relative to nuclear spin relaxation, and τ_e often limits the "memory" of the system more than the effect of random motions, particularly in macromolecules (i.e. $\tau_e \ll \tau_c$ where $\tau_e \approx T_{1e} = T_{2e}$). The fastest electron relaxation times ($<10^{-12}$ s) effectively decouple the electronic spin from that of nearby nuclei, so that the nuclear spins detect a static net magnetization due to the Boltzmann excess of the averaged electronic spin, but no discrete electronic spin states. The interaction of this apparently static magnetic dipole with nuclear spins is modulated by molecular motion as is any other dipole–dipole interaction, and can lead to what is called **Curie relaxation**. Curie contributions to nuclear spin relaxation are given by:

$$T_{1(Curie)}^{-1} = \frac{2}{5}\left(\frac{\mu_0}{4\pi}\right)^2 \frac{\omega_I^2 g_e^4 \mu_B^4 S^2 (S+1)^2}{(3kT)^2 r^6}\left(\frac{3\tau_c}{1+\omega_I^2\tau_c^2}\right)$$

$$T_{2(Curie)}^{-1} = \frac{1}{5}\left(\frac{\mu_0}{4\pi}\right)^2 \frac{\omega_I^2 g_e^4 \mu_B^4 S^2 (S+1)^2}{(3kT)^2 r^6}\left(4\tau_c + \frac{3\tau_c}{1+\omega_I^2\tau_c^2}\right) \qquad (11.26)$$

Note that the $T_{1(Curie)}^{-1}$ expression will decrease across the $\omega_c\tau_c$ transition as τ_c increases (i.e. $T_{1(Curie)}^{-1}$ is **dispersive**) in the same manner as it does for diamagnetic molecules. In most cases, the Curie contribution to T_1 relaxation for macromolecules in solution can be ignored. However, $T_{2(Curie)}^{-1}$ does not decrease with increasing τ_c, as it contains the $4\tau_c$ term, and so Curie broadening becomes more important at higher fields for slowly tumbling molecules, particularly for high γ spins such as ^{1}H.

Slower electronic relaxation times (10^{-9}–10^{-11} s) couple more effectively to nearby nuclear spins (they provide spectral density in the appropriate range for nuclear spin transitions), and unpaired electron spins are usually the dominant factor in the relaxation of nuclear spins within ~8 Å of the paramagnetic center in such cases. For those nuclei that interact with the electronic spin only by dipolar interactions, the relaxation is analogous to nuclear dd relaxation. As such, the relaxation behavior is a function of the transition frequencies of the spins and the spectral density function appropriate for the local environment. The correlation time that determines the spectral density near the paramagnetic center (and hence the relaxation behavior of nearby nuclear spins) is given by:

$$\tau_c^{-1} = \tau_r^{-1} + \tau_e^{-1} + \tau_m^{-1} \tag{11.27}$$

where τ_r^{-1} is the rotational correlation time, τ_e^{-1} is the electronic spin relaxation time, and τ_m^{-1} is the lifetime of any chemical exchange process that might occur involving the electron or the nuclei. The Bloembergen–Solomon equations (Reference 7) modified for electron–nuclear spin dipolar interactions are:

$$T_{1(dd)}^{-1} = \frac{2}{15}\left(\frac{\mu_0}{4\pi}\right)^2 \frac{g_e^2 \mu_B^2 \, {}^2\gamma_I^2 S(S+1)}{r^6}\left(\frac{7\tau_c}{1+\omega_e^2\tau_c^2} + \frac{3\tau_c}{1+\omega_I^2\tau_c^2}\right)$$

$$T_{2(dd)}^{-1} = \frac{1}{15}\left(\frac{\mu_0}{4\pi}\right)^2 \frac{g_e^2 \mu_B^2 \, {}^2\gamma_I^2 S(S+1)}{r^6}\left(4\tau_c + \frac{13\tau_c}{1+\omega_e^2\tau_c^2} + \frac{3\tau_c}{1+\omega_I^2\tau_c^2}\right) \tag{11.28}$$

where $|\omega_e + \omega_I| \cong |\omega_e - \omega_I| \cong \omega_e$, as the electronic transition frequency is much higher than the nuclear frequency. Because the relaxation exhibits an r^{-6} dependence on the distance between the nuclear spin and the electronic spin, the relaxation behavior of the nuclear spin is often used as a measure of the distance between the two spins, and can be incorporated into structural calculations. This assumes that the electronic spin is a point magnetic dipole, which works well enough when the nucleus is relatively far from the unpaired electron. However, if the electronic spin is delocalized (over ligands to a metal center, for example), the point dipole approximation breaks down, especially for spins close to the metal center, and the paramagnetic relaxation becomes much less precise as a structural restraint.

If the nuclear spin exhibits a hyperfine coupling A to the electronic spin, the modulation of this coupling also causes relaxation. The contributions due to Fermi contact interactions are given by:

$$T_{1(contact)}^{-1} = \frac{2}{3}\frac{A^2}{2}\frac{\tau_e}{1+(\omega_I - \omega_e)^2\tau_e^2}$$

$$T_{2(contact)}^{-1} = \frac{1}{3}\frac{A^2}{2}\left(\frac{\tau_e}{1+(\omega_I - \omega_e)^2\tau_e^2} + \tau_e\right) \tag{11.29}$$

Note that at long electronic relaxation times τ_e, the contribution of contact interactions to T_1 relaxation is minimal, but remains important for T_2.

Relaxation and the density matrix

Despite the number of equations presented in this chapter, we have still only dealt with relaxation interpretively rather than dealing with it at a quantum mechanical level. In Chapter 5, we discussed the use of the density matrix to describe evolution of coherence and populations of spin ensembles. (Recall that the product operator formalism is a shorthand simplification of the density matrix formalism for the case of two weakly coupled spins.) In order to allow relaxation to be integrated into the density matrix, Redfield (Reference 8) developed a formalism for relaxation in operator form (the **relaxation superoperator**) to use with density matrix calculations of spectral properties (observables). Appendix B deals with the Redfield formalism for those who wish more detail on this aspect of relaxation.

Problems

11.1 Show that the curves in Figure 11.2 are calculated from the special case where $\omega_H = \omega_X$ and dipolar relaxation is dominant. Assume that the value of $d^2/4 = 1.05 \times 10^9$.

11.2 What is the ratio of the spectral density due to a motion with a correlation time of 10^{-9} s to that at 10^{-12} s at the following frequencies (in units of radians): 0, 10^3, 10^7, 10^9, 10^{12}?

11.3 For which of the following does the extreme narrowing condition hold? (A) A ^{1}H, ^{15}N spin system at 11.74 T with a $\tau_c = 10^{-10}$ s? (assume minimal chemical shift anisotropy) . (B) A ^{1}H, ^{1}H spin system at 11.74 T, with $\tau_c = 10^{-12}$ s? (C) A ^{1}H, ^{19}F spin system at 11.74 T, also with a $\tau_c = 10^{-12}$ s?

11.4 The chloroform line in Figure 11.14 shows a distinct triplet structure due to ^{2}H,^{13}C coupling, rather than a relaxation-broadened singlet, even though ^{2}H is a quadrupolar nucleus. What is the maximum ^{2}H relaxation rate (T_2^*) that would still allow the triplet to be observed? The spectrum was obtained using an 11.74 T magnet.

11.5 Show that the magnetization in each experiment in Figure 11.7 is in the appropriate form for the relaxation measurement desired during the relaxation period T.

11.6 Using the information given in the caption for Figure 11.9, show how the exchange rate constant given for electron self-exchange in putidaredoxin was calculated.

11.7 In the sample described in Figure 11.13, the ^{11}B nucleus of BH_4^- has an observed R_1 relaxation rate constant of 0.355 s^{-1}, while ^{10}B has an observed R_1 rate constant of 0.191 s^{-1}. The nuclear quadrupole moments of ^{10}B and ^{11}B are known quantities (Q (^{11}B) = 0.04065 barns, Q (^{10}B) = 0.08472 barns), although the quadrupolar coupling constants χ are not. The gyromagnetic ratios of the two boron spins are 2.8740×10^7 rad T^{-1} s^{-1} for ^{10}B and 8.5794×10^7 rad T^{-1} s^{-1} for ^{11}B. Given that the two boron nuclides are in identical environments, using Equations 11.21, 11.22 and the following expressions:

$$R_{1dd(I)} = \frac{1}{3}(2\gamma_I\gamma_S)^2 S(S+1)\tau_c \sum_S r_{IS}^{-6} \quad \text{(dd relaxation of spin } I \text{ due to the}$$

attached ^{1}H(S))

$$R_{1Q(I)} = 3\pi^2(2I+3)(1+\eta^2/3)[10I^2(2I-1)]^{-2}\chi_I^2\tau_c \quad \text{(quadrupolar relaxation of}$$

spin I due to electric field gradient coupling, with η being an asymmetry parameter constant for BH_4^-)

$\chi = eQ(eq_{zz})/h$ (the relation between quadrupolar coupling constant and quadrupole moment, where eq_{zz} is the electric field gradient)

calculate the expected relative contributions of dd and quadrupolar relaxation R_{1dd} and R_{1q} for the two boron nuclides in BH_4^-, given that the observed $R_{1(obs)}$ values are 0.355 s^{-1} (^{11}B) and 0.191 s^{-1} (^{10}B). Recall that ^{10}B is a spin 3 nucleus, while ^{11}B is spin 3/2. (Hint: two equations with two independent variables are required.)

11.8   Given an experimental paramagnetic relaxation rate constant for a particular nuclear spin, what other information would one need and what assumptions are required in order to calculate a distance between the unpaired electron and the nuclear spin?

References

1.   G. Lipari and A. Szabo, *J. Am. Chem. Soc.* 104, 4546 (1982).
2.   D. E. Woessner, *J. Chem. Phys.* 37, 647–654 (1962).
3.   J. W. Peng and G. Wagner, *J. Magn. Reson.* 98, 308 (1992).
4.   O. Millet *et al.*, *J. Am. Chem. Soc.* 122, 2867–2877 (2000).
5.   R. J. Kurland and B. R. McGarvey, *J. Magn. Reson.* 2, 286 (1970).
6.   B. F. Volkman *et al.*, *J. Am. Chem. Soc.* 121, 4677–4683 (1999).
7.   N. Bloembergen, *J. Chem. Phys.* 27, 572–573 (1957); I. Solomon, *Phys. Rev.* 99, 559–565 (1955).
8.   A. Redfield, *Adv. Magn. Reson.* 1, 1–31 (1965).

12

DIFFUSION, IMAGING, AND FLOW

To the nonscientist, "magnetic resonance" means medical imaging (**magnetic resonance imaging**, or **MRI**). In fact, the number of commercially installed MRI units in the world (estimated at >20 000 in 2001, and probably closer to 30 000 in 2004) dwarfs the number of magnets used for NMR spectroscopy. Although the same basic principles apply to both MRI and NMR, it is not always immediately obvious to a practitioner of one how the other works. In this chapter, we will discuss the basics of MRI (Reference 1), as well as the related topics of diffusion and flow measurement.

Magnetic field inhomogeneity, $T_{2(macro)}$ and diffusion measurement by NMR

As noted in Chapter 4, the line width of an NMR resonance reflects both the intrinsic contributions of spin–spin interactions, what we will call $T_{2(micro)}$, and the inhomogeneity of the magnetic field over the volume of the NMR sample, represented by $T_{2(macro)}$. More precisely, this means:

$$T_{2(obs)}^{-1} = T_{2(macro)}^{-1} + T_{2(micro)}^{-1} \qquad (12.1)$$

An estimate of $T_{2(macro)}$ is obtained from the magnetic field inhomogeneity $B_0(r)$ by $T_{2(macro)}^{-1} = \gamma(B_0(r))$, where r denotes position in the sample. One of the advantages of

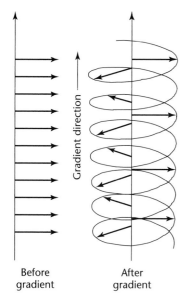

Figure 12.1 Field gradient spatial phase encoding. Prior to application of a field gradient (left side of figure), all isochromats are aligned, with the same phase relationship. After application of a linear gradient along the z axis shown, the phase is advanced for each isochromat by an amount $\phi = \gamma p G_z z t$ (Equation 12.3).

the spin-echo pulse sequence is to remove the effects of inhomogeneous broadening from the observed T_2 relaxation rate of a signal. However, as an introduction to imaging, let us consider what happens when an axial linear field gradient $\Delta B / \Delta z$ is applied over the length of the NMR sample. As shown in Figure 12.1, the application of a pulsed field gradient imposes a positionally determined phase advance onto the chemical shift evolution of spins that would otherwise precess at the same Larmor frequency. This results in the division of the sample into a series of isochromats along the direction of the field gradient (**z** in this case). One can think of this imprinted phase advance as a type of **spatial encoding**, with the magnitude of the phase advance of a spin determined by the position of the spin along the z axis of the sample. The application of a π pulse followed by a refocusing gradient of the same intensity and duration (or alternatively a gradient of equal but opposite direction without the π pulse, as in a gradient-echo experiment (see Figure 7.20) "de-codes" the spatial encoding by removing the gradient-imposed phase shift. The signal intensity detected after refocusing should be that which would be expected if a delay time equal to the time required for the two gradients were present, i.e. whatever the $T_{2(micro)}$ of the sample would dictate. However, complete signal recovery after the gradient echo depends upon each spin remaining at the same position throughout the sequence, i.e. the absence of diffusion between the encoding and decoding gradients. Diffusion of a spin across isochromats between the encoding and decoding gradient results in a less-than-perfect refocusing of the isochromats, giving a smaller-than-expected signal after refocusing. Signals from spins that diffuse more rapidly (small molecules) will be attenuated by the gradient sequence more than those due to large molecules, providing a means of measuring linear diffusion coefficients by NMR. The possibility of measuring diffusion by NMR was first discussed by Hahn in the same paper in which he described the spin-echo experiment (Reference 2). However, until the development of pulsed field gradients, the experiment was relatively impractical, as static gradients present during signal acquisition result in line broadening.

The first description of practical diffusion measurement by NMR using pulsed field gradients was by Stejskal and Tanner (Reference 3). The Stejskal–Tanner (ST) experiment is essentially a $\pi/2$ pulse followed by a spin echo in which the encoding and decoding gradients are separated by a diffusion delay Δ. The diffusion coefficient D is related to the observed attenuation of signal intensity by:

$$I \propto I_0 \exp\left[-\gamma^2 G^2 \tau_g^2 D \left(\Delta - \tau_g / 3 \right) \right] \qquad (12.2)$$

where G is gradient strength and τ_g is the duration of the gradient.

As the coherences of interest are transverse during Δ, the ST experiment is limited in its application to molecules with relatively long T_2 relaxation times. A useful variation of the ST experiment for measuring diffusion of larger molecules makes use of the **stimulated echo**, also described originally by Hahn (Reference 2). The stimulated echo experiment consists of three $\pi/2$ pulses, in which the second pulse places the coherence along $-z$, where it is unaffected by T_2 processes. The third pulse returns the coherence to the transverse axis, where the echo is detected after a time equal to the spacing between the first two pulses. In the stimulated echo version of the ST sequence (Figure 12.2), a spatial encoding gradient is applied after the first $\pi/2$ pulse. The second $\pi/2$ pulse stores one component of the spatially encoded coherence along $-z$. Here, the spatially encoded coherence is subject only to T_1 effects, and only after the diffusion delay is complete is the magnetization returned to the transverse plane and the decoding gradient applied. An example of diffusion measurement using the stimulated echo sequence, as well as the methods used for extracting diffusion coefficients from the data, are given in Problem 12.2.

Basic imaging concepts: phase and frequency encoding of position in a macroscopic sample

Usually, medical MRI uses the ^{1}H signal of water for generating images. This takes advantage of the abundance of water protons in most living tissues, and as we will see, makes it possible to characterize flow and diffusion using relaxation-based techniques. However, given enough time and signal density, any NMR active nucleus can be used for imaging purposes. The application of a linear-field gradient along the laboratory z axis results in a positionally determined shift $\Delta\omega$ of the detected signal given by Equation 7.29:

$$\omega_{obs} = \omega_0 + \gamma B_g = \omega_0 + \gamma G_z z = \omega_0 + \Delta\omega$$

$$\Delta\omega = \gamma G_z z$$

where G_z is the field gradient given in units of change in field strength change per unit displacement along the z axis and z is the displacement along the z axis. Based on this, the presence of a field gradient *during* acquisition of an NMR signal results in **frequency encoding** of the position of spins in the sample. One sees a spreading of the

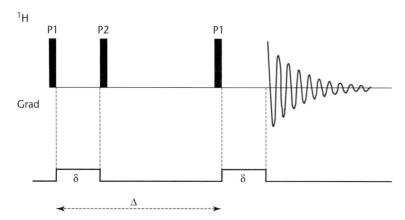

Figure 12.2 Stimulated echo sequence for diffusion measurement. All pulses are $\pi/2$. The delay Δ is the diffusion time, and δ is the length of the gradient pulse (see Equation 12.2). The phase cycling is the same as for a NOESY experiment: P1, *x*, –*x*; P2, *x*; P3, *x*, *x*, –*x*, –*x*; receiver phase, *x*, –*x*, –*x*, *x*.

NMR signal over a range of frequencies and, if one knows the magnitude and direction of the gradient, it is simple to replace the frequency axis with a distance-from-center measurement (see Figure 12.3). In this way, the frequency encoding of the NMR signal can be used to directly obtain a one-dimensional image of the sample.

Note that the implicit assumption in frequency encoding is that the inherent chemical shifts of all of the spins used in the imaging experiment are the same (usually that of bulk water). Since almost all of the proton density is in the form of bulk water in most tissues, this is not a bad assumption. However, for fatty tissues, the signal due to lipid (C–H) protons is a significant contributor, and will result in a spatial offset in the FT-derived image.

If the gradient is applied to transverse (single-quantum) coherence for a period of time and then turned off, the coherence is dephased into a series of isochromats due to a phase shift ϕ, the magnitude of which is determined by the position of the isochromat in the sample and the intensity and duration of the gradient:

$$\phi = \gamma p G_z z t \tag{12.3}$$

where p is the order of the coherence (assumed to be –1 for the moment) and the other variables are as described above. This positionally dependent phase shift, called **phase encoding**, also provides positional information, and if used in combination with frequency encoding provides the basis of MRI.

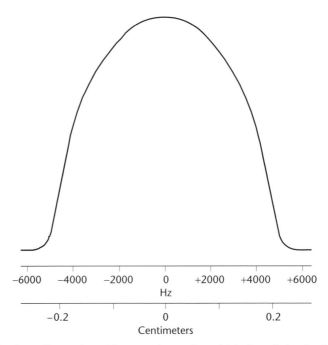

Figure 12.3 One-dimensional image along the width (*x* axis in the laboratory frame) of water in a standard NMR tube. Image was obtained by applying a 5.6 G/cm along the *x* axis during acquisition of a ^{1}H FID. The tube diameter is 0.42 cm.

Spatially selective pulses

The ability of magnetic field gradients to divide a sample into spatially resolved sectors (isochromats) is critical to the imaging experiment. Equally important is the ability to use shaped RF pulses in combination with field gradients to provide spatially selective RF pulses to excite particular regions of the sample. The principle of spatially selective excitation is straightforward. A shaped RF pulse designed to selectively excite spins with a particular frequency offset $\Delta\omega$ is applied while a gradient is present. Only those spins with the appropriate frequency offset due to the field gradient will be excited by the pulse. This provides spatially selective excitation in a narrow "slice" along the direction of the gradient.

A commonly used selective pulse for imaging is the **sinc** pulse (Figure 12.4), which has a time-modulated amplitude of the form:

$$B(t) \propto \left[\sin(a_0 t)/a_0 t \right] \tag{12.4}$$

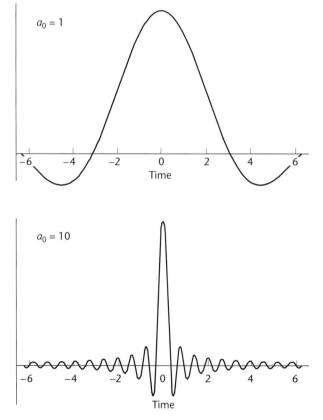

Figure 12.4 Sinc pulse described by Equation 12.4. The sinc pulse is the time domain transform of a delta function, and the width of the excitation envelope of the pulse is given by the constant a_0. The larger a_0 is, the shorter (and less selective) the pulse. (Top) sinc pulse with $a_0 = 1$, plotted versus time from -2π to 2π. (Bottom) $a_0 = 10$, over the same time.

The Fourier transform of the sinc function in the frequency domain is a step function around a_0 such that:

$$|\omega| < a_0, B(\omega) = A$$
$$|\omega| > a_0, B(\omega) = 0 \tag{12.5}$$

The frequency term a_0 is a constant chosen by the user in order to determine how narrow the excitation window of the pulse will be. The excitation will occur at the carrier frequency of the pulse, which can be modulated by pulse phase shift incrementation in order to generate off-carrier pulses.

The ability to use a combination of a gradient and a selective pulse is the first step in the imaging experiment, called **slice selection**. A "slice" of a three-dimensional object is a thin section in which transverse magnetization is generated.

Spatial equivalents of NMR parameters

As shown in Figure 12.3, a one-dimensional image (i.e. an image of a sample along one axis in the laboratory frame) essentially replaces the frequency domain obtained upon FT with its spatial equivalent, i.e. a position along the imaged axis corresponding to a particular frequency in the spectrum. The two-dimensional image familiar from MRI is equivalent to a two-dimensional NMR experiment (or, more properly, a plane from a three-dimensional experiment) in which frequency is replaced by a spatial dimension along both axes. As such, concepts used in FT-NMR such as spectral width, digital resolution and foldover have precise equivalents in imaging.

The spectral width, which is the range of frequencies that are sampled within the Nyquist limits, is replaced by the **field-of-view** (L), the length over which signal is measured in a particular dimension. Recall that, if signal intensity is sampled in a series of discrete measurements (as a series of time points) each of length $DW = 1/(2v_{max})$, that the range of frequencies accurately sampled is given by $v_{max} = |v_A - v_r|$, where v_r is the carrier frequency and v_A is the frequency of the resonance furthest from v_r at the edge of the spectrum. In terms of the encoding gradient G and the field-of-view L, the dwell time DW required to correctly sample the signal is given by:

$$DW = \frac{2\pi}{\gamma G L} \tag{12.6}$$

Every other basic parameter of NMR (resolution, signal-to-noise ratio, digital resolution) has an equivalent in MRI. **Contrast** within an image (the difference between regions of different intensity in the image) is a function of signal-to-noise ratio. Resolution (the ability to distinguish fine structure within the image) is a somewhat more complicated function of the gradient strength, relaxation rates and the digital resolution available for the image. The **point-spread function** (**PSF**, Equation 12.7) describes the image of a single point (i.e. a delta function):

$$PSF(x) = \frac{\exp(-i\pi x/L)\sin(\pi N x/L)}{N\sin(\pi x/L)} \qquad (12.7)$$

where N is the number of digitized points, x is the position of the point, and L is the field of view. The resolution dx along the x dimension is then given by:

$$dx = \frac{1}{PSF(0)} \int_{-L/2}^{L/2} PSF(x)\,dx \qquad (12.8)$$

Basic two-dimensional imaging sequences

The basic imaging sequence must encode image information by phase and frequency modulation. The **gradient-echo** imaging sequence (Figure 12.5) provides both of these, and is the basis for many imaging experiments. The gradient-echo sequence begins with a slice-selective excitation pulse that generates transverse coherence in a desired slice of the sample (along the z axis). This is followed by a phase-encoding gradient that is incremented over the series of acquisitions from a maximum negative to a maximum positive gradient strength. This has analogy with

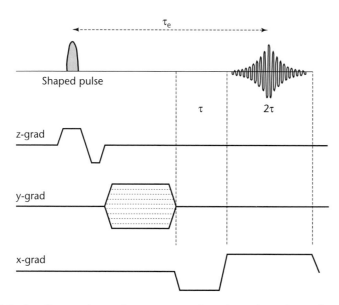

Figure 12.5 Gradient-echo pulse sequence for planar imaging. The z gradient is applied in combination with a shaped pulse in order to obtain slice-selective excitation, followed by a refocusing gradient. The y gradient is applied with a different value at each increment from $-G_y$ to $+G_y$ in order to obtain phase encoding. The x gradient is applied prior to acquisition for a time τ and then the gradient is reversed for a time 2τ, with the gradient echo reaching a maximum at the middle of the reverse gradient. This provides the frequency encoding during acquisition of the signal. The total echo time τ_e from the selective pulse to maximum of the echo is usually on the order of a few milliseconds.

the t_1 frequency-labeling period of a two-dimensional NMR experiment. However, in the imaging experiment, there is no time incrementation that would result in frequency encoding. Only the phase advance due to the gradient is incremented. The last step is the application of a gradient echo in which the dephasing gradient is applied for half as long as the refocusing gradient. The result is a signal maximum at half-way through the refocusing gradient. It is this echo that is recorded as a time-domain signal.

One disadvantage of the gradient-echo experiment is that T_2^* (including effects from field inhomogeneities) dominates the relaxation of signals. As a result, the acquisition time must be of very short duration, limiting the digital resolution available for the experiment. An alternative is the spin-echo experiment. In this experiment, a refocusing slice-selective π pulse is applied half-way between the excitation pulse and the maximum of the observed echo (that is still generated by a gradient for frequency encoding). This removes the effects of field inhomogeneities from the observed signal.

k-Space

As in NMR, the signal that is acquired in an imaging experiment is a complex time-domain signal made up of heterodyned (i.e. subtracted from the carrier frequency) frequencies arising from all of the excited spins in the sample. Assuming that slice-selective excitation is used, the desired image will be a two-dimensional array of small volumes (**voxels**) with dimensions Δx, Δy, Δz, where x and y are the dimensions of the array and z is the axis along which slice selection occurs (i.e. Δz is the width of the selected slice of the sample). As in two-dimensional NMR, the number of voxels is determined by the number of complex points acquired in each dimension. If 256 complex points are acquired per acquisition (the directly observed "t_2" frequency-encoded dimension), and phase-encoding is incremented in the indirectly-detected ("t_1") dimension in 256 steps, an array of 256×256 voxels is generated. As one might guess, zero-filling, linear prediction and other varieties of data massaging can be used for processing that change the number of points in the final image. The number of points that are present upon FT determine the number of points in the transformed image.

Prior to FT in either dimension, the data array from a gradient-echo experiment such as that shown in Figure 12.5 consists of a series of phase-modulated echos in which the center of the echo represents the maximum refocusing of isochromats by the gradient that is applied during the acquisition (no frequency modulation from the carrier) (see Figure 12.6a). Since all of the signals in the sample will contribute to this part of the echo (i.e. everything is refocused) it will provide most of the intensity to the image. As the echo tails, there is rapid dephasing of signals from regions of the sample distant from each other, since these are at quite different frequencies. However, signals from adjacent regions of the sample diphase more slowly, and only give rise to detectable interference patterns in the "tail" of the echo. It is here that most of the information for fine structure is contained. Emphasis of the outside features of the echo results in improved resolution, while emphasis of the more intense central features of the echo gives improved signal-to-noise ratio. This is precisely analogous to an FID, where emphasizing the longest-lived components improves spectral resolution at the expense of signal-to-noise ratio.

One can think of the two-dimensional plot of raw data from a gradient-echo experiment as a plot of "spatial frequencies" (Figure 12.6a). At the center of the plot (zero-frequency in both dimensions) all signaling voxels contribute. This represents the maximum of the echo in both dimensions. Near the center of the echo, interference from signals arising from features that are far apart from each other in space (they differ in frequency enough that they go out of phase quickly) in the imaged object gives rise to the low-frequency interference patterns. There is an inverse relationship between the observed frequency and the distance between voxels contributing to the interference. This is an example of **reciprocal space**, and has analogy with the reciprocal space used in X-ray crystallography. Physicists call such spaces **k-spaces**, and the raw time-domain data from an imaging experiment is a plot of the time evolution of signal in k-space. Figure 12.6a provides an example of raw k-space data. Figure 12.6b shows the same data set after transformation in the frequency-encoded (directly detected) dimension, and Figure 12.6c shows the completed image after FT in the phase-encoded dimension.

Contrast and contrast agents, relaxation, and flow

The contrast available in an image is a function of the available signal as well as the number of signal-averaged acquisitions that can be obtained in a given amount of time. As living subjects tend to move, the signal averaging that is possible is inherently limited: the sample (the patient) will not remain in precisely the same position for long, so a long signal averaging time is not possible. (In fact, a sudden increase in resolution of an ongoing imaging experiment is cause for some concern, see Problem 12.6.)

The signal intensity I (and therefore contrast) in a spin-echo imaging experiment depends on the number of spins N contributing to the signal, as well as the experiment time (i.e. the inverse of the repetition rate), τ_e, the echo time (see Figure 12.7) and the intrinsic relaxation rates of the spins. The dependence is given approximately by:

$$I(t) \propto N \exp\left(-\tau_e/T_2\right)\left(1 + \exp\left(-\tau_{ex}/T_1 - 2\exp\right) - \left(\tau_{ex} - \tau_e\right)/2/T_1\right) \quad (12.9)$$

As shown by this expression, both transverse and longitudinal relaxation rates will affect the contrast in an image, and intrinsic differences between these rates for spins in different types of tissues can be used to some advantage in identifying different types of tissue. In **T_1-weighted contrast** images, echo times τ_e are kept sufficiently short and repetition rates sufficiently high (short τ_{ex}, total experiment time) so that signals from areas of the image with short intrinsic T_1 values are emphasized. For **T_2-weighted contrast** the opposite is true. Repetition rates must be sufficiently slow that differences in T_1 do not affect the signal averaging, and the echo time is set to a value approximately equal to the shortest T_2 of interest. Finally, the density of reporting spins in a tissue can be used to enhance contrast. Obviously,

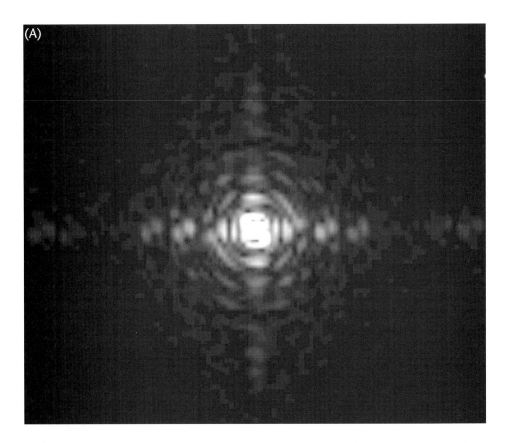

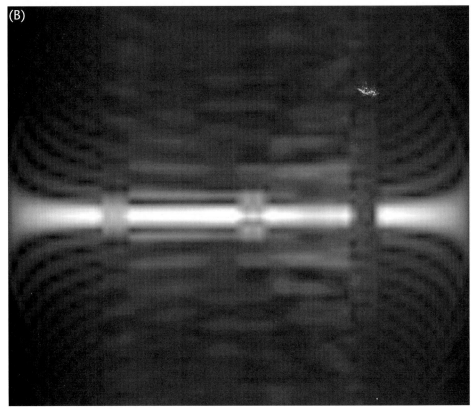

Figure 12.6 (A) Two-dimensional *k*-space data set (raw data) from a gradient-echo experiment. Figure courtesy of R. Rycyna, Bruker Biospin Inc. (B) Gradient echo imaging data set shown in (A) after transformation in the directly detected (frequency-encoded) dimension. Figure courtesy of R. Rycyna, Bruker Biospin Inc. (C) Completely transformed image from gradient echo imaging data set shown in (A). Figure courtesy of R. Rycyna, Bruker Biospin Inc.

blood and fluids have higher concentrations of ^{1}H spins than do more solid tissues, while bone and connective tissue have the least.

Besides providing basic contrast in an image, relaxation-weighted contrast can be used to determine regions of higher blood flow. Blood flow has the effect of refreshing the sample during the experiment. The previous sample is flushed and a new sample brought in during the course of the experiment. This means that the Boltzmann population difference is at a maximum and more signal can be obtained per unit time, especially in T_1-weighted experiments. This is of particular use when looking for necrotic tissue, where blood flow is restricted, and therefore will show up in the image as dimmer than healthy tissue with good blood supply. Spins in tumors often have longer relaxation times than nearby tissue, and this can also be used for image contrast (Reference 4).

A commonly used method of improving contrast is to introduce relaxation agents, often chelated paramagnetic metal ions, into the tissue to be imaged. This increases

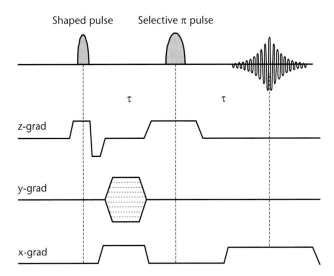

Figure 12.7 Spin-echo pulse sequence for planar imaging. The refocusing of magnetization by the spin echo that is generated by the slice-selective excitation and π pulses prior to frequency encoding improves signal by removing effects of field inhomogeneities (T_2^*) from the signal.

the relaxation rates of spins that interact with the metal and increases the available signal. One of the most common contrast agents is the complex of Gd(III) with the chelating agent diethylenetriaminepentaacetic acid (DTPA). Because rare-earth metal ions can have high coordination numbers, there are open coordination sites on the metal complex to which water can bind. Protons on bound water are relaxed by the paramagnetism of the metal ion, and upon exchange with bulk solvent, provide a fresh source of signal. The design of such agents attempts to enhance the exchange rates of bound water while still maintaining the very stable complex with the chelate that prevents the otherwise toxic gadolinium from injuring tissues.

One intriguing development that has occurred over the past 10 years is the use of **hyperpolarized noble gases** to provide contrast in MRI. Both ^{129}Xe and ^{3}He are noble gases with spins of 1/2. Because noble gas atoms have closed shell configurations and spherical electronic distributions, they do not interact strongly with their environment and for that reason can have extremely long T_1 values. For ^{129}Xe, a T_1 of ~20 min is observed under appropriate conditions. For ^{3}He, relaxation times can be on the order of many hours. Under normal circumstances this would render them useless as contrast agents (or even targets for spectroscopy). However, this slow relaxation behavior has one benefit: if one can generate a high degree of polarization via polarization transfer from another spin, it is possible to bamboozle Boltzmann and generate highly polarized states that are extremely long-lived.

Circularly polarized laser light can be used to selectively generate a highly spin-polarized excited state of the unpaired valence electron of rubidium atoms in the gas phase. If ^{129}Xe or ^{3}He is placed in contact with the rubidium atoms, electron–nuclear coupling between the rubidium valence electron and the noble-gas nuclear spin will transfer the electron polarization to the nucleus (a process known as **dynamic nuclear polarization**). The noble gas can be polarized to as high as a 50% excess of one nuclear spin orientation by this method (as opposed to the usual miniscule Boltzmann difference). After separation from the rubidium, the hyperpolarized gas can be administered to the patient via inhalation or other methods, and the polarization of the noble gas spins directly observed by appropriate pulse sequences. Such methods hold particular promise for applications in lung imaging or other regions of the body where there is low water content and ^{1}H imaging and contrast agents are less useful.

Rapid-scan MRI: echo-planar imaging and one-shot methods

As anybody who had a medical MRI scan in the early days of the technique could tell you, the most difficult part of the procedure was remaining motionless in a cramped enclosed space (i.e. the bore of the magnet) while the image was obtained. As anybody who has run a three-dimensional NMR experiment knows, it can take a long time (measured in hours or days) to acquire a three-dimensional data set: the problem is parallel for a whole body or sectional image, which is essentially a three-

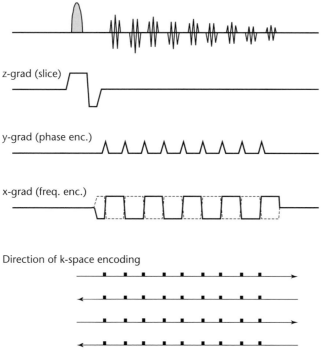

Figure 12.8 Echo-planar imaging sequence for acquisition of multiple steps of frequency encoding (multiple lines in *k*-space). Incrementation in the phase-encoded dimension (*y* axis) is accomplished by constant-sign "chirp" *y* gradients placed between successive frequency-encoding gradients. The frequency-encoding gradient along the *x* axis is reversed for each acquisition, so that successive lines in *k*-space are encoded in opposite directions (see lower diagram).

dimensional data set in reciprocal space. Thus, it was imperative to develop faster ways of sampling in reciprocal space if MRI was to be competitive with other imaging methods.

One of the most widely used rapid-scan methods is **echo-planar imaging** (**EPI**). EPI is based on the gradient-echo experiment, but incorporates multiple gradient echos and multiple phase-encoding gradients to obtain more than one row in k-space (more than one phase-encoding) in a single acquisition (Figure 12.8). This is functionally equivalent to acquiring data for more than one t_1 value in a two-dimensional NMR experiment in a single scan. In the best case, an entire image can be obtained in a single scan (so-called **one-shot** MRI). NMR spectroscopic equivalents of EPI have been described in recent years that allow a two-dimensional experiment to be acquired with a single scan.

Problems

12.1 A ^1H NMR sample that is 150 mm long gives a line width of 14 Hz at half-height in a 11.74 T field, but yields a T_2 by CPMG measurement of 1 s. Estimate the magnetic field inhomogeneity over the sample.

12.2 (a) Show that the phase cycling for the stimulated echo sequence shown in Figure 12.2 is appropriate for the desired coherence transfer pathway. (b) Three compounds, tetrabutylsilane, tetrabutylammonium chloride and chloroform, are present in the same NMR sample. The stimulated echo-diffusion experiment (Figure 12.2) gives the following normalized intensities (I/I_0) as a function of gradient strength (implicit in K^2 where $K = -\gamma G \tau_s$, see Equation 12.2) for a 57 ms delay Δ and 5 ms gradients (τ_g) for ^1H signals of the three compounds:

$K^2(\Delta-\tau_g 3))/$ 10^{-11} s/m^2	Tetrabutylsilane	Tetrabutylammonium chloride	Tetramethylsilane
0	1	1	1
1	0.55	0.81	0.90
2	0.36	0.67	0.77
3	0.25	0.49	0.67
4	0.12	0.36	0.57
5	0.07	0.30	0.48
6	0.04	0.26	0.45
7	–	0.17	0.35
8	–	0.12	0.27
9	–	0.10	0.24
10	–	0.07	0.20

Calculate the diffusion coefficients of the three compounds in m^2/s. Speculate on the reason for the different values of the constants.

12.3 Given a gradient intensity of 15 G/cm along the z axis (the long axis of the NMR tube) during signal acquisition, predict the ^1H chemical shift range (in Hz) for a water sample that extends 120 mm along the z axis in a susceptibility-matched NMR tube (that is, the tube has a flat bottom and a flat plug at the top, so there is no air-sample interface.) The tube and plug are constructed of material with magnetic susceptibility similar to that of water, and so there is little perturbation at the sample-tube interfaces. Sketch the spectral appearance that you expect. How would that change if a standard NMR tube (air at the top and a rounded bottom) is used? Repeat using an x-axis gradient (no sample spinning) of 30 G/cm.

12.4 A sinc pulse is often used to generate a selective excitation over a narrow region of a spectrum. As with any pulse, the excitation envelope (SW) is inversely proportional to the duration of the pulse. For a sinc pulse, this can be estimated as SW=$4/t_p$, where t_p is the pulse length. Calculate the width of the slice excited by a 3.2 ms sinc pulse in the presence of a field gradient of 1 G/cm for a ^1H image in a 3 T magnetic field.

12.5 Calculate the phase shift relative to the carrier frequency due to a 1 G/cm gradient applied for 50 ms to a proton at 20 mm to the positive side of the gradient.

12.6 A whole-body imaging experiment was being performed on a sedated mouse in which multiple scans were acquired at each slice in the image. A paramagnetic contrast agent was administered intravenously into the blood stream. While imaging slices through the thorax, resolution suddenly improved from one slice to the next, while the contrast got worse. Speculate on a reason for these observations.

12.7 An image of brain tissue is desired in which the intrinsic T_1 and T_2 values and proton densities of different areas of the brain are as follows:

Brain area	T_1 (s)	T_2 (s)	^1H density (relative)
Cerebrospinal fluid	2.5	0.3	1.0
Gray matter	0.9	0.1	0.7
White matter	0.7	0.1	0.6
Edema (stroke)	1.2	0.2	0.9
Tumor	1.5	0.1	0.9

Suggest a series of imaging experiments that could be used to determine the nature of an abnormality in (a) gray matter and (b) white matter of the brain.

References

1. P. C. Lauterbur, *Nature* 242, 190 (1973).
2. E. L. Hahn, *Phys. Rev.* 80, 580 (1950).
3. E. O. Stejskal and J. E. Tanner, *J. Chem. Phys.* 42, 288 (1965).
4. R. Damadian, *Science* 171, 1151 (1971).

APPENDIX A
TIME-DEPENDENT
PERTURBATIONS

The time-dependent Schrödinger equation and superposition states

The wave description of a stationary state of a quantum system is described in terms of a standing wave of a form $\varsigma(s)$ where s is a generalized position variable. A standing wave does not vary with time. When one considers the behavior of a system under the influence of a time-dependent perturbation such as the application of an RF field, the correct wave description of the system must also include a time-dependent term. By assuming that the time-dependent and time-independent portions of the wave description are separable, the complete wave description would be as follows:

$$\psi(s,t) = \varsigma(s)\varphi(t) \tag{A.1}$$

The appropriate wave descriptions of the system would be a series of functions $\Psi(s,t)$ that are eigenfunctions of the Hamiltonian operator $\hat{\mathbf{E}}$ in the time-dependent version of the Schrödinger equation:

$$\hat{\mathbf{E}}(s)\psi(s,t) = i\hbar\left(\frac{d\psi(s,t)}{dt}\right)_s \tag{A.2}$$

If the Hamiltonian operator does not have an explicit time dependence, substitution of Expresssion A.1 into Equation A.2 gives:

$$\varphi(t)\hat{\mathbf{E}}(s)\varsigma(s) = i\hbar\varsigma(s)\left(\frac{d\varphi(t)}{dt}\right)_s \tag{A.3}$$

This expression can be rearranged by dividing both sides by $\Psi(s,t) = \psi(s)\varphi(t)$ to put all the time-dependent terms on one side and all of the spatially dependent terms on the other:

$$\frac{1}{\varsigma(s)}\hat{\mathbf{E}}(s)\varsigma(s) = \frac{i\hbar}{\varphi(t)}\frac{d\varphi(t)}{dt} \tag{A.4}$$

However, the only way in which two expressions that are functions of two independent variables can be equal is if they both equal a constant; in this case, that constant is E, the energy of the state. Thus, we get:

$$\frac{i\hbar}{\varsigma(s)}\hat{\mathbf{E}}(s)\varsigma(s) = E = \frac{i\hbar}{\varphi(t)}\frac{d\varphi(t)}{dt} \tag{A.5}$$

Rearrangement of the left-hand side of A.5 with respect to E just gives the time-independent Schrödinger equation described in Chapter 1. We can solve the differential equation given in the right-hand side of A.5, which has the general solution:

$$\varphi(t) = \varphi(0)\exp\left[\frac{-iEt}{\hbar}\right] \tag{A.6}$$

Now we get the full form of the wave description of the state:

$$\psi(s,t) = \varsigma(s)\varphi(0)\exp\left[\frac{-iEt}{\hbar}\right] \tag{A.7}$$

Hilbert space, eigenvectors, and superposition of states

The set of wave functions ψ_i, which describe all of the possible stationary states of a quantum oscillator, are usually a complete set of eigenfunctions of an operator $\hat{\mathbf{R}}$ representing an observable which is a constant of a particular stationary state, such as energy or angular momentum. The ψ_i can be represented as a set of orthonormal vectors (called **eigenvectors** of the operator $\hat{\mathbf{R}}$ for which they constitute a set of eigenfunctions) in an abstract N-dimensional space called **Hilbert space**, where N is the number of different possible states of the system. Each "axis" in Hilbert space is orthogonal to all of the others, and is represented by one of the eigenvectors of $\hat{\mathbf{R}}$. The dot-product of two different eigenvectors of $\hat{\mathbf{R}}$ in Hilbert space will be zero ($\psi_i \cdot \psi_j = 0$, $i \neq j$); the dot product of an eigenvector of $\hat{\mathbf{R}}$ with itself will be unity

$(\psi_i \cdot \psi_j = 1, i = j)$. This is, of course, the definition of orthonormality, and is exactly analogous to three unit vectors representing the $\bar{x}$, $\bar{y}$ and $\bar{z}$ axes of Cartesian coordinate space. The dot product of any two of the three unit vectors in Cartesian space will be zero, and that of any one with itself will be unity. Furthermore, the $\bar{x}$, $\bar{y}$ and $\bar{z}$ axes form the **basis set** for Cartesian coordinate space, and any vector in Cartesian space can be represented as a linear combination of this basis set. By analogy, the complete set of eigenvectors of the operator $\hat{\mathbf{R}}$ describing the stationary states of the system form a basis set in Hilbert space; that is, any state of the system, even those "in between" stationary states (i.e. time-dependent states), may be described by a linear combination of elements of the basis set (eigenvectors of $\hat{\mathbf{R}}$). Mathematically, this is shown as:

$$\Psi_j = \sum_i^N c_i(t)\psi_i \qquad (A.8)$$

where Ψ_j is the description of an arbitrary state of the system, and N is the number of eigenstates of the system (and also the dimensionality of the Hilbert space). For example, as a quantum oscillator undergoes a transition from a stationary state described by ψ_i to another described by ψ_j, the "in-between" state is represented by:

$$\Psi(t) = c_i(t)\psi_i + c_j(t)\psi_j \qquad (A.9)$$

Perturbation theory: time-dependent perturbations of the Hamiltonian

Perhaps the simplest way to look at spectroscopic transitions using quantum theory is to take the solutions to the time-independent form of the Schrödinger equation and add a small time-dependent perturbation of the Hamiltonian operator. Then, the probability that this perturbation will cause a transition between two stationary states is determined. The modified Hamiltonian operator $\hat{\mathbf{E}}_{eff}$ has the form:

$$\hat{\mathbf{E}}_{eff} = \hat{\mathbf{E}} + \hat{\mathbf{E}}'(s,t) \qquad (A.10)$$

where $\hat{\mathbf{E}}$ is the time-independent portion of Hamiltonian operator used to determine the energies of the stationary states of the system, and is modified by a small time-dependent contribution $\hat{\mathbf{E}}'(s,t)$ that could be the result of an oscillating RF field. For this approach to be effective, $\hat{\mathbf{E}}'(s,t)$ must be much smaller than $\hat{\mathbf{E}}$, i.e. it is a "perturbation", not a large change. We can assume that there exists a complete set of N orthonormal time-independent eigenfunctions of $\hat{\mathbf{E}}$, $\varsigma_i(s)$, which describe a Hilbert space of dimension N, N being the total number of stationary states of the system. (It is important to remember that N does not have to be finite, although it is for Zeeman splitting.) We can form linear combinations with these eigenfunctions, represented as a $1 \times N$ matrix $\bar{\mathbf{V}}$, to generate time-dependent wave descriptions of the form:

$$\Psi(s,t) = \sum_i^N b_i(t)\varsigma_i(s) \tag{A.11}$$

where the $b_i(t)$ are time-dependent coefficients. These coefficients can also be represented by a $1 \times N$ matrix, $\bar{\mathbf{b}}$. This renders A.11 in the form:

$$\Psi(s,t) = \bar{\mathbf{V}}\bar{\mathbf{b}} \tag{A.12}$$

Note that this is not precisely the form that we encountered in the previous section, but we will later recall that:

$$\Psi(s,t) = \sum_i^N b_i(t)\varsigma_i(s) = \sum_i^N \varsigma_i(s)c_i(t)\exp\left[-\frac{iE_it}{\hbar}\right] \tag{A.13}$$

that is similar in form to Equation A.11, with the $\exp(i\xi_j)$ assumed to be included in $\varsigma_j(s)$. However, for the moment we will continue to use the description in Equation A.13. Replacing Equations A.12 and A.13 into the time-dependent form of the Schrödinger equation, we get:

$$i\hbar\bar{\mathbf{V}}\frac{\partial}{\partial t}\bar{\mathbf{b}} = \left(\hat{\mathbf{E}} + \hat{\mathbf{E}}'(s,t)\right)\bar{\mathbf{V}}\bar{\mathbf{b}} \tag{A.14}$$

Recalling that $\bar{\mathbf{V}}$ is a vector representing eigenfunctions of $\hat{\mathbf{E}}$, operation by the time-independent portion of the Hamiltonian operator $\hat{\mathbf{E}}$ on the right gives:

$$i\hbar\bar{\mathbf{V}}\frac{\partial}{\partial t}\bar{\mathbf{b}} = \left(\hat{\mathbf{E}} + \hat{\mathbf{E}}'(s,t)\right)\bar{\mathbf{V}}\bar{\mathbf{b}} = \bar{E}\bar{\mathbf{V}}\bar{\mathbf{b}} + \hat{\mathbf{E}}'(s,t)\bar{\mathbf{V}}\bar{\mathbf{b}} \tag{A.15}$$

where $\bar{E}$ is a $1 \times N$ vector containing the eigenvalues E_i of $\hat{\mathbf{E}}$ operating on $\bar{\mathbf{V}}$. Then, we use the orthonormality requirement to simplify the expression, multiplying both sides of the equation by the complex conjugate of $\bar{\mathbf{V}}$, $\bar{\mathbf{V}}*$, and integrating both sides of the expression over all space:

$$i\hbar\frac{\partial}{\partial t}\bar{\mathbf{b}} = \bar{E}\bar{\mathbf{b}} + \bar{E}'\bar{\mathbf{b}} \tag{A.16}$$

where the $N \times N$ matrix $\bar{E}'$ contains i,j elements of the form:

$$\int_0^\infty \varsigma_i(s) * \hat{\mathbf{E}}'(s,t)\varsigma_j(s)ds = E'_{ij} \tag{A.17}$$

We can now replace the $b_i(t)$ with the form from Equation A.13:

$$b_i(t) = c_i(t)\exp\left[-\frac{iE_it}{\hbar}\right] \tag{A.18}$$

which yields, after differentiation of the left-hand side of the expression, for each value of j:

$$i\hbar\left(\frac{-iE_j}{\hbar}\right)\exp\left(-\frac{iE_i t}{\hbar}\right)c_j(t)+\exp\left(-\frac{iE_i t}{\hbar}\right)\frac{\partial c_j(t)}{\partial t}=$$
$$E_j c_j(t)\exp\left[-\frac{iE_j t}{\hbar}\right]+\sum_i^N E'_{ji}c_i(t)\exp\left[-\frac{iE_i t}{\hbar}\right]\qquad(A.19)$$

Note that the first terms on both sides of Equation A.19 are the same, and so can be cancelled. Also, we note that since the $c_i(t)$ are functions only of time, the partial derivative terms can be replaced by the total derivative, represented by $c'_i(t)$, yielding:

$$i\hbar\exp\left(-\frac{iE_j t}{\hbar}\right)c'_j(t)=\sum_i^N E'_{ji}c_i(t)\exp\left[-\frac{iE_i t}{\hbar}\right]\qquad(A.20)$$

Rearranging this expression, we get:

$$c'_j(t)=\frac{1}{i\hbar}\exp\left[\frac{iE_j t}{\hbar}\right]\sum_i^N E'_{ji}c_i(t)\exp\left[-\frac{iE_i t}{\hbar}\right]\qquad(A.21)$$

which can be written in terms of a transition frequency, ω_{ji}, between the jth and ith states by using the relationship $\omega_{ji}=(E_j-E_i)/\hbar$:

$$c'_j(t)=\frac{1}{i}\sum_i^N E'_{ji}c_i(t)\exp\left(-i\omega_{ji}t\right)\qquad(A.22)$$

We now have an expression for the rate of change for the time-dependent coefficients that determine the contribution of particular stationary-state wave functions to the overall description of the state in terms of the transition frequencies and the off-diagonal terms of $\bar{\bar{E}}'$.

If we now expand the wave description of the state of the system in terms of the $c_i(t)$, we get, as above:

$$\Psi(s,t)=\sum_i^N \varsigma_i(s)c_i(t)\exp\left[-\frac{iE_i t}{\hbar}\right]\qquad(A.23)$$

The complex square of $\Psi(s,t)$, $\Psi*\Psi$, will, by orthogonality, give nonzero terms when $i=j$, and normalization requires that the sums of the complex squares of the $c_i(t)$ be unity (i.e. the system has to be in *some* state). Thus, the complex square of the $c_i(t)$, $c_i*(t)\cdot c_i(t)$, gives the probability of finding the system in the ith state at time t. If there is no perturbation, the system will remain in the starting state (call it state j), and $c_j*(t) \cdot c_j(t) = 1$, while all other $c_i*(t) \cdot c_i(t) = 0, i \neq j$. If a time-dependent perturbation is present, the possibility of a transition to another state exists, and at a later time t', the probability that the system will occupy the new state (call it state k), will increase; i.e. $c_k*(t') \cdot c_k(') \neq 0$.

In order to evaluate the $c_i(t)$, we use Equation A.22, and assume that at t close to 0, the $c_i(0) \approx 0, i \neq j$, where j is the starting state. Thus, if we are considering a transition to state k,

$$c_k'(t) \approx \frac{1}{i} E_{kj}' c_k(t) \exp\left(-i\omega_{kj}t\right) \tag{A.24}$$

Integration of this expression yields first-order values for $c_i(t)$:

$$c_k(t') \approx \frac{1}{i} \int_0^{t'} E_{kj}' \exp\left(-i\omega_{kj}t\right) dt \tag{A.25}$$

In order to correctly evaluate this integral, it is important to remember that the cross-term for the Hamiltonian, E_{kj}', also contains a time dependence, and it is this time dependence which will determine the appearance of the NMR spectrum.

Semiclassical interactions between EMR and quantum oscillators using perturbation theory

From Chapter 6, we recall that the time-dependent portion of the Hamiltonian operator due to the application of an RF field along the x axis (with the static field $\vec{B}_0$ along $\bar{z}$), is given by:

$$\hat{\mathbf{E}}'(t) = -\gamma\hbar\hat{\mathbf{I}} \cdot \vec{B}_1 \cdot 2\cos\omega t = -\gamma\hbar\hat{\mathbf{I}}_x B_1 2\cos\omega t \tag{A.26}$$

The factor of two in the amplitude is used for convenience in Equation A.26, since we need to use the equality:

$$\cos\omega t = \frac{\exp(i\omega t) + \exp(-i\omega t)}{2} \tag{A.27}$$

Using the bra-ket notation, the $E_{kj}' = \langle \varsigma_k| -\gamma\hbar\hat{\mathbf{I}}_x B_1 2\cos\omega t |\varsigma_j\rangle = -\gamma\hbar B_1 2\cos\omega t \langle \varsigma_k|\hat{\mathbf{I}}_x|\varsigma_j\rangle$, and replacing this into Equation A.25, we get:

$$c_k(t') \approx -\frac{\gamma B_1}{i} \left\langle \varsigma_k | \hat{\mathbf{I}}_x | \varsigma_j \right\rangle \int_0^{t'} \left(\exp(i\omega t) + \exp(-i\omega t)\right)\exp\left(-i\omega_{kj}t\right) dt \tag{A.28}$$

$$c_k(t') \approx -\frac{\gamma B_1}{i} \left\langle \varsigma_k | \hat{\mathbf{I}}_x | \varsigma_j \right\rangle \int_0^{t'} \left[\exp\left(-i\left(\omega_{kj}+\omega\right)t\right) + \exp\left(i\left(\omega_{kj}+\omega\right)t\right)\right] dt \tag{A.29}$$

And integration yields:

$$c_k(t') \approx -\gamma B_1 \left\langle \varsigma_k \mid \hat{\mathbf{I}}_x \mid \varsigma_j \right\rangle \left[\frac{\exp\left[i\left(\omega_{kj} - \omega\right)t\right]}{\omega_{kj} - \omega} + \frac{\exp\left[i\left(\omega_{kj} + \omega\right)t\right]}{\omega_{kj} + \omega} \right] \quad (A.30)$$

Equation A.30 is then evaluated from $t = 0 \rightarrow t = t'$ to give:

$$c_k(t') \approx -\gamma B_1 \left\langle \varsigma_k \mid \hat{\mathbf{I}}_x \mid \varsigma_j \right\rangle \left[\frac{\exp\left[i\left(\omega_{kj} - \omega\right)t'\right] - 1}{\omega_{kj} - \omega} + \frac{\exp\left[i\left(\omega_{kj} + \omega\right)t'\right] - 1}{\omega_{kj} + \omega} \right] \quad (A.31)$$

Far from resonance ($\omega_{jk} \neq \omega$) both terms in Equation A.31 will be small, but close to resonance ($\omega_{jk} \approx \omega$) the first term will predominate, and Equation A.31 simplifies to:

$$c_k(t') \approx -\gamma B_1 \left\langle \varsigma_k \mid \hat{\mathbf{I}}_x \mid \varsigma_j \right\rangle \left[\frac{\exp\left[i\left(\omega_{kj} - \omega\right)t'\right] - 1}{\omega_{kj} - \omega} \right] \quad (A.32)$$

The absolute value of the square of Equation A.32 will yield the probability of transition from state j to state k:

$$P_{kj}(\omega, t) = \mid c_k(t) \mid^2 \approx -\gamma^2 B_1^2 \left\langle \varsigma_k \mid \hat{\mathbf{I}}_x \mid \varsigma_j \right\rangle^2 \left[\frac{\exp\left[i\left(\omega_{kj} - \omega\right)t'\right] - 1}{\omega_{kj} - \omega} \right] \left[\frac{\exp\left[-i\left(\omega_{kj} - \omega\right)t'\right] - 1}{\omega_{kj} - \omega} \right]$$

$$(A.33)$$

Using the identity $\sin^2(z/2) = -(\exp(iz) - 2 + \exp(-iz))/4$, Equation A.33 can be further simplified to the real expression:

$$P_{kj}(\omega, t) = \gamma^2 B_1^2 \left\langle \varsigma_k \mid \hat{\mathbf{I}}_x \mid \varsigma_j \right\rangle^2 \left[\frac{\sin^2((\omega_{kj} - \omega)t/2)}{4(\omega_{kj} - \omega)^2} \right] \quad (A.34)$$

The line width narrows as the time over which the probability is measured increases. Integration of Equation A.34 over all frequencies will give the rate constant for the transition. Multiplying both numerator and denominator by $(t/2)^2$, the integral becomes:

$$k_{kj}(\omega, t) = \frac{\gamma^2 B_1^2 t^2}{16} \left\langle \varsigma_k \mid \hat{\mathbf{I}}_x \mid \varsigma_j \right\rangle^2 \int_0^\infty \left[\frac{\sin^2((\omega_{kj} - \omega)t/2)}{((\omega_{kj} - \omega)t/2)^2} \right] d\omega \quad (A.35)$$

and remembering that $\omega = 2\pi v$ we can further rearrange the expression:

$$k_{kj}(v,t) = \frac{\gamma^2 B_1^2 t}{8} \left\langle \varsigma_k \mid \hat{\mathbf{I}}_x \mid \varsigma_j \right\rangle^2 \int_0^\infty \left[\frac{\sin^2((v_{kj} - v)\pi t)}{((v_{kj} - v)\pi t)^2} \right] \pi t dv \tag{A.36}$$

A further simplification is possible when the definite integral relationship:

$$\int_{-\infty}^\infty \left[\frac{\sin^2 u}{u^2} \right] du = \pi \tag{A.37}$$

If $u = \pi(v_{kj} - v)t$ and $du = -\pi t dv$, we can use A.37 to arrive at:

$$k_{kj}(t) = \frac{\gamma^2 \pi B_1^2 t}{8} \left\langle \varsigma_k \mid \hat{\mathbf{I}}_x \mid \varsigma_j \right\rangle^2 \tag{A.38}$$

which is the rate constant as a function of time for a transition to occur from state j to state k in the presence of the RF field of strength B_1 fixed along the x axis of the rotating frame.

APPENDIX B
DENSITY MATRIX FORMALISM
AND THE RELAXATION
SUPERMATRIX

A density matrix description of the ^{1}H, ^{15}N HMQC experiment

The density matrix is a concise representation of a quantum mechanical ensemble, and is used to predict the behavior of the ensemble under the influence of time-dependent perturbations. This includes the generation and loss of coherences between stationary states and population changes of those states as the result of time-dependent perturbations. For a system consisting of N discrete eigenstates, the density matrix contains N^2 elements.

As an example, we will consider the evolution of the density matrix in the course of the ^{1}H, ^{15}N HMQC experiment (Figure 8.17). The coupling between the ^{1}H and ^{15}N spins is much smaller than the differences in their transition energies (always the case for nuclei with different γ values); the products of the uncoupled basis functions provide a suitable basis set for construction of the coupled spin system after normalization. This basis set can be represented as a series of kets:

$$| \alpha\alpha >, | \alpha\beta >, | \beta\alpha >, | \beta\beta > \qquad (B.1)$$

The expectation value of any appropriately constructed operator representing an observable of the system $\hat{\mathbf{O}}$ for each discrete state is given by:

$$< O_{ijkl} > = < ij \, | \, \hat{\mathbf{O}} \, | \, kl > \qquad (B.2)$$

where the bras $<ij|$ are the complex conjugates of the corresponding ket. This is identical to the treatment that was described in Chapter 4.

However, the operation shown in B.2 only allows one to calculate the expectation value for $\hat{O}$ in one state of the system, not the overall expectation value for the entire system. In order to determine this, one needs to know the relative populations of all states of the system, and determine the average expectation value $\langle \bar{O} \rangle$ from the population-weighted values of $\langle O_{ijkl} \rangle$.

In order to accomplish this, we need a wave representation of the ensemble. This can be of the form:

$$\psi = c_1 | \beta\beta \rangle + c_2 | \alpha\beta \rangle + c_3 | \beta\alpha \rangle + c_4 | \alpha\alpha \rangle \qquad (B.3)$$

where the constants c_i must be normalized so that $\sum_n c*_i c_i = 1$ over all N states (i.e.

the system has to be in *some* state, and this is reflected by a net unit probability). The student should be aware that the wave representation of the ensemble shown in Equation B.3 is similar to that for a single system in a superposition of states.

In the chosen example of a two-spin ^{1}H, ^{15}N-coupled spin system, $N = 4$. In the absence of time-dependent perturbations, the expectation value of the ensemble $\langle \bar{O} \rangle$ would be equal to the $\langle O_{ijkl} \rangle$ for each state weighted by the probability of the system being in state ij given by $c*_i c_j$, where we now replace the double subscript on the constants with the single subscripts from Equation B.3. Recall from Appendix A that:

$$\Psi(s,t) = \sum_i^N \varsigma_i(s) c_i(t) \exp\left[-\frac{iE_i t}{\hbar} \right] \qquad (A.23)$$

The elements of the density matrix $\bar{\sigma}$ for a given ensemble are defined as $\sigma_{ij} = \langle c_i c_j* \rangle$. In the absence of any external perturbation that results in the mixing of states, the off-diagonal elements are zero, since the complex exponent term (which changes sign in the complex conjugate) gives:

$$c_i(t) c_j*(t) \exp i\left[-\frac{E_i - E_j}{\hbar} \right] \qquad (B.4)$$

Integration of this term with respect to time introduces an arbitrary phase factor, which, in the absence of any coherent perturbation at frequency $\omega_{ij} = (E_i - E_j)\hbar$, can have any value between 0 and 2π, and thus averages to zero.

On the other hand, the diagonal elements are nonzero at equilibrium. Due to the orthonormality restrictions on the basis set used to generate Equation B.3, the diagonal elements $c*_i c_i$ represent the equilibrium populations of the four states of the two-spin system that form the basis set shown in B.3. Thus, the density matrix for an NH pair at equilibrium would have the form:

$$\begin{bmatrix} P_1 & 0 & 0 & 0 \\ 0 & P_2 & 0 & 0 \\ 0 & 0 & P_3 & 0 \\ 0 & 0 & 0 & P_4 \end{bmatrix} \tag{B.5}$$

arranged so that the highest energy state is $P_1 = \langle c_1 c_1 * \rangle$ and the lowest energy state is $P_4 = \langle c_4 c_4 * \rangle$. Based on the gyromagnetic ratios of ^{15}N and ^1H, the Boltzmann distributions of populations between the four states should be:

$$\begin{bmatrix} P_1 & 0 & 0 & 0 \\ 0 & P_1+p & 0 & 0 \\ 0 & 0 & P_1+10p & 0 \\ 0 & 0 & 0 & P_1+11p \end{bmatrix} \tag{B.6}$$

where p is the Boltzmann excess of an isolated ^{15}N population at equilibrium. By moving the constants out, B.6 can be shown as the sum of two matrices:

$$P_1 \begin{bmatrix} 1 & 0 & 0 & 0 \\ 0 & 1 & 0 & 0 \\ 0 & 0 & 1 & 0 \\ 0 & 0 & 0 & 1 \end{bmatrix} + p \begin{bmatrix} 0 & 0 & 0 & 0 \\ 0 & 1 & 0 & 0 \\ 0 & 0 & 10 & 0 \\ 0 & 0 & 0 & 11 \end{bmatrix} \tag{B.7}$$

The first matrix is the identity matrix, and is invariant to any transformations that we will need to perform. The second matrix represents the population differences between the states, as we expected, and the weighting factor p can be ignored (just realizing that it is always present), so the representation of the density matrix at equilibrium for an NH pair is:

$$\bar{\sigma}_{eq} = \begin{bmatrix} 0 & 0 & 0 & 0 \\ 0 & 1 & 0 & 0 \\ 0 & 0 & 10 & 0 \\ 0 & 0 & 0 & 11 \end{bmatrix} \tag{B.8}$$

RF pulses

RF pulses result in the time-dependent mixing of stationary states that in turn give rise to nonzero off-diagonal elements of the density operator matrix that represent coherences. Although the wave equation B.3 is assumed to be an eigenfunction of the time-independent Hamiltonian $\hat{E}_0$, this is not the case in the presence of the time-dependent Hamiltonian $\hat{E}_{eff}(t) = \hat{E}_0 + \hat{E}'(t)$, where $E'(t)$ is a small perturbation. In order to determine the time evolution of $\bar{\sigma}$ in this situation, a more general equation is used, the quantum mechanical equivalent of the classical Liouville equation:

$$\frac{d\sigma}{dt} = \frac{i}{\hbar}\left[\sigma, \hat{E}_{eff}(t)\right] \tag{B.9}$$

where the term in brackets is the commutator between the density operator and the Hamiltonian operator. Integration of B.9 with respect to time has the following solution:

$$\sigma(t) = \exp\left(\frac{-iE_{eff}t}{\hbar}\right)\sigma(0)\exp\left(\frac{iE_{eff}t}{\hbar}\right) \tag{B.10}$$

In the laboratory frame of reference, $E_{eff}(t)$ contains both the Zeeman term E_0 and the oscillating RF field $E'(t)$. In the rotating frame, however, this simplifies to the operator for a pulse along a defined axis $\beta\hat{\mathbf{I}}_\varphi$ with β being the tip angle (which implicitly defines the time, since the tip angle is determined by the frequency of the B_1 field and the duration of the pulse) and the axis indicated by the subscript φ (see Equation 6.25). Placing this into Equation B.9 yields:

$$\sigma(t) = e^{-i\beta\hat{\mathbf{I}}_\varphi}\sigma(0)e^{i\beta\hat{\mathbf{I}}_\varphi} \tag{B.11}$$

Note that this is the same form as was given for the rotation operator that represents a pulse applied in the product operator formalism (Chapter 6), in which the rotation operator representing a pulse was given by $\hat{R}_\varphi^{-1}\hat{I}\hat{R}_\varphi$. In fact, Equation 6.25 is the form for the pulse rotation operator for a single spin. For a matrix representation of the rotation operator for a two-spin IS system, we need to expand the expression $R_\varphi = e^{i\beta\hat{\mathbf{I}}_\varphi}$. In the case of an **x** pulse (φ = x) with a tip angle β, the series expansion B.12 applies:

$$R_{\beta x} = e^{i\beta\hat{\mathbf{I}}_x} \cong 1 + i\beta I_x + \frac{(i\beta)^2 I_x^2}{2!} + \frac{(i\beta)^3 I_x^3}{3!} + \frac{(i\beta)^4 I_x^4}{4!} + \cdots \tag{B.12}$$

Using the Pauli spin matrix representation of $\hat{\mathbf{I}}_x$, it is simple to show that:

$$\hat{\mathbf{I}}_x^n = \frac{1}{2^n}[1], n = even$$

$$\hat{\mathbf{I}}_x^n = \frac{2}{2^n}\hat{\mathbf{I}}_x, n = odd \tag{B.13}$$

where [1] is the identity matrix. Using these values in the expansion shown in Equation B.12, and then using the relationship between complex exponents and trigonometric relationships, Equation B.12 becomes:

$$R_{\beta x}\left[1-\frac{(\beta/2)^2}{2!}+\frac{(\beta/2)^4}{4!}-\cdots\right]\times\begin{bmatrix}1 & 0\\0 & 1\end{bmatrix}+i\left[\beta/2-\frac{(\beta/2)^3}{3!}+\frac{(\beta/2)^5}{5!}-\cdots\right]\times\begin{bmatrix}0 & 1\\1 & 0\end{bmatrix}$$

(B.14)

$$R_{\beta x}=\cos\beta/2\begin{bmatrix}1 & 0\\0 & 1\end{bmatrix}+i\sin\beta/2\begin{bmatrix}0 & 1\\1 & 0\end{bmatrix}=\begin{bmatrix}\cos\beta/2 & i\sin\beta/2\\i\sin\beta/2 & \cos\beta/2\end{bmatrix}$$ (B.15)

For the two-spin system, the corresponding matrix for **x** pulses that connect states 1 to 2 and 3 to 4 (in this case, ^{15}N pulses) is:

$$R_{\beta x}(^{15}N)=\begin{bmatrix}\cos\beta/2 & i\sin\beta/2 & 0 & 0\\i\sin\beta/2 & \cos\beta/2 & 0 & 0\\0 & 0 & \cos\beta/2 & i\sin\beta/2\\0 & 0 & i\sin\beta/2 & \cos\beta/2\end{bmatrix}$$ (B.16)

The inverse (or reciprocal) matrix, defined by $R\times R^{-1}=[1]$, is obtained by standard matrix algebra methods. Because the rotation operators are Hermitian, this is obtained by transposition and taking the complex conjugate. So the inverse rotation operator has the form :

$$R^{-1}_{\beta x}(^{15}N)=\begin{bmatrix}\cos\beta/2 & -i\sin\beta/2 & 0 & 0\\-i\sin\beta/2 & \cos\beta/2 & 0 & 0\\0 & 0 & \cos\beta/2 & -i\sin\beta/2\\0 & 0 & -i\sin\beta/2 & \cos\beta/2\end{bmatrix}$$ (B.17)

For a 90° (x) pulse, this gives the following results:

$$R_{90x}(^{15}N)=\frac{1}{\sqrt{2}}\begin{bmatrix}1 & -i & 0 & 0\\-i & 1 & 0 & 0\\0 & 0 & 1 & -i\\0 & 0 & -i & 1\end{bmatrix}$$

$$R^{-1}_{90x}(^{15}N)=\frac{1}{\sqrt{2}}\begin{bmatrix}1 & i & 0 & 0\\i & 1 & 0 & 0\\0 & 0 & 1 & i\\0 & 0 & i & 1\end{bmatrix}$$ (B.18)

The corresponding matrix for a ^{1}H pulse on the same two-spin system (connecting states 1 to 3 and 2 to 4) is:

$$R_{\beta x}(^1\text{H}) = \begin{bmatrix} \cos\beta/2 & 0 & i\sin\beta/2 & 0 \\ 0 & \cos\beta/2 & 0 & i\sin\beta/2 \\ i\sin\beta/2 & 0 & \cos\beta/2 & 0 \\ 0 & i\sin\beta/2 & 0 & \cos\beta/2 \end{bmatrix} \quad\text{(B.19)}$$

For a 90° (x) pulse on ^{1}H, this gives:

$$R_{90x}(^1\text{H}) = \frac{1}{\sqrt{2}} \begin{bmatrix} 1 & 0 & i & 0 \\ 0 & 1 & 0 & i \\ i & 0 & 1 & 0 \\ 0 & i & 0 & 1 \end{bmatrix}$$

$$R^{-1}_{90x}(^1\text{H}) = \frac{1}{\sqrt{2}} \begin{bmatrix} 1 & 0 & -i & 0 \\ 0 & 1 & 0 & -i \\ -i & 0 & 1 & 0 \\ 0 & -i & 0 & 1 \end{bmatrix} \quad\text{(B.20)}$$

Application of a ^{1}H 90° (x) pulse (the first pulse in an HMQC experiment) to the equilibrium system represented by Equation B.8, one obtains:

$$R^{-1}_{90x} \times \sigma_{eq} \times R_{90x} = \frac{1}{2} \begin{bmatrix} 1 & 0 & -i & 0 \\ 0 & 1 & 0 & -i \\ -i & 0 & 1 & 0 \\ 0 & -i & 0 & 1 \end{bmatrix} \times \begin{bmatrix} 0 & 0 & 0 & 0 \\ 0 & 1 & 0 & 0 \\ 0 & 0 & 10 & 0 \\ 0 & 0 & 0 & 11 \end{bmatrix} \times \begin{bmatrix} 1 & 0 & i & 0 \\ 0 & 1 & 0 & i \\ i & 0 & 1 & 0 \\ 0 & i & 0 & 1 \end{bmatrix} =$$

$$\begin{bmatrix} 5 & 0 & -5i & 0 \\ 0 & 6 & 0 & -5i \\ 5i & 0 & 5 & 0 \\ 0 & 5i & 0 & 6 \end{bmatrix}$$

$$\text{(B.21)}$$

Note that the product matrix is a linear combination of a series of population operators (along the diagonal) indicating that state populations have been perturbed, as well as off-diagonal elements that correspond to nonzero elements in the matrix representation of the Cartesian operator $\hat{\mathbf{I}}_y$ (see Equation 6.9).

Time evolution of the density matrix with chemical shift and coupling

Off-diagonal elements of the density matrix that are generated by pulses evolve according to the general expression:

$$\sigma_{ij}(t) = \sigma_{ij}(0)\exp(-i\omega_{ij}t) \qquad (\text{B.22})$$

One can see that this is identical to the time evolution of single-element operators described in Chapter 6. The transition frequency ω_{ij} given by:

$$\omega_{ij} = \left(E_i - E_j\right)/ \qquad (\text{B.23})$$

In the rotating frame, the transition energies are those shown in Figure 5.1, with the signs of the couplings adjusted for the fact that the signs of the gyromagnetic ratios of ^{15}N and ^1H are opposite. Thus:

$$\sigma_{13}(t) = \sigma_{13}(0)\exp(-i(\Omega_H - \pi J_{NH})t)$$
$$\sigma_{24}(t) = \sigma_{24}(0)\exp(-i(\Omega_H + \pi J_{NH})t) \qquad (\text{B.24})$$

Evolution during the $\tau = 1/2J$ delay gives both chemical shift and coupling to these terms, so after converting the complex exponential relating to coupling evolution to sin and cos equivalents, combining terms and reconverting to complex exponentials, one gets:

$$\bar{\sigma}(\tau) = \begin{bmatrix} 5 & 0 & 5\exp\left(-i\Omega_H\tau\right) & 0 \\ 0 & 6 & 0 & -5\exp\left(-i\Omega_H\tau\right) \\ 5\exp\left(i\Omega_H\tau\right) & 0 & 5 & 0 \\ 0 & -5\exp\left(i\Omega_H\tau\right) & 0 & 6 \end{bmatrix}$$

$$(\text{B.25})$$

In the HMQC experiment, this evolution is followed by a ^{15}N 90° (x) pulse, which yields the following matrix:

$$\bar{\sigma}(2) = \frac{1}{2}\begin{bmatrix} 11 & i & 0 & -10i\exp\left(-i\Omega_H\tau\right) \\ -i & 11 & 10i\exp\left(-i\Omega_H\tau\right) & 0 \\ 0 & -10i\exp\left(i\Omega_H\tau\right) & 11 & i \\ 10i\exp\left(i\Omega_H\tau\right) & 0 & -i & 11 \end{bmatrix}$$

$$(\text{B.26})$$

Note that the off-diagonal elements representing DQ (elements 1,4 and 4,1) evolution between ^1H and ^{15}N are now nonzero, as are the ZQ (2,3 and 3,2) and ^{15}N SQ transitions.

After evolution for a time $t_1/2$, the matrix will have the following appearance (ignoring the evolution of the single-quantum ^{15}N operators, which will not be detected).

$$\sigma(t_1/2) = \frac{1}{2} \begin{bmatrix} 11 & i & 0 & -10i\exp[-i\Omega_H\tau - i(\Omega_H+\Omega_N)t_1/2] \\ -i & 11 & 10i\exp[-i\Omega_H\tau - i(\Omega_H-\Omega_N)t_1/2] & 0 \\ 0 & -10i\exp[i\Omega_H\tau + i(\Omega_H-\Omega_N)t_1/2] & 11 & i \\ 10i\exp[i\Omega_H\tau + i(\Omega_H+\Omega_N)t_1/2] & 0 & -i & 11 \end{bmatrix}$$

(B.27)

The ^{1}H π pulse in the middle of the t_1 evolution is represented by the following rotation operators:

$$R_{180x}(^1H) = \begin{bmatrix} 0 & 0 & i & 0 \\ 0 & 0 & 0 & i \\ i & 0 & 0 & 0 \\ 0 & i & 0 & 0 \end{bmatrix}$$

$$R^{-1}_{180x}(^1H) = \begin{bmatrix} 0 & 0 & -i & 0 \\ 0 & 0 & 0 & -i \\ -i & 0 & 0 & 0 \\ 0 & -i & 0 & 0 \end{bmatrix}$$

(B.28)

This yields the following matrix immediately after the π pulse:

$$\sigma(t_1/2) = \frac{1}{2} \begin{bmatrix} 11 & i & 0 & -10i\exp[-i\Omega_H\tau + i(\Omega_H-\Omega_N)t_1/2] \\ -i & 11 & 10i\exp[i\Omega_H\tau - i(\Omega_H+\Omega_N)t_1/2] & 0 \\ 0 & -10i\exp[-i\Omega_H\tau - i(\Omega_H+\Omega_N)t_1/2] & 11 & i \\ 10i\exp[-i\Omega_H\tau - i(\Omega_H-\Omega_N)t_1/2] & 0 & -i & 11 \end{bmatrix}$$

(B.29)

Note that the net sign of evolution of the ^{1}H chemical-shift term changes for each of the four antidiagonal terms, while the net sign of the ^{15}N shift evolution does not. At the end of the t_1 period, the ^{1}H shift portion of the evolution has refocused, while the ^{15}N has not. This leaves:

$$\sigma(t_1/2) = \frac{1}{2} \begin{bmatrix} 11 & i & 0 & -10i\exp[-i\Omega_H\tau - i\Omega_N t_1] \\ -i & 11 & 10i\exp[i\Omega_H\tau + i\Omega_N t_1] & 0 \\ 0 & -10i\exp[-i\Omega_H\tau - i\Omega_N t_1] & 11 & i \\ 10i\exp[-i\Omega_H\tau + i\Omega_N t_1] & 0 & -i & 11 \end{bmatrix}$$

(B.30)

At this point, the last pulse, a ^{15}N $\pi/2$ (x) pulse is applied:

$$R^{-1}_{90x}(^{15}N) \times \sigma(t_1) \times R_{90x}(^{15}N) = \begin{bmatrix} 6 & 0 & -5/2A & -5/2B \\ 0 & 5 & -5/2B & 5/2A \\ -5/2A* & -5/2B* & 6 & 0 \\ -5/2B* & 5/2A* & 0 & 5 \end{bmatrix}$$

$$A = \exp[i\Omega_H\tau]\exp[-i\Omega_N t_1] + \exp[i\Omega_H\tau]\exp[i\Omega_N t_1]$$
$$= \exp[i\Omega_H\tau][(\cos\Omega_N t_1 - i\sin\Omega_N t_1) + (\cos\Omega_N t_1 + i\sin\Omega_N t_1)]$$
$$= 2\exp[i\Omega_H\tau]\cos\Omega_N t_1$$
$$B = \exp[i\Omega_H\tau]\exp[-i\Omega_N t_1] - \exp[i\Omega_H\tau]\exp[i\Omega_N t_1] \tag{B.31}$$

Of these terms, only the 1,3 and 2,4 elements and their complex conjugates will develop into observable coherence on ^1H that is detected. All of these terms are modulated as the cosine of Ω_N, the chemical shift of ^{15}N, during t_1. They also retain the modulation due to the chemical shift of ^1H that evolved during the $\tau = 1/2J$ period. This is refocused during the final $\tau = 1/2J$ period: note that the signs of the exponential are opposite to that of the expected evolution on ^1H during τ, and so these terms will go to unity at the end of the refocusing τ period (no chemical shift evolution on ^1H during the pulse sequence).

The treatment above is concerned only with the evolution of coupled ^1H and ^{15}N spins. The evolution of ^1H spins not coupled to ^{15}N will require a separate (albeit simpler) treatment. Either phase cycling or gradient coherence selection is required to eliminate those signals.

Semiclassical relaxation theory and the Redfield relaxation matrix

The other consideration in the time evolution of the density matrix is relaxation. The Bloch equations can be used to provide a complete description of the time-dependent behavior of a single spin, including line width of the resulting resonance in the frequency domain. This classical treatment can be extended to any ensemble of uncoupled spins. However, when treating a coupled two-spin system using the time-independent version of the Schrödinger equation to predict resonance positions and intensities, we did not include relaxation and so line widths were not determined. In order to include relaxation in the density matrix representation of a coupled spin system, one needs to specify all of the individual contributions that connect state populations to coherences, and one coherence to another. Redfield introduced a statistical mechanical formalism to analyze more complex spin systems that accomplishes this (Reference 1). The Redfield relaxation matrix approach can be used to analyze the relaxation behavior of coupled systems, including the details of correlation effects such as those responsible for the different line widths observed for individual lines in multiplets. This type of effect was discussed in Chapter 9 in relation to the TROSY experiment.

For a system containing j discrete stationary states, the equilibrium density matrix $\hat{\sigma}(0)$ is a diagonal (i.e. off-diagonal elements are zero) $j \times j$ matrix with diagonal elements σ_{ii} representing the normalized Boltzmann populations of the stationary states of the system, such that $\sum_j \sigma_{ii} = 1$ over all j states of the system. Off-diagonal elements represent coherences between states, and become nonzero in the presence of an oscillating RF field close to the transition frequency that is represented by the connected elements. For two states α and α' (represented by matrix elements $\sigma_{\alpha\alpha}$ and $\sigma_{\alpha'\alpha'}$), this is a time variation in the off-diagonal element connecting the two states $\sigma_{\alpha\alpha'}$ corresponding to $\exp(i\omega_{\alpha\alpha'}t)$. The time dependence of the off-diagonal terms in the absence of relaxation will be of the form:

$$\sigma_{\alpha\alpha'} \propto \exp(i\omega_{\alpha\alpha'}t)$$
$$d\sigma_{\alpha\alpha'}/dt = i\omega_{\alpha\alpha'}\sigma_{\alpha\alpha'} \qquad \text{(B.32)}$$

In order to include relaxation in the time evolution of a given density matrix element, the connections between that element and all of the other elements of the density matrix must be explicitly included. This would give a term of the form:

$$\sum_{\beta\beta'} P_{\alpha\alpha'\beta\beta'}\sigma_{\beta\beta'} \qquad \text{(B.32)}$$

where $P_{\alpha\alpha'\beta\beta'}$ are the elements of the relaxation matrix P that connect $\sigma_{\alpha\alpha'}$ to the other elements of the density matrix. (The Greek letter P is used to distinguish the relaxation matrix from the rotation matrix used for describing pulses in the previous section.) This would give the following expression for time evolution of the density matrix element:

$$d\sigma_{\alpha\alpha'}/dt = i\omega_{\alpha\alpha'}\sigma_{\alpha\alpha'} + \sum_{\beta\beta'} P_{\alpha\alpha'\beta\beta'}\sigma_{\beta\beta'} \qquad \text{(B.33)}$$

Compare this form to that of the Bloch equations (Equations 5.12–5.14).

For a coupled two-spin system, there are four stationary states, and hence the density matrix is a 4×4 matrix (16 elements):

$$\hat{\sigma} = \begin{vmatrix} \alpha\alpha & \alpha'\alpha & \beta\alpha & \beta'\alpha \\ \alpha\alpha' & \alpha'\alpha' & \beta\alpha' & \beta'\alpha' \\ \alpha\beta & \alpha'\beta & \beta\beta & \beta'\beta \\ \alpha\beta' & \alpha'\beta' & \beta\beta' & \beta'\beta' \end{vmatrix} \qquad \text{(B.34)}$$

The corresponding relaxation matrix would require a 16×16 (264-element) representation in order to connect each element in the density matrix to every other element (see Expression B.1). Despite this apparent complexity, the relaxation matrix is not as formidable as it would first appear. Many of the elements can be grasped intuitively. For example, off-diagonal element $P_{\alpha\alpha\beta\beta}$ is the probability per unit time that a transition from state β to state α will occur, with the corresponding element $R_{\beta\beta\alpha\alpha}$ represents the opposite transition probability. The two terms differ by a

Boltzmann factor that insures that when equilibrium is reached, no further net population changes will occur. The diagonal elements of $\hat{\hat{R}}$ connect each element in the density matrix with itself. In the case of populations, such as $P_{\alpha\alpha\alpha\alpha}$, the diagonal element represents the rate at which state α is depleted to all other states, i.e. $-P_{\alpha\alpha\alpha\alpha} = \sum_{\beta} P_{\beta\beta\alpha\alpha}$. Diagonal elements involving single-quantum transitions, such as $R_{\alpha\alpha'\alpha\alpha'}$ represent the tendency of random spin flips (zero-quantum transitions) to occur in coupled oscillators that have the same natural frequency. This gives the following form to Equation B.31 for such transitions:

$$\sigma_{\alpha\alpha'} \propto \exp(i\omega_{\alpha\alpha'} - P_{\alpha\alpha'\alpha\alpha'})t \qquad (B.35)$$

Note that this is simply the expression for damped harmonic oscillator, and the term $P_{\alpha\alpha'\alpha\alpha'}$ will contribute to the natural line width of the resonance line corresponding to that transition. An important corollary to this is that transition frequencies that are different from $\omega_{\alpha\alpha'}$ will not couple effectively to that transition and will not result in relaxation, i.e. $\left(\omega_{\alpha\alpha'} - \omega_{\beta\beta'}\right) > P_{\alpha\alpha'\beta\beta'}$. This also effectively removes relaxation interactions between different levels of coherence, rendering those elements of the Redfield matrix that couple nondegenerate coherences effectively zero. For this reason, the Redfield matrix (B.35) has the appearance of a kite, with off-diagonal elements that cross-relax coherences of different frequencies zero, and is often called the "Redfield kite".

$$
\hat{\hat{P}} =
\begin{bmatrix}
\alpha\alpha\alpha\alpha & \alpha'\alpha'\alpha\alpha & \beta\beta\alpha\alpha & \beta'\beta'\alpha\alpha & \cdot & \cdot & \cdot & \cdot & \cdot & \cdot & \cdot & \cdot & \cdot & \cdot \\
\alpha\alpha\alpha'\alpha' & \alpha'\alpha'\alpha'\alpha' & \beta\beta\alpha'\alpha' & \beta'\beta'\alpha'\alpha' & \cdot & \cdot & \cdot & \cdot & \cdot & \cdot & \cdot & \cdot & \cdot & \cdot \\
\alpha\alpha\beta\beta & \alpha'\alpha'\beta\beta & \beta\beta\beta\beta & \beta'\beta'\beta\beta & \cdot & \cdot & \cdot & \cdot & \cdot & \cdot & \cdot & \cdot & \cdot & \cdot \\
\alpha\alpha\beta'\beta' & \alpha'\alpha'\beta'\beta' & \beta\beta\beta'\beta' & \beta'\beta'\beta'\beta' & \cdot & \cdot & \cdot & \cdot & \cdot & \cdot & \cdot & \cdot & \cdot & \cdot \\
\cdot & \cdot & \cdot & \cdot & \alpha\alpha'\alpha\alpha' & \cdot & \cdot & \cdot & \cdot & \cdot & \cdot & \cdot & \cdot & \cdot \\
\cdot & \cdot & \cdot & \cdot & \cdot & \alpha'\alpha\alpha'\alpha & \cdot & \cdot & \cdot & \cdot & \cdot & \cdot & \cdot & \cdot \\
\cdot & \cdot & \cdot & \cdot & \cdot & \cdot & \beta\beta'\beta\beta' & \cdot & \cdot & \cdot & \cdot & \cdot & \cdot & \cdot \\
\cdot & \cdot & \cdot & \cdot & \cdot & \cdot & \cdot & \beta'\beta\beta'\beta & \cdot & \cdot & \cdot & \cdot & \cdot & \cdot \\
\cdot & \cdot & \cdot & \cdot & \cdot & \cdot & \cdot & \cdot & \alpha\beta\alpha\beta & \cdot & \cdot & \cdot & \cdot & \cdot \\
\cdot & \cdot & \cdot & \cdot & \cdot & \cdot & \cdot & \cdot & \cdot & \alpha'\beta\alpha'\beta & \cdot & \cdot & \cdot & \cdot \\
\cdot & \cdot & \cdot & \cdot & \cdot & \cdot & \cdot & \cdot & \cdot & \cdot & \alpha\beta'\alpha\beta' & \cdot & \cdot & \cdot \\
\cdot & \cdot & \cdot & \cdot & \cdot & \cdot & \cdot & \cdot & \cdot & \cdot & \cdot & \alpha'\beta'\alpha'\beta' & \cdot & \cdot \\
\cdot & \cdot & \cdot & \cdot & \cdot & \cdot & \cdot & \cdot & \cdot & \cdot & \cdot & \cdot & \beta\alpha\beta\alpha & \cdot \\
\cdot & \cdot & \cdot & \cdot & \cdot & \cdot & \cdot & \cdot & \cdot & \cdot & \cdot & \cdot & \cdot & \beta'\alpha\beta'\alpha \\
\cdot & \cdot & \cdot & \cdot & \cdot & \cdot & \cdot & \cdot & \cdot & \cdot & \cdot & \cdot & \cdot & \cdot \\
\cdot & \cdot & \cdot & \cdot & \cdot & \cdot & \cdot & \cdot & \cdot & \cdot & \cdot & \cdot & \cdot & \cdot \\
\end{bmatrix}
$$

$$\beta\alpha'\beta\alpha' \qquad \beta'\alpha'\beta'\alpha'$$

$$(\text{B.37})$$

INDEX